Pesticide Regulation Handbook

Third Edition

McKenna & Cuneo

Partners
John D. Conner, Jr.
Lawrence S. Ebner
Stanley W. Landfair
Charles A. O'Connor, III
Kenneth W. Weinstein

Associates
Wayne T. Ault
Susan K. Boushell
Thomas B. Johnston
Gregory J. Madden
Michael R. Neilson
Monica A. Wallach

Technology Sciences Group Inc.

David J. Brookman, Ph.D.
Jeffrey H. Driver, Dr.P.H., M.T., C.L.S.
Jack D. Early, Ph.D.
James F. Hobson, Ph.D., D.A.B.T.
Edwin L. Johnson, M.A., MPA
Andrew P. Jovanovich, Ph.D., MBA
Arthur D. Lawyer, Ph.D.
Alain F. Pelfrene, M.D., Ph.D.
Robert R. Stewart, Ph.D.
Gary K. Whitmyre, M.A., D.A.B.T.

AF718834

McGraw-Hill, Inc.

New York St. Louis San Francisco Auckland Bogotá
Caracas Lisbon London Madrid Mexico Milan
Montreal New Delhi Paris San Juan São Paulo
Singapore Sydney Tokyo Toronto

Executive Enterprises Publications Co., Inc., New York, New York

KF
3959
.P47
1993

Copyright © 1993 by McKenna and Cuneo. All rights reserved. Printed in the United States of America. Except as permitted under the United States Copyright Act of 1976, no part of this publication may be reproduced or distributed in any form or by any means, or stored in a data base or retrieval system, without the prior written permission of the publisher.

1 2 3 4 5 6 7 8 9 0 HAL/HAL 9 8 7 6 5 4 3 2

ISBN 0-07-546322-9

The sponsoring editor for this book was Gail F. Nalven, and the production supervisor was Thomas G. Kowalczyk.

Printed and bound by Arcata Graphics/Halliday.

This publication is designed to provide accurate and authoritative information regarding its subject matter. It is sold with the understanding that the publisher is not engaged in rendering legal, accounting, or other professional service. If legal advice or other expert assistance is required, the services of a competent professional person should be sought.—*From a Declaration of Principles Jointly Adopted by a Committee of the American Bar Association and a Committee of Publishers.*

WITHDRAWN

McKENNA & CUNEO

McKenna & Cuneo is an international law firm with over 230 attorneys in Washington, D.C., Los Angeles, San Francisco, San Diego, Denver and Brussels, Belgium. For more than three decades, the firm has advised and represented pesticide producers, users and their trade associations on many of the most important pesticide regulatory issues. **McKenna & Cuneo** also is active in the fields of environmental regulation and litigation, toxic torts and product liability, government contracts, international trade and business, food and drug law, and other areas.

TECHNOLOGY SCIENCES GROUP INC.

Technology Sciences Group, a subsidiary of McKenna & Cuneo, specializes in registering new pesticides with EPA and state authorities, and in assisting in the resolution of pesticide regulatory problems. The company also offers unique expertise in registration of biological products. In the chemical regulatory area, **Technology Sciences Group** offers consulting services in all facets of TSCA compliance. Technology Sciences Group is comprised of Pesticide, Chemicals, Toxicology, Legislative Affairs, and State Affairs Divisions.

ACKNOWLEDGEMENT

The authors wish to thank Robert A. Howe and the firm's other legal assistants for their efforts in cite checking, proofreading and indexing the manuscript. We also acknowledge the research assistance of Suzanne W. King and Richard S. Moskowitz, and the efforts of Mary C. Sonosky, legislative specialist, in compiling the text of FIFRA as amended. The authors also thank Robert M. Rosenberg, Director of Government Affairs, National Pest Control Association, for reviewing the chapter on Regulation of Pesticide Applicators.

PREFACE TO THE THIRD EDITION

Since 1987 when we first revised our original *Pesticide Regulation Handbook*, the pesticide manufacturing and user communities have experienced numerous significant developments, including most notably the 1988 amendments to FIFRA. These amendments added a new § 4 to FIFRA, launching EPA's ambitious program to complete the reregistration of all pesticides within nine years. In addition, since 1987 EPA has demonstrated an increasing commitment to aggressive enforcement of FIFRA through civil and even criminal initiatives. The role of pesticides in food safety is the continuing object of scientific, regulatory and legal debate. But during this time, states (most frequently California) and even local communities have not been content to assume a back seat in the field of pesticide regulation. Also, with distressing frequency, pesticides are the subject of toxic tort suits. The only constant feature is change. In this Third Edition, we have attempted to highlight these changes, while attempting to meet our original goal — to provide a comprehensive yet readable reference for understanding pesticide regulation.

As always, we invite inquiries and comments about this edition or pesticide regulation in general.

THE AUTHORS

January 1991

CONTENTS

Chapter 9

Chapter 10

Chapter 11

Chapter 12

Chapter 13

Chapter 17

Chapter 1
EVOLUTION OF PESTICIDE REGULATION

Pesticides are chemical or biological substances intended to control weeds, insects, fungi, rodents, bacteria, and other pests. They protect food crops and livestock, control household pests, promote agricultural productivity, and protect public health. Because they can affect living organisms, pesticides are highly regulated to ensure that their use will be safe to man and the environment.

Pesticides are regulated by the U.S. Environmental Protection Agency (EPA) primarily under the Federal Insecticide, Fungicide, and Rodenticide Act (FIFRA), 7 U.S.C. §§ 136-136y (1988), as amended by the Food, Agriculture, Conservation, and Trade Act of 1990, Pub. L. No. 101-624, 104 Stat. 3359 (Nov. 28, 1990), and the Federal Food, Drug, and Cosmetic Act (FFDCA), 21 U.S.C. §§ 301-393 (1988). FIFRA requires pesticides to be registered by EPA and authorizes the Agency to prescribe conditions for their use. The FFDCA requires the Agency to establish maximum acceptable levels of pesticide residues in foods. Both laws require EPA to weigh the benefits derived from the use of a pesticide against any risks that it may pose to public health and the environment, when establishing the conditions for a pesticide's use.

EPA has registered about 600 pesticide active ingredients, which are used to produce approximately 24,000 pesticide products. More than a billion pounds of pesticides are used each year, with about four-fifths of the production going to agriculture.

PESTICIDE REGULATION BEFORE 1972

In the early twentieth century, agricultural chemicals consisted mainly of insecticides and fungicides. These products contained inorganic compounds as their active ingredients, principally arsenicals such as lead arsenate and Paris green (copper aceto-arsenite), and also copper, mercury, and zinc, to name a few.

Federal regulation of pesticides started with the Insecticide Act of 1910, which made it unlawful to manufacture any insecticide or fungicide that was "adulterated or misbranded."

Ch. 191, 36 Stat. 331 (1910). By setting standards for these products and declaring adulterated any products that deviated from these standards, the Insecticide Act was designed to protect consumers from substandard or fraudulent insecticides and fungicides. It provided for inspections, seizure of adulterated or misbranded products, and prosecution of violations as misdemeanors. The Insecticide Act, however, was essentially a labeling statute; it did not require registration or establish significant safety standards for pesticides. It remained the federal law for thirty-seven years.

By World War II, synthetic organic pesticides had emerged, demonstrated great effectiveness, and received rapid market acceptance. Organochlorine insecticides such as aldrin, dieldrin, chlordane, heptachlor, lindane, endrin, toxaphene, and DDT were highly effective in controlling insects at very low dosages.

In 1947, Congress replaced the Insecticide Act with the more comprehensive FIFRA. Pub. L. No. 80-104, 61 Stat. 163 (1947). FIFRA required that pesticides be registered by the Secretary of Agriculture before their sale or distribution in interstate or foreign commerce. States were left to regulate intrastate sales of pesticides. The 1947 Act required warnings on the labels of highly toxic pesticides and instructions for their safe use.

Despite the enhanced powers of the U.S. Department of Agriculture (USDA), the 1947 FIFRA remained fundamentally a labeling statute. If the labeling was not adequate to protect the public, the product was legally misbranded and subject to seizure and cancellation. In enforcement proceedings USDA had the burden of proving that the product was misbranded, and the Department had no power to deny or cancel a registration that apparently failed to comply with the provisions of the Act. The Secretary of Agriculture could only notify the applicant or the registrant that it should take certain corrective measures. The applicant or registrant could refuse to comply with the Secretary's recommendation and still obtain "registration under protest."

Congress ended the practice of protest registration in 1964 and authorized the Secretary of Agriculture to deny or cancel a registration. Pub. L. No. 88-305, 78 Stat. 190

(1964). The 1964 amendment also authorized the Secretary to suspend a registration immediately if necessary to prevent an imminent hazard to the public.

In 1954, Congress enacted the Miller Amendment to FFDCA, which required that a maximum acceptable level (tolerance) be established for pesticide residues in foods and animal feed. FFDCA § 408, Pub. L. No. 518, 68 Stat. 511 (1954). In 1958, Congress enacted the Food Additives Amendment to the FFDCA, which regulated pesticide residues in processed foods. FFDCA § 409, Pub. L. No. 85-929, 72 Stat. 1785 (1958). It required that, if a pesticide residue in a processed food exceeded the tolerance for the related raw agricultural commodity, a food additive regulation must be established. These two statutory provisions have remained essentially unchanged since their enactment, but still remain the focus of ongoing Congressional debate over food safety.

On December 2, 1970, President Nixon's Reorganization Plan No. 3 created EPA. Reorg. Plan No. 3 of 1970, 40 C.F.R. pt. 1 (1970), *reprinted in* 5 U.S.C. app. at 1132-37 (1982), *and in* 84 Stat. 2086 (1970). Under the Reorganization Plan, EPA assumed the pesticide regulatory functions of USDA and the Food and Drug Administration (FDA) under FIFRA and the FFDCA. 35 Fed. Reg. 15,623 (1970).

1972 AMENDMENT TO FIFRA

The Federal Environmental Pesticide Control Act of 1972 (FEPCA) completely overhauled FIFRA and significantly increased the authority of EPA to regulate pesticides. Pub. L. No. 92-516, 86 Stat. 973 (1972). FEPCA is the backbone of the current FIFRA.

FEPCA retained the basic provisions of the 1947 FIFRA: Registration is still the centerpiece of regulation; labeling is defined broadly; and adulterated and misbranded products are unlawful. But FEPCA added a new statutory standard for registration: A pesticide could be registered only if it did not cause "unreasonable adverse effects" on human health or the environment. 86 Stat. 980-81. Under FEPCA, EPA was authorized to classify pesticides into general- and/or restricted-use categories; to regulate pesticide use; to require federal registration of pesticides sold intrastate; to register establishments that produce pesticides; to require pesticide producers to maintain records necessary for enforcement of FIFRA; to conduct inspections of establishments that produce or hold pesticides for sale; to issue "stop sale, use, or removal" orders against persons holding illegal pesticides; to seek federal district court orders condemning and seizing illegal pesticides; and to enlist the cooperation of the states in enforcing FIFRA.

FIFRA still prohibits substandard or inadequately labeled pesticides (as in the 1910 Act) and requires registration (as in the 1947 FIFRA), but FEPCA otherwise transformed FIFRA in almost all respects. Importantly, it directed the Administrator to reregister all currently registered pesticides within five years. Congress relaxed this overly ambitious timetable in 1978 when it admonished the Administrator to accomplish reregistration "in the most expeditious manner practicable." Pub. L. No. 95-396, 92 Stat. 819, 827 (1978). In 1988, frustrated with EPA delays in reregistering existing pesticides, Congress set a specific statutory timetable and a new funding mechanism for accomplishing reregistration.

Since the passage of FEPCA in 1972, Congress has amended FIFRA five times — in 1975, 1978, 1980, 1988, and 1990 — each time with the intention of fine-tuning the statute. Congress considered but failed to adopt extensive changes to FIFRA in 1982 and 1986, and many issues not addressed in the 1988 amendments, such as export controls, food safety and ground water contamination, continue to prompt legislative initiatives.

1975 AMENDMENT TO FIFRA

In 1975, Congress tried to counterbalance a perceived environmental bias in EPA's decisionmaking on pesticides and to respond to "the often-stated concern that EPA has not adequately considered the impact of its actions on agriculture" and that "more effective consultation is necessary." S. Rep. No. 452, 94th Cong., 1st Sess. 8 (1975). The 1975 amendment to FIFRA required EPA to give the Secretary of Agriculture advance notice of proposals to cancel or change the classification of a pesticide, and of proposed regulations; directed EPA to consider the impact on agriculture before initiating cancellation proceedings; and established the FIFRA Scientific Advisory Panel (SAP) to review proposed cancellation actions and proposed regulations. Pub. L. No. 94-140, 89 Stat. 751 (1975). Among other things, the 1975 amendment also tried to clarify the scope and effective date of FIFRA § 3(c)(1)(D), which was enacted by FEPCA in 1972 in order to protect trade secret data and to compensate pesticide data submitters for EPA's use of their test data in registering another company's product.

1978 AMENDMENT TO FIFRA

In 1978, Congress substantially revised FIFRA by enacting the Federal Pesticide Act of 1978, Pub. L. No. 95-396, 92 Stat. 819. Congress revamped § 3(c)(1)(D) entirely by granting data submitters the right to ten years of exclusive use of their data concerning new active ingredients, removing most trade secret protection for health and safety data, and transferring the responsibility for adjudicating data compensation from EPA to private arbitrators.

In other important provisions, the 1978 amendment granted conditional registration authority to allow EPA to process the backlog of applications for registration despite the absence of full supporting data accompanying the applications; authorized interim administrative review (or Special Review), but only on the basis of "a validated test or other significant evidence raising prudent concerns of unreasonable adverse risk to man or to the environment," 92 Stat. 826; prohibited disclosure of pesticide data to foreign or multinational pesticide producers; subjected pesticides and devices intended for export to FIFRA's misbranding provisions; redefined "use inconsistent with the label" under FIFRA § 2(ee), 92 Stat. 819, in order to permit certain departures from the label specifications; and required the Administrator to recognize the distinction between agricultural and non-agricultural pesticides in processing registrations and in setting registration standards and guidelines.

1980 AMENDMENT TO FIFRA

In 1980, Congress added several provisions to FIFRA § 25. Pub. L. No. 96-539, 94 Stat. 3194 (1980). As amended in 1980, FIFRA § 25(a)(4) provides for a bicameral congressional veto of EPA's prospective rules and regulations under FIFRA;[1] FIFRA § 25(d) authorizes SAP to create its own subpanels and requires the Administrator to solicit comments from SAP on suspension (as it already does on cancellation) actions; and FIFRA § 25(e) requires the Administrator to specify procedures for peer review of the design, protocols, and conduct of major scientific studies performed by or under the auspices of EPA, or relied on by EPA in initiating changes in classification, suspensions, or cancellations of pesticides.

1988 AMENDMENT TO FIFRA

In response to concerns that EPA's pace in reregistering pesticides was "exceedingly slow," Congress enacted the Federal Insecticide, Fungicide, and Rodenticide Act Amendments of 1988, Pub. L. No. 100-532, 102 Stat. 2654. H.R. Rep. No. 939, 100th Cong., 2d Sess. 28, *reprinted in* 1988 U.S. Code Cong. & Admin. News 3474, 3477. The 1988 FIFRA amendment established a five-phase reregistration program for all active ingredients first registered prior to November 1, 1984, and a schedule for completing reregistration within three to nine years. To fund the cost of reregistration, Congress imposed reregistration fees of up to $150,000 per active ingredient.

In other changes, Congress created a "fast track" registration of end-use products for "me-too" registrants (those registering products similar to an already registered pesticide); required prior Congressional approval before EPA could indemnify registrants (except for end users) holding cancelled or suspended pesticides; authorized EPA to regulate the

---

1 The bicameral veto of § 25(a)(4) appears to be an unconstitutional violation of the separation of powers provisions of Article I of the Constitution under the broad holding of the U.S. Supreme Court in *Immigration & Naturalization Service v. Chadha*, 462 U.S. 919 (1983).

storage, disposal and transportation of suspended or cancelled pesticides, and directed EPA to reimburse some or all of the associated storage costs; established new data and labeling requirements for storage and disposal; and mandated EPA regulation of pesticide containers to promote safe storage and disposal.

The 1988 amendment also broadened the class of regulated entities subject to FIFRA recordkeeping requirements. Under the prior law, EPA could prescribe such requirements only for producers. Now, recordkeeping requirements apply to all registrants and applicants for registration as well. The 1988 amendment increased EPA's inspection and enforcement authority, and also increased the maximum fine for criminal FIFRA violations from $25,000 to $50,000.

1990 AMENDMENT TO FIFRA

The Food, Agriculture, Conservation, and Trade Act of 1990, Pub. L. No. 101-624, secs. 1491-1496, 104 Stat. 3359 (Nov. 28, 1990) (commonly referred to as the Farm Bill), contains five amendments to FIFRA. Pursuant to these amendments, certified applicators of restricted use pesticides must maintain adequate records. Registrants of minor use pesticides need not submit field residue data from geographic areas where the pesticide will not be registered for use. The amendments also vest the Administrator with discretion to reduce or waive a reregistration fee if it would "significantly reduce the availability of the pesticide." In addition, the Act amends FIFRA § 6(f), concerning voluntary cancellation of a pesticide registration, in order to provide for public notice and comment upon a registrant's application for voluntary cancellation, and imposes several additional requirements pertaining to the voluntary cancellation of minor use pesticides.

1989-90 PROPOSED AMENDMENTS TO FIFRA

In February 1989, a mass media blitz by national environmental interest groups propelled the issues of pesticide safety back into public consciousness. Alar residues in apples and their purported risks to children dominated media attention. Members of the House and Senate committees with jurisdiction over FIFRA once again began introducing legislation to amend not only FIFRA, but also the tolerance-setting provisions of FFDCA.

The central themes of the major bills introduced during the 101st Congress were food safety, export restrictions, and ground water contamination. Although these bills made substantial progress in committee, they lost momentum during the celebrated debates over the federal budget - except for that of export controls. The House and Senate agriculture committees introduced export controls as a 1990 Farm Bill amendment designed to prevent the so-called "circle of poison" (whereby hazardous pesticides exported for use abroad return to the United States in the form of unsafe or unregulated residues in food). Ultimately, however, this amendment on pesticide exports also was defeated.

The proposed amendment would have revised § 17 of FIFRA to prohibit the export of pesticides which are not registered or which are not subject to a food tolerance. The legislation would have prohibited the export of a pesticide unless it met the following four requirements: (1) the product had a registration or food tolerance; (2) the exporter had complied with all the prior informed consent (PIC) requirements; (3) the product label was written in the official language of the receiving country; and (4) pesticide manufacturers and exporters had satisfied the transportation, storage and disposal requirements of § 19 of FIFRA. Undoubtedly this pesticide export proposal as well as the food safety and ground water contamination initiatives will reemerge during the 102d Congress.

Chapter 2
ADMINISTRATION OF EPA'S PESTICIDE PROGRAM

The U.S. Environmental Protection Agency (referred to in this book as "EPA" or "the Agency") is headed by an Administrator and Deputy Administrator, who are appointed by the President with the advice and consent of the Senate. They oversee the ten Assistant Administrators in charge of EPA's various program and operational offices and the ten Regional Administrators. See EPA organization chart, figure 2-1, which appears at the end of this chapter; 40 C.F.R. pt. 1. Assistant and Regional Administrators also are presidential appointees.

The authority to regulate pesticides that FIFRA grants to the EPA Administrator has been delegated within the Agency to the Assistant Administrator for Pesticides and Toxic Substances, who has overall responsibility for implementing the pesticide program. The Assistant Administrator, however, has delegated day-to-day operation of the program to the Director of the Office of Pesticide Programs (OPP). The Assistant Administrator also oversees the Office of Toxic Substances (OTS) and the Office of Compliance Monitoring (OCM), which, along with OPP and several specialized staffs, constitute EPA's Office of Pesticides and Toxic Substances (OPTS). See OPTS organization chart, figure 2-2.

EPA headquarters is located at 401 M Street, S.W., Washington, D.C. 20460, in a commercial development known as Waterside Mall. The central offices for OPTS, OCM and OTS, and the Office of General Counsel, Pesticides Division, are located in Waterside Mall, but OPP is located across the Potomac River in Crystal Mall #2, 1921 Jefferson Davis Highway, Arlington, Virginia 22202 (tel. (703) 557-7090). Although EPA periodically publishes a directory that lists the telephone numbers of EPA professional staff down to the branch level, telephone numbers assigned to Agency personnel change frequently. An EPA Personnel Locator can be reached by calling (202) 382-2090 between 8:30 a.m. and 4:30 p.m.

eastern time on weekdays; for best results the caller should know the exact name of the office or person he or she is seeking.

OFFICE OF PESTICIDE PROGRAMS

OPP has an annual budget of approximately $84 million. In 1988 OPP reorganized into seven divisions: (1) Registration Division (RD); (2) Special Review and Reregistration Division (SRRD); (3) Health Effects Division (HED);[2] (4) Environmental Fate and Effects Division (EFED); (5) Field Operations Division (FOD); (6) Program Management and Support Division (PMSD); and (7) Biological and Economic Analysis Division (BEAD). See OPP organization chart, figure 2-3. The functions and responsibilities of each of these divisions are described to some extent in the EPA Delegations Manual, and are described in more detail in Pesticide Registration (PR) Notice 88-8 (Oct. 3, 1988).

Most pesticide registrants' primary contact with OPP is through OPP's Registration Division. That division coordinates new pesticide registrations, amendments to existing registrations, and related activities. The newly created Special Review and Reregistration Division manages all Special Reviews, and is responsible for supervising the reregistration and data call-in programs. The Health Effects, Environmental Fate and Effects, and Biological and Economic Analysis Divisions provide technical support to the Registration and Special Review/Reregistration Divisions and are composed largely of staff scientists and economists. The Program Management and Support Division performs administrative functions such as preparation of OPP's budget and design and operation of computerized data bases. The new Field Operations Division is responsible for implementing OPP's actions, policies and programs in the field (i.e., in the EPA regions throughout the United

[2] The "new" HED should not be confused with the "old" HED, or Hazard Evaluation Division. Prior to the reorganization, the "old" HED encompassed what are now the Health Effects Division and the Environmental Fate and Effects Division.

States), and for managing certain information services, notably OPP's Public Docket and the OPP Freedom of Information Act program.

Registration Division

OPP's Registration Division (tel. (703) 557-7760) is headed by a Director and Deputy Director and consists primarily of four branches: Insecticide-Rodenticide Branch, Fungicide-Herbicide Branch, Antimicrobial (formerly Disinfectants) Branch, and Registration Support Branch. Each branch is headed by a chief and is divided into teams and/or sections. See Registration Division organization chart, figure 2-4.

The chiefs of the three product-oriented branches designate a product manager (PM) for each active ingredient falling within the purview of the branch. The PM coordinates all routine registration activity relating to that active ingredient, such as applications to register new products, formulations, or uses. This includes serving as the principal contact at OPP for all applicants or registrants of the pesticide, except for reregistration issues (see below). The PMs receive technical advice on concerns such as toxicology, exposure and benefits associated with a given application or amendment from a product team composed of technical experts from RD and the other OPP Divisions. A registrant can ascertain the identity of the PM for a given product by contacting RD or the branch chief for the branch that handles the particular type of product in question.

In July 1985 EPA issued a policy regarding communications between pesticide registrants and OPP staff (PR Notice 85-4 (July 30, 1985)). Under this policy all communications must be directed to or coordinated by the PM. Meetings with the PM or his staff must be scheduled in advance; registrants can no longer just "drop in" for a visit. In addition, meetings must be requested by letter or telephone, and the subject of the meeting identified. A PM may deny a request for a meeting if he feels that the proposed agenda does not warrant discussion, the purpose of the discussion is not sufficiently clear, or insufficient notice has been given. PR 85-4 also applies to contacts with the designated Review Manager for Special Reviews.

The Registration Support Branch handles administrative, processing, and support functions for the other three Registration Division branches. The Branch's "front-end processing unit" receives all registration applications and petitions for tolerances and assigns them to the appropriate product management branch.

The remainder of the Branch is separated into five sections. The Product Chemistry Section reviews and evaluates product chemistry data and clears inert ingredients from requirements for tolerances, when appropriate, and also implements OPP's strategy for regulation of inert ingredients. See Chapter 3. The Precautionary Review Section reviews acute toxicity studies and recommends precautionary labeling requirements to the product management branches. The Administrative Processing Section handles all administrative functions such as transfers and withdrawals of registrations, analyzes RD's program targets and accomplishments, and acts as liaison on automatic data processing systems for the Branch and Division. The Section also maintains the Data Submitters List, which lists each registrant who has submitted data in support of a registration of a particular active ingredient. The Project Coordination Section develops OPP regulations, PR Notices, and policy statements. The Emergency Response and Minor Use Section evaluates requests by states for emergency permission to use a pesticide on a crop that is not on the label. See Chapter 3. This Section also processes petitions for tolerances in minor use crops that are submitted to OPP under the IR-4 program.

In addition to the four branches, the RD maintains a separate Program Coordination Staff that is responsible for coordinating the mailing of PR Notices, which provide guidance to registrants on various policy matters relating to registration of pesticides. PR Notices can be initiated by RD or the OPP Director's Office. In addition, the Program Coordination Staff will provide interested members of the public with RD's own "Standard Operating Procedures" (SOPs), which are internal guidelines for operation of the Division's business. Pesticide Assessment Guidelines, which provide technical guidance regarding compliance with the data requirements for registration set forth at 40 C.F.R. part 158, are discussed in Chapter 3.

Special Review and Reregistration Division

The relatively new Special Review and Reregistration Division (SRRD) (tel. (703) 557-7760) was created from the Special Review Branch of RD that existed throughout the 1980s. It is a reformation of the old Special Review Division (SPRD) that existed during the late 1970s. In addition to Special Reviews, SRRD has assumed the reregistration responsibilities formerly conducted by the Registration Division Product Managers.

Within the new SRRD, the Special Review Branch coordinates Special Reviews, which is EPA's term for the process of reviewing on a high priority basis the risks and benefits of previously registered pesticides. See Chapter 8. A Review Manager is assigned to coordinate each review and oversee a team of technical staff who assemble risk and benefits data and make detailed risk and benefits assessments on the pesticide under review. The team normally includes staff from the Health Effects, Environmental Fate and Effects, and Biological and Economic Analysis Divisions.

The Reregistration Branch is responsible for reregistering "List A" pesticides (i.e., products containing active ingredients that were the subject of registration standards prior to December 24, 1988, when the 1988 FIFRA amendments were enacted). See Chapter 4. The Reregistration Branch is organized into two sections. One manages reregistration of fungicides and herbicides and the other the insecticides, rodenticides and antimicrobials. In each section, a chemical review manager is assigned to manage the reregistration process for each reregistration "case," which may involve more than one active ingredient. See SRRD Reregistration Branch, Reregistration Plan for Chemicals on List A (Mar. 1990).

The Generic Chemical Support Branch is responsible for reregistration, in accordance with FIFRA § 4, of pesticides in Lists B, C and D (i.e, the lists of active ingredients EPA has created to comply with FIFRA § 4(c)(2)), and also manages OPP's special data call-in program. See Special Review and Reregistration Division organizational chart, figure 2-5.

Health Effects Division

The Health Effects Division (HED) (tel. (703) 557-7351) reviews scientific data on the toxicological effects of pesticides and the potential rates of human and/or domestic animal exposure to individual pesticides. From these data, HED develops assessments of the potential risk of pesticide products to humans and domestic animals. These risk assessments are used to support registration, special review and reregistration decisions. The new HED also provides clerical and administrative support for the FIFRA Scientific Advisory Panel (SAP). See page 18 and Health Effects Division organizational chart, figure 2-6.

HED contains two Toxicology Branches, one devoted to insecticides and rodenticides and one to herbicides, fungicides and anti-microbials. These branches review data from acute and subacute toxicology studies, long-term chronic feeding studies, and oncogenicity, mutagenicity, and teratogenicity/reproductive effects studies.

The Dietary Exposure Branch estimates human exposure to pesticides in food by reviewing plant and animal metabolism data and data regarding the nature and levels of residues in treated commodities, foods, and feeds. The Non-Dietary Exposure Branch estimates occupational exposure to pesticides, i.e., exposure of pesticide mixers, loaders and applicators, and exposure of consumers to residential-use pesticides, including lawn care products.

The major responsibility of the Science Analysis and Coordination Branch (SACB) is creation of risk assessments from the toxicological data provided by the toxicology branches and the exposure data from the exposure branches. SACB also oversees the Dietary Risk Evaluation System (formerly the Tolerance Assessment System) that is used to set the "reference dose" for a given pesticide residue. EPA considers the reference dose (formerly called the acceptable daily intake (ADI)) when the Agency sets tolerance levels for pesticides in foods. See Chapter 5. SACB also provides peer review for proposed classifications of carcinogens and developmental toxicants.

Environmental Fate and Effects Division

The Environmental Fate and Effects Division (EFED) (tel. (703) 557-7695) reviews scientific data on the effects of pesticides on animals other than humans and domestic pets and on the fate of pesticides when they are released into the environment as a result of application or disposal or by other means. EFED writes assessments of the risk to non-target wildlife and has lead responsibility for OPP's efforts in the areas of biotechnology, pesticide monitoring, quality assurance, groundwater protection, pesticide disposal, and endangered species protection. See Environmental Fate and Effects Division organizational chart, figure 2-7.

The Environmental Fate and Groundwater Branch profiles the environmental fate of pesticides by examining dissipation, photolysis, hydrolysis, metabolism, leaching, and accumulation studies. The Ecological Effects Branch evaluates wildlife and aquatic organism data from toxicity studies on birds and fish, and develops assessments of the hazards pesticides pose to non-target fish, other aquatic organisms, and/or wildlife.

Field Operations Division

This new division (tel. (703) 557-7092) is responsible for ensuring that the policies and programs OPP develops are effectively implemented throughout the nation. To that end, FOD has absorbed several functions that previously had been the responsibility of various OPP or OPTS offices. For example, the Occupational Safety Branch of FOD encompasses the old Pesticides Farm Safety Staff once located in PMSD, and the Applicator Certification and Training Branch performs the functions of a former OPTS office by the same name. The Integrated Pest Management Program, formerly of the Benefits and Use Division of OPP, has been transferred to FOD's Program Communication Branch.

The Public Information Branch of FOD is responsible for OPP's Freedom of Information Act programs, which once were located in PMSD. The Public Information Branch also is responsible for OPP's Public Docket.

OPP's Public Docket actually consists of separate dockets for: (1) rulemakings; (2) Special Reviews; and (3) registration standards processes. Rulemaking dockets contain public comments submitted to EPA in response to *Federal Register* notices that solicit such comments. Special Review and registration standards dockets contain minutes of meetings between EPA and members of the public and communications with persons outside the government. OPP publishes a monthly index of these documents.

EPA regulations require OPP to create a public special review docket whenever EPA first notifies registrants of a pesticide that it is considering issuance of a Notice of Special Review. 40 C.F.R. § 154.15. EPA regulations require OPP to create a public registration standard docket for a pesticide active ingredient when the Agency begins review of data for a standard on that ingredient or when EPA publishes a notice that states that the Agency will develop a registration standard for that active ingredient. EPA generally publishes such schedules annually.

The FOD Division Director also supervises three special assistants, the Regional Liaison, the Agricultural Liaison, and the State Liaison. These assistants exchange information on EPA policies and programs with the EPA regional offices, the U.S. Department of Agriculture, and state agencies in charge of pesticide programs.

Program Management and Support Division

Most pesticide registrants will have few, if any, contacts with the new PMSD (tel. (703) 557-2440) now that the Pesticide Farm Safety Staff, Freedom of Information Act, and Public Docket functions have moved to FOD. The remaining responsibilities of PMSD are mostly internal, and involve program planning, budget formulation and execution, program evaluation, administrative support, and internal management of information and systems.

Biological and Economic Analysis Division

The Biological and Economic Analysis Division (BEAD) (tel. (703) 557-0500) collects and analyzes data on pesticide use and the benefits associated with such use.

The Biological Analysis Branch provides scientific and technical analyses relating to the efficacy and field performance of pesticides. These analyses are used when OPP estimates the benefits of a pesticide during a Special Review, and are taken into consideration when OPP processes requests for emergency exemptions under FIFRA § 18. The Branch also provides comprehensive use data assessment to predict frequency and duration of worker exposure. The Economic Analysis Branch assesses the economic benefits of particular pesticides and their alternatives in support of Special Reviews, cancellation or suspension, and other regulatory actions.

The Analytical Chemistry Branch validates the analytical methods for parent compounds and metabolites that are submitted by registrants to support pesticide tolerances in foods. The Branch also validates the analytical methods for parent compounds, inert ingredients and impurities that registration applicants submit in support of registrations of technical or formulated products. The Branch provides pesticide technical reference standards to enforcement agencies in the states and assists EPA regional and state enforcement laboratories. In addition, the Branch provides support for special projects such as the National Pesticide Wellwater Survey.

The Branch maintains two laboratories, the Analytical Chemistry Laboratory in Beltsville, Maryland, and the Environmental Chemistry Laboratory in Bay St. Louis, Mississippi.

SCIENTIFIC ADVISORY PANEL

FIFRA § 25(d) directs the EPA Administrator to create a Scientific Advisory Panel consisting of seven members appointed by the Administrator from a list of twelve nominees, six nominated by the National Institutes of Health and six by the National Science

Foundation. The Administrator is to utilize a system of staggered terms of appointment, and members of the Panel must be selected on the basis of their professional qualifications to assess the effects of the impact of pesticides on health and the environment. To the extent feasible to ensure multidisciplinary representation, the Administrator must appoint representatives from the disciplines of toxicology, pathology, environmental biology, and related sciences. When vacancies occur on the Panel due to expiration of a term, resignation, or any other reason, the Administrator must select a replacement from a group of four nominees, two submitted by each of the nominating entities.

EPA must submit to SAP for comment as to the impact on health and the environment any action proposed in notices of intent to cancel registration or to change classifications of a pesticide issued under FIFRA § 6(b). See Chapter 8. The Administrator also must submit to SAP for comment any proposed or final form of regulations issued under FIFRA. Any action the Administrator takes to suspend the registration of a pesticide under FIFRA § 6(c) also must be submitted to SAP.

SAP also performs peer review with respect to the design, protocols and conduct of major scientific studies conducted under FIFRA by EPA or by any other federal agency, state, institution, or individual under a grant, contract, or cooperative agreement from or with EPA.

The Administrator can use nominees to SAP from a current list maintained by SAP to peer review the results of any scientific studies relied upon by EPA during actions relating to a change in classification, suspension, or cancellation of a pesticide.

OFFICE OF COMPLIANCE MONITORING

Pesticide registrants also may have contact with the Office of Compliance Monitoring (OCM) (tel. (202) 382-3807) within OPTS.

The Compliance Monitoring Staff became the Office of Compliance Monitoring under a final rule effective June 20, 1985. 50 Fed. Reg. 26,721 (1985). With this change, OCM also assumed a parallel position, along with the Office of Toxic Substances and the

Office of Pesticide Programs, as one of the three separate offices within OPTS. See OPTS organization chart, figure 2-2. OCM, headed by a Chief Executive Officer, directs EPA's compliance programs for pesticides and toxic substances.

OCM initiates civil administrative complaints and other administrative orders in enforcement cases involving issues of first impression or overriding national significance, or in cases involving alleged violations in more than one EPA region. OCM is assisted in this regard by attorneys from the Pesticides and Toxic Substances Division of EPA's Office of General Counsel (see below) and EPA's Office of Enforcement (OE). OCM also provides guidance to Regional Administrators and Regional Counsel regarding pesticide enforcement activities. In addition to carrying out FIFRA-related enforcement activities, OCM coordinates with the Office of Toxic Substances (OTS) in conducting regulatory and compliance programs under the Toxic Substances Control Act (TSCA) and participates in development of regulations under TSCA.

OCM has three divisions: the Compliance Division, the Policy and Grants Division, and the Laboratory Data Integrity Program. The Compliance Division develops national compliance monitoring procedures under both FIFRA and TSCA, including procedures for controlling imminent hazards involving pesticides and toxic substances. The Division provides liaison with EPA's National Enforcement Investigations Center; provides inspector training programs for the regions and states; promulgates internal guidelines for preparation of enforcement cases; provides case development support for headquarters and regional enforcement cases; evaluates headquarters, regional, and state case development efforts; and monitors compliance with data call-in deadlines.

The Policy and Grants Division develops overall compliance monitoring policies and strategies, proposed legislation, and guidance to the states and the regions. It also sets policy and procedures for the FIFRA and TSCA state cooperative enforcement agreement programs; conceives of new programs; assists the regions in implementing and managing state cooperative enforcement agreement programs; manages the national pesticide applicator certification and training program; and evaluates state certification plans for compliance with FIFRA.

The Laboratory Data Integrity Staff develops good laboratory practice compliance programs under both FIFRA and TSCA. The Staff provides compliance and data audit guidance, policies and procedures for the operation of FIFRA and TSCA laboratory compliance monitoring programs, and assesses the quality and effectiveness of these programs.

OFFICE OF GENERAL COUNSEL

EPA's Office of General Counsel (OGC) serves as in-house legal counsel to the Agency's various program offices, including OPP. Headed by a General Counsel and Deputy General Counsel, OGC is divided into divisions which parallel EPA's various program offices. Thus, OGC's Pesticides and Toxic Substances Division serves as legal counsel to OPTS. Its activities are coordinated by an Associate General Counsel for Pesticides and Toxic Substances. This Division of OGC is composed of approximately twenty attorneys. They advise the pesticide program staff on various matters of statutory and regulatory interpretation, and represent the Assistant Administrator for Pesticides and Toxic Substances in cancellation and suspension proceedings under § 6 of FIFRA. They also assist the Environment and Natural Resources Division of the U.S. Department of Justice in judicial proceedings relating to FIFRA and the pesticide program. Most routine FIFRA enforcement proceedings, however, are brought by EPA's Regional Administrators, who are represented in such proceedings by the Agency's Regional Counsel.

OFFICE OF ENFORCEMENT

The Office of Enforcement (OE) was formerly known as the Office of Enforcement and Compliance Monitoring (OECM), which caused some confusion between OECM and the Office of Compliance Monitoring, which is a program office under OPTS. OE is separate from OPTS and is headed by its own Assistant Administrator.

The Pesticides and Toxic Substances Division of OE provides attorneys who represent OCM when it brings an enforcement action under FIFRA. OE also provides guidance to EPA regional offices on questions relating to enforcement of FIFRA and other statutes.

EPA REGIONAL OFFICES

EPA's ten regional offices, each of which is headed by a Regional Administrator, play a substantial role in the administration of EPA's pesticide program. Each regional office has a separate branch that handles matters relating to pesticides and toxic substances. The following table lists the locations and phone numbers of EPA's regional offices:

REGION	CITY	DIRECT DIAL #
Region 1	Boston, MA	(617) 565-3715
Region 2	New York, NY	(212) 264-2657
Region 3	Philadelphia, PA	(215) 597-9800
Region 4	Atlanta, GA	(404) 347-4727
Region 5	Chicago, IL	(312) 353-2000
Region 6	Dallas, TX	(214) 655-6444
Region 7	Kansas City, KS	(913) 551-7000
Region 8	Denver, CO	(303) 293-1603
Region 9	San Francisco, CA	(415) 556-6322
Region 10	Seattle, WA	(206) 442-1200

The regional offices administer the EPA grants to states, which provide money for FIFRA enforcement and pesticide applicator certification and training programs. They supervise the writing of the grants, ensure that all activities described in the grant are completed, and prepare mid-year and year-end evaluations of activities under the grant. They also supervise facility inspection programs, which usually are carried out by a corresponding state agency under an EPA grant, and, in cooperation with these state agencies, investigate complaints of FIFRA violations. The regional offices also bring the vast majority of EPA enforcement actions under FIFRA. They also inform state agencies about new EPA headquarters programs or initiatives.

FIGURE 2-1

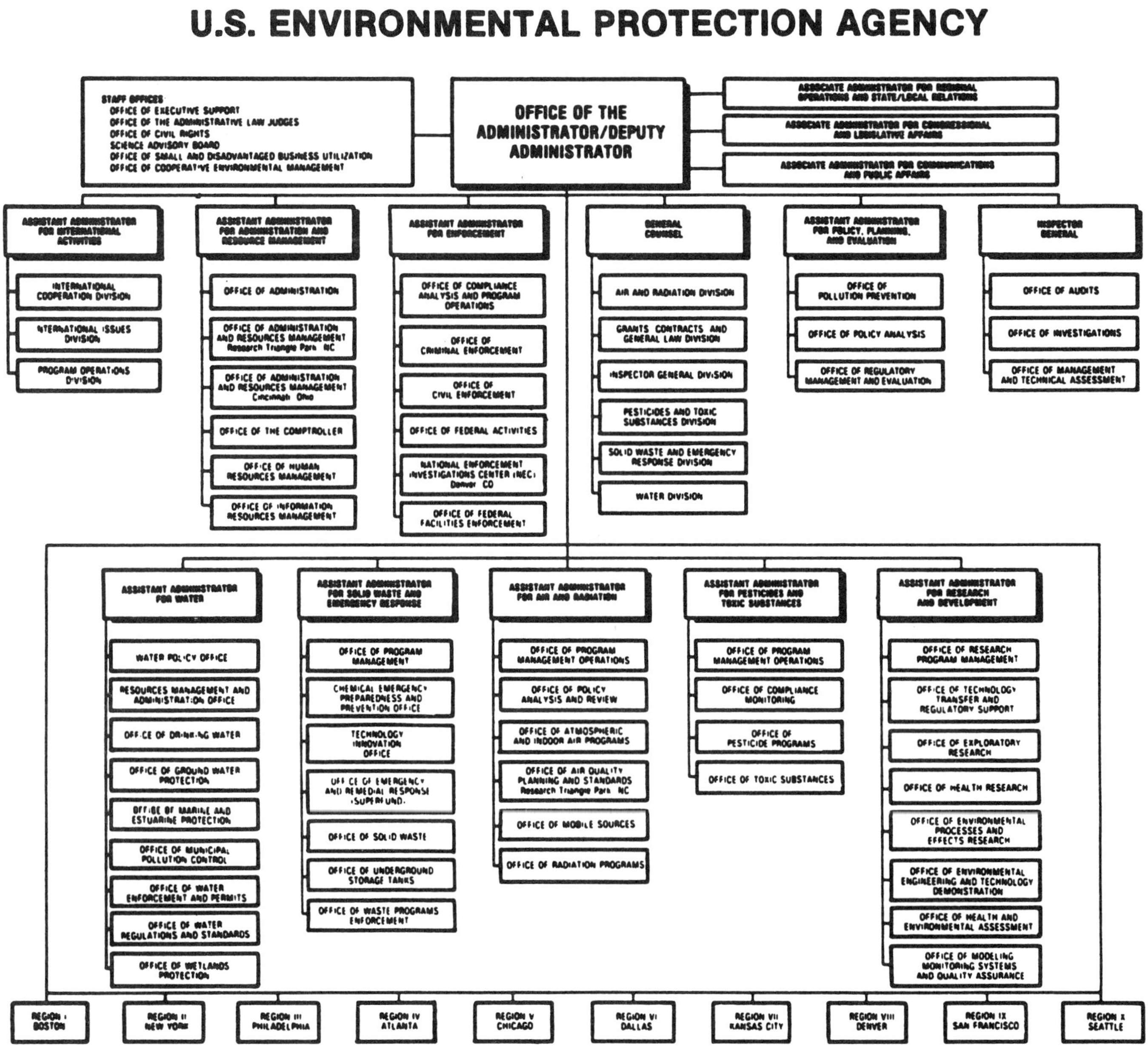
U.S. ENVIRONMENTAL PROTECTION AGENCY
OFFICE OF THE ADMINISTRATOR/DEPUTY ADMINISTRATOR

FIGURE 2-2

OFFICE OF PESTICIDES AND TOXIC SUBSTANCES ORGANIZATIONAL CHART

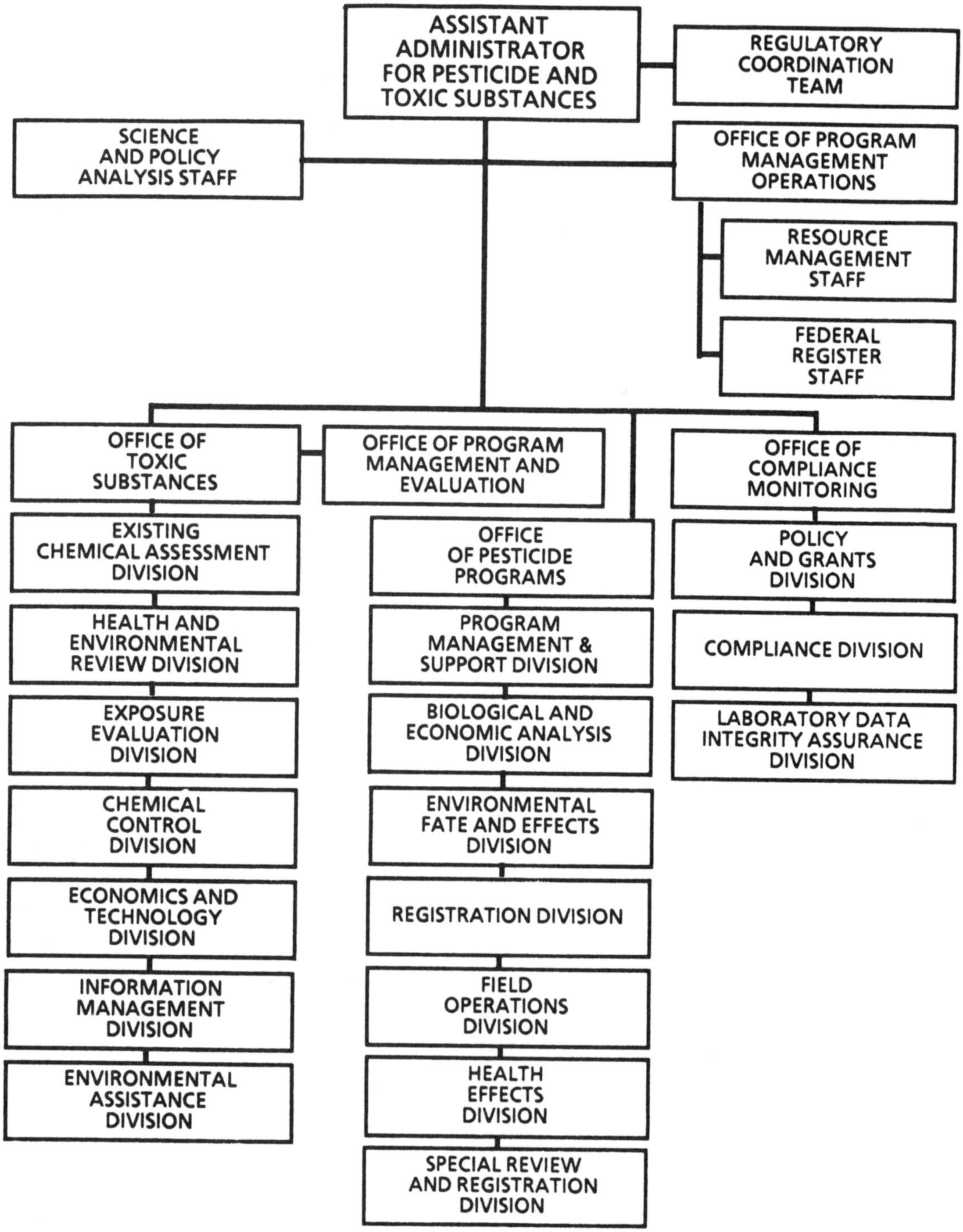

FIGURE 2-3

OFFICE OF PESTICIDE PROGRAMS
ORGANIZATIONAL CHART

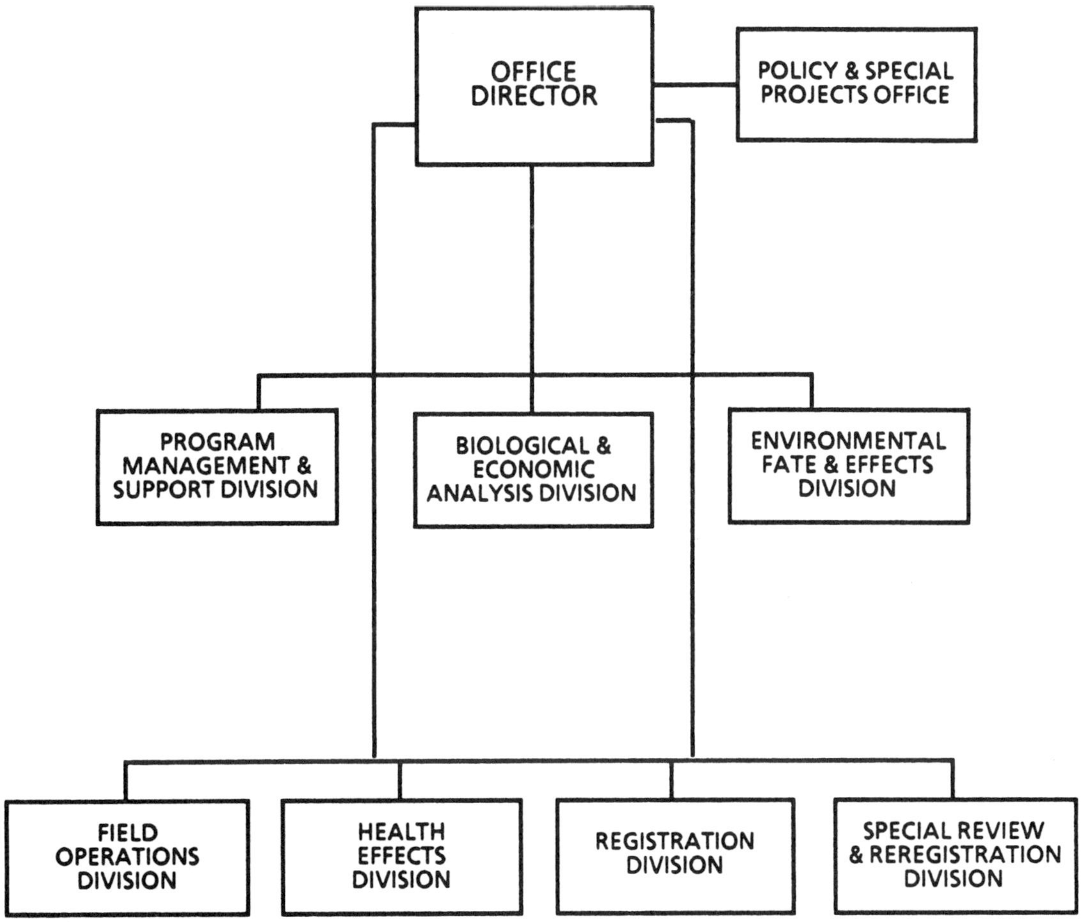

FIGURE 2-4

REGISTRATION DIVISION
ORGANIZATIONAL CHART

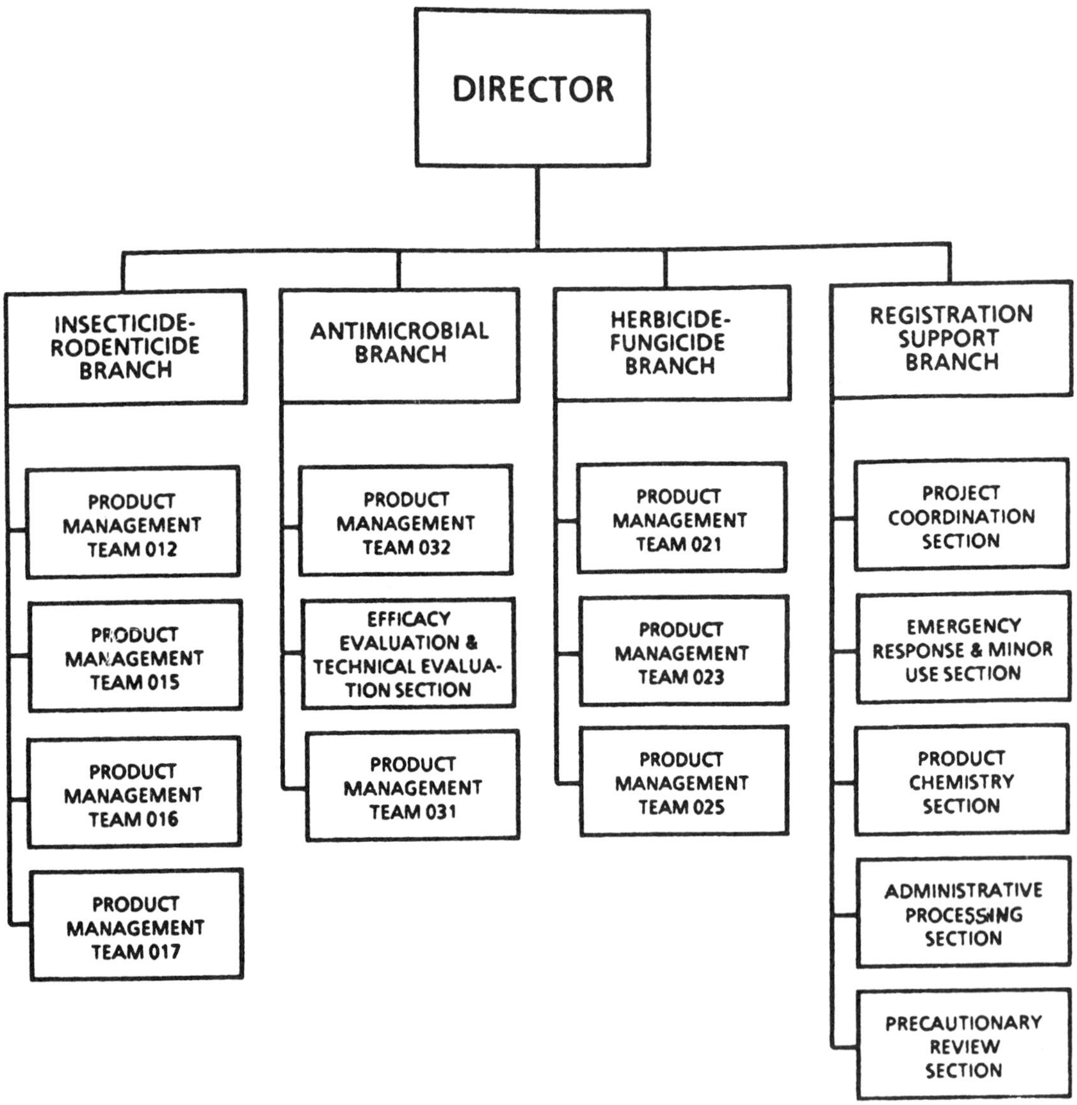

FIGURE 2-5

SPECIAL REVIEW & REREGISTRATION DIVISION
ORGANIZATIONAL CHART

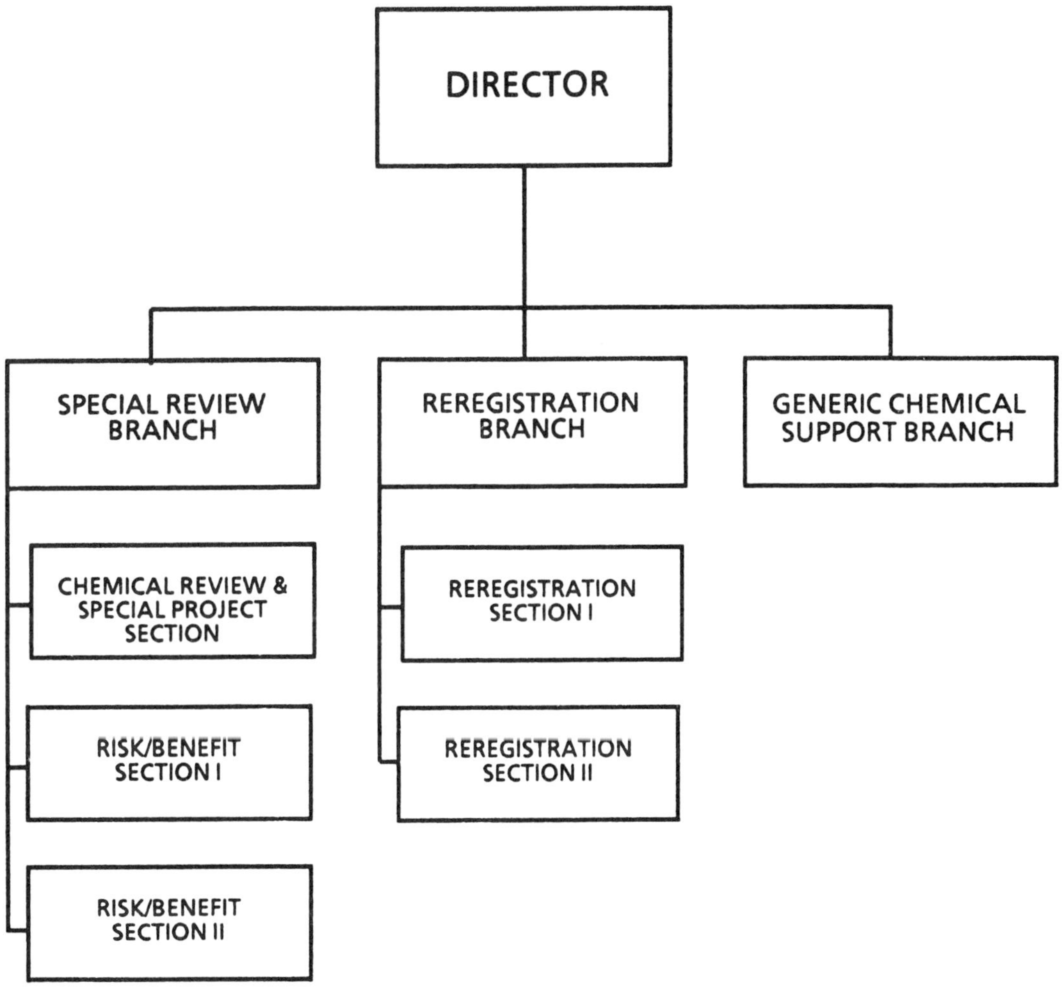

FIGURE 2-6

HEALTH EFFECTS DIVISION
ORGANIZATIONAL CHART

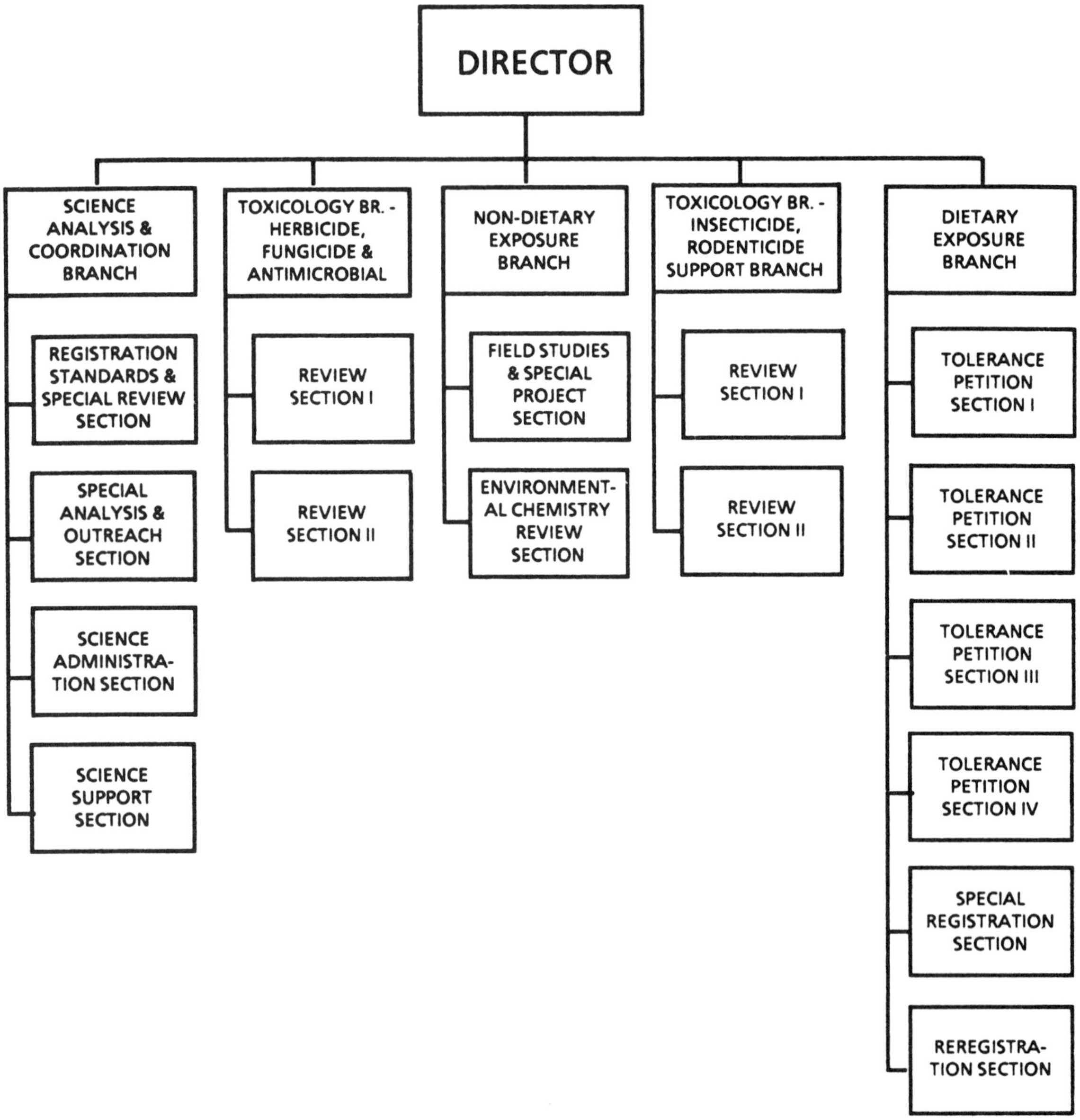

FIGURE 2-7

ENVIRONMENTAL FATE & EFFECTS DIVISION
ORGANIZATIONAL CHART

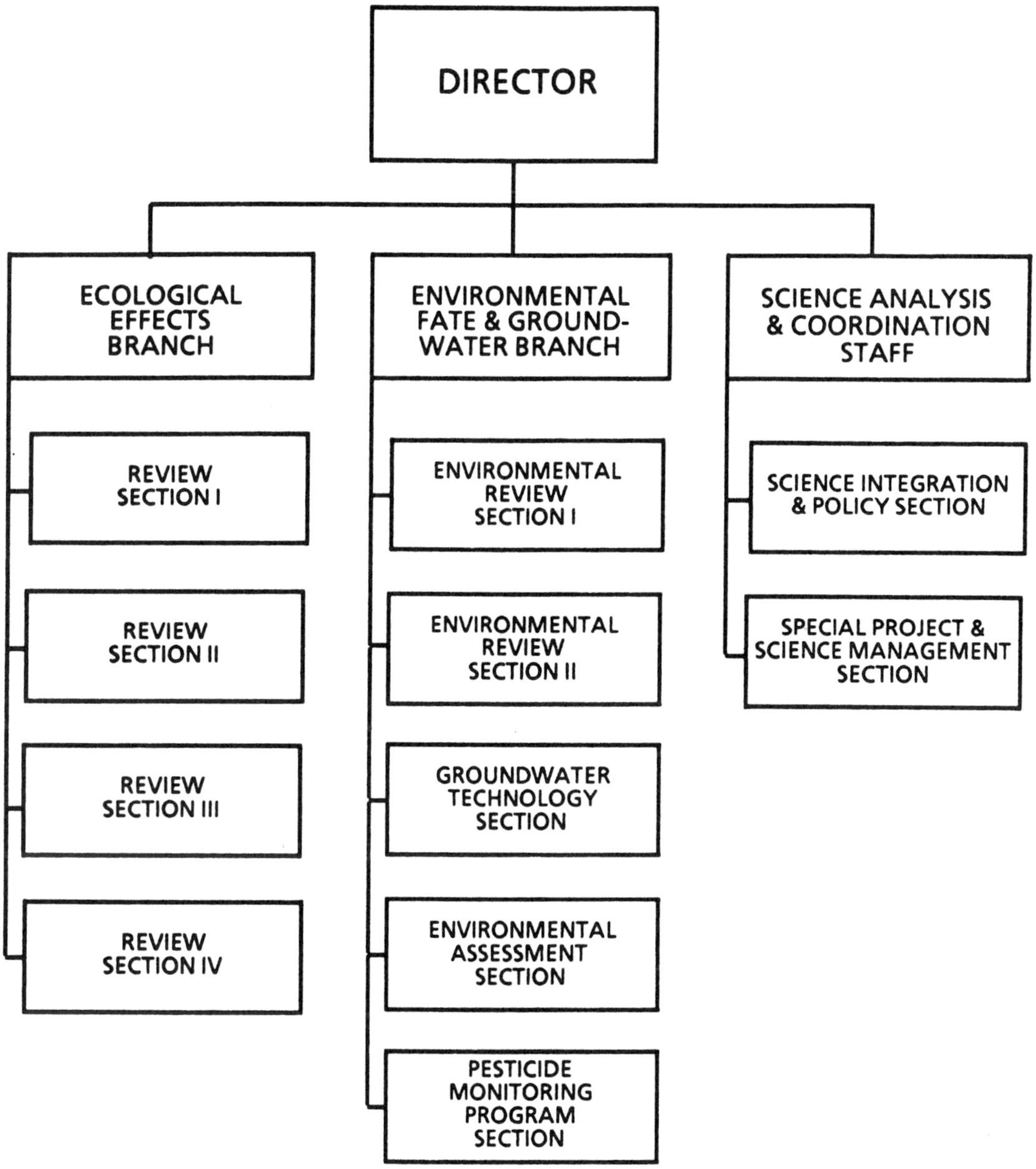

Chapter 3
PESTICIDE REGISTRATION

Registration is the heart of the federal program for regulating pesticides. Section 3 of FIFRA makes it unlawful for any person to sell or distribute any pesticide in any state without a registration. Thus, a registration is equivalent to a federal license to distribute or sell a pesticide in commerce. FIFRA § 3(c)(5) establishes the criteria under which EPA may approve an application for registration.

To implement the statutory criteria for registration, EPA has promulgated extensive regulations concerning pesticide registration procedures, data requirements, classification, labeling, and packaging, which are published at title 40 of the *Code of Federal Regulations*.[3] These provisions are discussed below.

PRODUCTS SUBJECT TO PESTICIDE REGISTRATION REQUIREMENTS

FIFRA regulates the manufacture, sale, delivery, and use of products defined under the Act as "pesticides." Under FIFRA § 2(u), a "pesticide" is any substance or mixture of substances intended: (a) for preventing, destroying, repelling, or mitigating any pest, or (b) for use as a plant regulator, defoliant, or dessicant. Specifically excluded from this definition are animal feeds containing pesticides and "new animal drugs," both of which fall under the jurisdiction of the Federal Food, Drug, and Cosmetic Act (FFDCA). (The requirements of the FFDCA are discussed in full in Chapter 5.)

"Pest" is defined in FIFRA § 2(t) as "(1) any insect, rodent, nematode, fungus, weed, or (2) any other form of terrestrial or aquatic plant or animal life or virus, bacteria, or other micro-organism (except viruses, bacteria, or other micro-organisms on or in living man or other living animals) which the Administrator declares to be a pest." EPA regulations

---

[3] Unless indicated otherwise, all references in this chapter to the *Code of Federal Regulations* are to title 40, and are cited, for example, in this abbreviated format: § 158.4

further define this term. Under § 152.5, the Agency will consider the following organisms to be "pests" when they exist under circumstances that make them "deleterious to man or the environment": (a) vertebrate animals other than man; (b) invertebrate animals (other than internal parasites of living man or other living animals), including but not limited to insects and other arthropods, nematodes, and mollusks such as slugs and snails; (c) plants growing where not wanted, including mosses, algae, liverworts, and all plants of higher orders, and plant parts such as roots; and (d) fungi, bacteria, viruses, and other micro-organisms, other than those on or in living man or other living animals and those on or in processed food or processed animal feed, beverages, and drugs (as defined in FFDCA § 201(g)(1)).

Determining Whether a Product Is a Pesticide

Under FIFRA § 2(u), the intended use of a material determines whether it will be considered a pesticide. EPA regulations further explain that a pesticide is "any substance (or mixture of substances) intended for a pesticidal purpose, i.e., use for the purpose of preventing, destroying, repelling, or mitigating any pest or use as a plant regulator, defoliant, or dessicant." § 152.15. To determine whether a substance is "intended for a pesticidal purpose," the Agency has established a three-part test. EPA will consider whether:

> (a) [t]he person who distributes or sells the substance claims, states, or implies (by labeling or otherwise):
>
> > (1) [t]hat the substance (either by itself or in combination with any other substance) can or should be used as a pesticide; or
> >
> > (2) [t]hat the substance consists of or contains an active ingredient and that it can be used to manufacture a pesticide; or

(b) [t]he substance consists of or contains one or more active ingredients and has no significant commercially valuable use as distributed or sold other than (1) use for a pesticidal purpose (by itself or in combination with any other substance), or (2) use for manufacture of a pesticide; or

(c) [t]he person who distributes or sells the substance has actual or constructive knowledge that the substances will be used, or is intended to be used, for a pesticidal purpose.

§ 152.15. No single part of this test, or any combination of them, is determinative. Rather, EPA will consider all three factors as a whole.

Products That Are Not Pesticides

EPA regulations also enumerate certain products that the Agency considers are *not* pesticides because they are not intended for use against pests. These include products labeled and intended for use *only* to control fungi, bacteria, viruses, or other microorganisms in man or animals, and products labeled and intended for use on invertebrate parasites, or nematodes in living man or animals. § 152.8.

Similarly, there are some products intended only to aid in the growth of desirable plants which EPA does not consider to be pesticides, and thus are not subject to regulation under the Act. Such products include fertilizers that do not otherwise contain pesticides; plant nutrient products; plant inoculant products consisting of microorganisms applied to the plant or soil for the purpose of enhancing the availability or uptake of plant nutrients; and soil amendment products added to the soil for the purpose of improving soil characteristics. § 152.8.

Lastly, EPA has listed certain products which it excludes from the definition of pesticides because they are not deemed to be used for a pesticidal effect. Provided that their labeling does not bear any pesticidal claims, the Agency excludes deodorizers, bleach, and cleaning agents; products that are intended for use only to attract pests for survey or detection purposes and do not contain toxicants; and products intended to repel pests only by providing a physical barrier against pest access (such as pruning paints used on trees) and which contain no toxicants. § 152.10.

Exemptions from Registration Requirements

Although FIFRA § 3 requires the registration of virtually every pesticide that is distributed or sold in commerce, another section of the Act, FIFRA § 25(b)(1), allows EPA to establish exemptions for pesticides that are (1) adequately regulated by another federal agency, or (2) of a character not requiring FIFRA regulation. In addition, FIFRA § 3 itself permits the transfer of certain pesticide products without the benefit of registration, even though a registration for the product is otherwise required, where the transfer (1) is from one registered establishment to another, or (2) pursuant to the terms of an experimental use permit. The EPA criteria for implementing these exemptions are set forth in detail in regulations appearing at 40 C.F.R. part 152, subpart B.

Exemptions for Pesticides Regulated by Another Federal Agency. EPA has specifically exempted from the registration requirements all biological control agents, except eucaryotic microorganisms (including protozoa, algae and fungi), procaryotic microorganisms (including bacteria), and viruses. § 152.20(a). The Agency also has exempted certain human drugs subject to regulation under the FFDCA. § 152.20(b).

Exemptions for Pesticides of a Character Not Requiring FIFRA Regulation. EPA has exempted from the registration requirements articles or substances that contain or are treated with a pesticide to protect the article or substance itself. Examples of such articles or substances include wood products treated with a pesticide to protect the wood against insects, and paint products treated with a pesticide to protect the paint coating from fungus or mold. § 152.25(a). The Agency also has exempted pheromones and pheromone traps; preservatives used for biological specimens; vitamin hormone products; and products consisting of foods which are used to attract pests and which contain no active ingredients. §§ 152.25(b)-(e).

Exemptions for Pesticides Transferred Between Registered Establishments. FIFRA § 3(b)(1) permits the transfer of an unregistered pesticide from one registered establishment to another, provided that both establishments are operated by the same producer, and that the purpose of the transfer is solely for packaging at the second establishment or for use as a constituent part of another pesticide produced at the second establishment.

In the past, under the Agency's "contract manufacturing" policy, EPA construed the language "operated by the same producer" broadly to allow the transfer of unregistered products between various combinations of companies that contracted for the manufacture, formulation, or packaging of a pesticide. EPA FIFRA Compliance Program Policy Compendium, No. 3.6 (May 10, 1982). EPA changed this policy substantially in 1987 through the issuance of Pesticide Registration (PR) Notice 87-7 (June 3, 1987). Under this "new" contract manufacturing policy, which was formalized in regulations promulgated in May 1988, EPA continues to permit the transfer of unregistered pesticides between registered establishments operated by the same producer. § 152.30(a). However, the Agency now construes the phrase "operated by the same producer" to mean that both establishments must be owned or leased by a single person or company, and to exclude establishments operated by different persons or companies under a contractual relationship. § 152.3(q).

An unregistered pesticide may be transferred between registered establishments *not* operated by the same producer only if (a) the transfer is solely for the purpose of further formulation, packaging or labeling into a product that is registered; (b) each active ingredient in the product at the time of the transfer is present in the product as a result of incorporation of either a registered product or a pesticide produced by the registrant of the final product; and (c) the product is properly labeled under the Act. § 152.30(b).

Thus, under the present regulations, a producer still may contract out for processing, packaging, or labeling. However, the producer may transfer its unregistered product to the contractors only if it is the registrant of the product ultimately distributed, and only if it owns the product during processing, packaging, or labeling operations. A producer no longer is permitted to manufacture a pesticide product under contract for another owner/registrant. Such a product must be registered by the producer prior to transfer to the owner/registrant, if the transfer represents a change in ownership of the product.

Exemptions for Pesticides Sold or Distributed Under Experimental Use Permits. Under FIFRA § 5, an unregistered product may be distributed or sold in accordance with terms established by EPA in an experimental use permit. Any product so transferred, however, must be labeled in accordance with all requirements for an experimental product. § 152.30(c). (The requirements for obtaining an experimental use permit are discussed below.)

In addition, three other provisions of FIFRA permit exemptions from the registration requirements. FIFRA § 17 allows for the transfer of unregistered pesticides for export. FIFRA § 18 permits EPA to exempt federal and state agencies from any provision of the Act, under emergency conditions. FIFRA § 19 permits the transfer of unregistered pesticides for purposes of disposal. EPA regulations implementing this provision appear at §§ 152.30(d)-(f). Finally, EPA has discretion to permit the distribution or sale of existing stocks of a pesticide whose registration is cancelled or suspended. § 152.30(g).

It also should be noted that FIFRA § 24(c) permits state governments to issue state registrations for additional uses of federally registered pesticides to meet "special local needs." Procedures and requirements for obtaining special local needs registrations are discussed in Chapter 13.

REGULATION OF DEVICES

Devices used to control or destroy pests are not subject to the registration requirements of FIFRA § 3. Nevertheless, devices are subject to regulatory requirements under numerous other sections of the Act. These include labeling and misbranding requirements under FIFRA §§ 2(p) and 2(q); registration and reporting requirements under FIFRA § 7; recordkeeping requirements under FIFRA § 8; inspection requirements under

FIFRA § 9; import and export restrictions under FIFRA § 17; and child resistant packaging requirements imposed pursuant to FIFRA § 25(c)(3).

A "device" is defined as "any instrument or contrivance (other than a firearm) which is intended for trapping, destroying, repelling, or mitigating any pest . . . but not including equipment used for the application of pesticides when sold separately therefrom." FIFRA § 2(h). If an article uses any physical or mechanical means to trap, destroy, repel or mitigate a pest, EPA will consider the article to be a device. If, on the other hand, the article incorporates any substance or mixture of substances for that purpose, then EPA will consider it to be a pesticide. 41 Fed. Reg. 51,065 (1976). Thus, for example, FIFRA does not regulate as devices pesticide application equipment such as crop dusters or hand sprayers, unless such equipment is sold in conjunction with a pesticide.

EPA excludes from regulation as devices such products as rat and mouse traps, fly swatters, tillage equipment for weed control, and fish traps. EPA considers that such devices do not require regulation under FIFRA because they depend for their effectiveness more upon the performance of the user than on the device itself. 40 Fed. Reg. 28,266 (1975). The Administrator, however, has expressly determined that EPA will regulate those devices for which pesticidal claims are made. These include, but are not limited to, devices with accompanying claims that they: (1) kill, inactivate, entrap, or suppress the growth of fungi, bacteria, or viruses in various sites (ultraviolet light systems, ultrasonic devices, ozone generators, and water and air filters unless they contain pesticides, in which case they are regulated as pesticides); (2) repel birds (high frequency sound generators, carbide cannons, foils, and rotating devices); (3) kill or entrap certain insects (black light traps, fly traps, electronic and heat screens, fly ribbons, and flypaper); or (4) repel certain mammals (mole thumpers, sound repellents, foils, and rotating devices). 40 Fed. Reg. 28,266 (1975).

Violations of any of the FIFRA regulatory provisions discussed above may result in enforcement action pursuant to FIFRA §§ 12, 13 or 14. For example, in *VRP Corp.*, IF&R Docket No. IX-267C, 1981 FIFRA Lexis 1 (EPA Dec. 28, 1981) (initial decision), the administrative law judge (ALJ) deduced from sales literature accompanying VRP's device that the device was "designed to eliminate and control infestations of rats and mice wherever they occur in farm buildings, residences, and field acreage." 1981 FIFRA Lexis at 4. During the administrative hearing of the case, the Agency staff produced evidence that the device produced an electromagnetic field which, beyond three meters, had strength "much less than the earth's magnetic field," and that a similar device had failed to interfere with eating, drinking or breeding habits of rats within a ten-foot area. Claims of efficacy for an electronic rodent repellent were found to be "false and misleading" under FIFRA § 12(a)(1)(F), rendering the product misbranded under FIFRA § 12.

EXPERIMENTAL USE PERMITS

To develop a pesticide product and prepare it for registration under FIFRA § 3, extensive scientific testing of the product's chemical properties, safety and efficacy usually must be performed. For agricultural products, this may include testing under actual field conditions using commercial application equipment.

To allow for the use of unregistered pesticides for testing purposes, EPA is empowered under FIFRA § 5 to issue experimental use permits (EUPs) establishing limited conditions for the transportation application and disposal of an unregistered test product. Products subject to an EUP may be sold or distributed only between participants in the test program and only for use under conditions provided by the EUP. § 172.2(b).

Exemptions from the EUP Requirement

An applicant must first ascertain whether an EUP is required for the particular testing desired. For example, if the purpose of testing a product is solely to determine a product's pesticidal value, toxicity, or other properties, and the producer or applicator does not expect to receive a benefit in pest control from its uses, EPA does not consider the product a pesticide within the meaning of FIFRA and the product is not subject to regulation under the Act.[4] § 172.3(a). Thus, no EUP is required. EPA will presume that testing is for this purpose and that no EUP is required under the following conditions:

- For land uses, if the test of a substance against a particular pest is performed on no more than ten acres of land, and food or feed crops resulting from the tests are destroyed or fed to research animals (unless a tolerance or exemption from tolerance exists for the test product). § 172.3(a)(1).
- For aquatic uses, if the test of a substance against a particular pest is performed on no more than one surface acre of water; water used in the test is not subsequently used for irrigation, drinking, or swimming; and fish and other aquatic organisms in the water are not used for food or feed (unless a tolerance or exemption from tolerance exists for the test product). § 172.3(a)(2).
- For tests conducted only on experimental animals, if the animals are not used for food or feed (unless a tolerance or exemption from tolerance exists for the test product). § 172.3(a)(3).

These exemptions are not exclusive. The Agency may exempt testing under other conditions at its discretion. § 172.3(b).

Application for an EUP

An application for an EUP may be submitted by any company or person wishing to accumulate information necessary to register a product under FIFRA. § 172.2(a). The applicant may be a potential registrant, an independent researcher or testing laboratory, or any similar agent or consultant of a manufacturer. The application must be submitted to

---

4 A chemical which is not considered a pesticide, and thus not subject to regulation under FIFRA, generally will be subject to different regulatory requirements under the Toxic Substances Control Act (TSCA). This Act and its requirements are discussed in further detail in Chapter 17, and in McKenna & Cuneo's TSCA Handbook (2d ed. 1990). Producers conducting experiments with genetically engineered pesticides also should note that EPA's regulatory requirements for these products are more stringent. These requirements are discussed further in Chapter 12.

the Registration Division, Office of Pesticide Programs, at EPA's Washington, D.C. headquarters office. Information required to be submitted with the application is described in regulations at § 172.4, and is also described below. The information and supporting data are submitted on EPA's standardized Form 8570-17, which is organized as follows:

Section A must include a data sheet giving the chemical and physical properties of the test chemical and a complete statement of the names and percentages by weight of each active and inert ingredient in the formulation to be shipped.

Section B must include a copy of the proposed experimental label. The minimum labeling requirements are set forth at § 172.6.

Section C must contain toxicity data, including LD5 values, eye and skin irritation data on the formulated product, and subacute, teratology (one species), mutagenicity, and possibly chronic and reproduction data on the active ingredient. Data on the product's toxicity to fish and wildlife may be included in Section C when appropriate.

Section D must contain residue data, including, when appropriate, data on: (1) food or feed commodities; (2) non-food crops such as tobacco; or (3) on foliage or other sites to which the product may be exposed and in which residues may pose a risk to man or the environment. Section D also must include a description of analytical methods used, a summary of the residue data acquired, and, when appropriate, environmental fate data.

Section E must contain product performance information demonstrating that the product is useful for the purposes proposed. Because the submission of efficacy data has been waived for all products except certain disinfectant and public health uses, Section E need not contain actual efficacy data but should contain a summary of the results of efficacy testing performed to date on the product.

Section F must contain a statement explaining whether a tolerance exists or is being requested, if the product is to be tested in a manner that may result in residues in food or feed. If a tolerance is being requested, the temporary tolerance petition must be provided with the EUP application. If all food or feed derived from the experimental program is to be destroyed or fed to experimental animals, a statement must be included to that effect.

Section G must include details concerning the proposed experimental program, including:

1. qualifications, names, addresses and telephone numbers of participants;

2. names of states in which the product will be used, along with the amount of active ingredient and acreage (or other appropriate measures) to be used in each state, and the names of states in which the pesticide may be shipped for further distribution;

3. details of the proposed program, including types of pests or organisms targeted; the crops, animals, surfaces, or sites to be treated; the geographical areas where the material is to be used; the use patterns, intended plot sizes, number of plots, number of replicates, and other test parameters to be used;

4. objectives of the EUP program, including a statement specifying the type of data to be collected and the intended gain from conducting the program;

5. justification for the quantity of active ingredient proposed to be used under the permit, including a statement specifying the various parameters used to determine the quantity of active ingredient;

6. a statement proposing a suitable duration for the EUP commensurate with the program objectives; and

7. details concerning the method of disposing of unused materials at the conclusion of the testing program.

EUP applications should be submitted to EPA as far in advance as possible of the first shipping date (preferably at least six months). Although FIFRA § 5(a) requires EPA to complete EUP reviews within 120 days, EPA rarely has been able to satisfy this statutory deadline.

Permit Conditions

EUPs usually remain valid for one year and are renewable upon request, if circumstances warrant renewal. § 172.5. A renewal application must be submitted before the EUP expires. § 172.9. EPA reserves the right to limit the quantity of pesticide to be used in the testing program, and may place such limitations on the permit as the Agency believes are necessary to protect the public health and environment. The Agency also may request that additional studies be performed during the permit period if the test product contains a new chemical or mixture of chemicals.

Program Surveillance and Reporting

After issuance of an EUP, the permittee becomes responsible for supervising and evaluating the testing performed and for immediately reporting to EPA any adverse effects that may be discovered. Although interim reports to the Agency are no longer required, a final report must be submitted to the Administrator within 180 days after the permit expires, unless an extension has been approved. The permittee must include in the report all data obtained during the program; a description of the disposition of any unused pesticides, pesticide containers, or affected food or feed; and details concerning shipment of the test pesticide. Although the permittee must maintain field notes, these need not be submitted to EPA unless requested. In addition, the permittee and participants in the program must hold open to inspection by authorized EPA representatives any premises involved in the testing. § 172.8.

Denial or Revocation of an EUP

EPA may deny an application for an EUP if the Agency determines that the EUP is not justified, would cause unreasonable adverse effects on the environment, or for any other reason provided for by law. EPA similarly may revoke an EUP if the Agency finds that its terms have been violated or are inadequate to avoid unreasonable adverse effects on the environment, or if new evidence demonstrates that a tolerance associated with the EUP is inadequate to protect public health. § 172.10.

If EPA denies an application for an EUP, or revokes an EUP after it is issued, the Agency must notify the applicant or permittee of the reason for the action. The applicant or permittee is entitled to an informal hearing to confer with the Agency regarding the matter, and to request reconsideration. § 172.10.

REGISTRATION CRITERIA

The 1972 amendments to FIFRA transformed the Act from a pesticide labeling statute into a comprehensive regulatory scheme whose overall purpose is to "protect man and the environment," and revamped the statutory criteria for registration. These criteria, set forth in FIFRA § 3(c)(5), focus upon whether a product causes "unreasonable adverse effects on the environment." [5] On the basis of these amendments, EPA increased substantially the amount and types of data required to be submitted in support of an application for registration.

Subsequent amendments to FIFRA in 1978 created a separate standard for conditional registration. Under FIFRA § 3(c)(7), pesticides that are supported by less than the full complement of data ordinarily required may be registered, provided that any data requirements deferred are met at the time that all similar products are reregistered. The criteria for an unconditional registration under FIFRA § 3(c)(5) remain unchanged.

The criteria for registration under FIFRA § 3(c)(5) and FIFRA § 3(c)(7) are interrelated: a conditional registration will not be granted unless it has been determined that a product does not cause unreasonable adverse effects on the environment.

---

[5] "Unreasonable adverse effects on the environment" is defined under FIFRA § 2(bb) as "any unreasonable risk to man or the environment, taking into account the economic, social, and environmental costs and benefits of the use of any pesticide."

Unconditional Registration

FIFRA § 3(c)(5) requires the Administrator to register a pesticide if he finds that: (1) its composition warrants its proposed claims; (2) its labeling complies with the Act; (3) it will perform its intended function without unreasonable adverse effects on the environment; and (4) it will not generally cause unreasonable adverse effects on the environment, when used in accordance with widespread and commonly recognized practice.

Under FIFRA § 3(c)(5)(A), a pesticide must perform in accordance with the claims made in the registration statement. EPA may not impose standards of effectiveness which differ from those claims. *S.L. Cowley & Sons Mfg. Co., Inc. v. EPA*, 615 F.2d 1312 (10th Cir. 1980). EPA's review of the effectiveness of pesticide products for registration purposes is limited. The Agency has determined that the effectiveness of most pesticides can be adequately regulated by the marketplace, and thus has waived under FIFRA § 3(c)(5) the automatic submission of efficacy data as a prerequisite to registration, except in the case of products intended for human health where pest control cannot be observed by the user, and for new uses of products previously identified as preventing unreasonable risks of adverse effects on the environment. Applicants are required, however, to generate efficiency data and have them readily available for submission, should EPA request such data. § 158.640(b)(1). (EPA's requirements for product performance data are discussed in further detail in Chapter 4.)

Under FIFRA § 3(c)(5)(B), a pesticide must comply with FIFRA's labeling requirements. FIFRA requires that the labeling of all registered products bear certain precautionary statements and directions for use. Labeling that fails to include this information or that is false or misleading is considered misbranded under FIFRA § 2(q) and will preclude registration.

Under FIFRA §§ 3(c)(5)(C) and (D), EPA must grant an application for registration if; (1) when used as intended (i.e., in accordance with the labeling and directions for use), and (2) when used in accordance with widespread and commonly recognized practice, the pesticide does not cause "unreasonable adverse effects on the environment." Under these criteria, EPA has the authority to deny registration if a pesticide is generally misused with the result that it causes unreasonable adverse effects, even if it does not cause unreasonable adverse effects when used as directed. FIFRA § 2(bb) defines "unreasonable adverse effects on the environment" to mean "any unreasonable risk to man or the environment, taking into account the economic, social, and environmental costs and benefits of the use of any pesticide." Under this definition, EPA must balance the risks and benefits of a pesticide when considering whether to approve its registration. EPA's balancing of risks and benefits is discussed in further detail in Chapter 9.

Section 3(c)(5) also prohibits EPA from considering the "essentiality" of a pesticide in determining whether it should be registered, or from registering one pesticide in preference to another when both satisfy the registration criteria. However, the availability of an alternative product that poses less risk may reduce the benefits of a pesticide and alter the risk/benefit calculation.

Conditional Registration

Section 3(c)(7) authorizes EPA to issue conditional registrations for "generic" (often referred to as "me-too" or "follow-on") products, new uses of currently registered products, and new active ingredients. The provision was enacted to eliminate inequities that arose as a result of the increased data requirements under FIFRA § 3(c)(5) between applicants for new registrations and those who already held registrations for the same products. After the § 3(c)(5) criteria were enacted, more data were required to support a new registration than existed in the data base for registrations already granted. A "double standard" thus arose under which registrants of products already registered could continue marketing without satisfying the new data requirements, but applicants for new registrations of the same product could not obtain EPA approval until they met all of the new data requirements. Moreover, virtually all applications for registration of generic products, for new uses of existing products, and for new active ingredients were denied because the new data requirements had not been met.

FIFRA § 3(c)(7) gives EPA discretion to issue conditional regulations in the following three circumstances:

Generic Products. Under FIFRA § 3(c)(7)(A), EPA may conditionally register or conditionally amend the registration of a product if:

(1) the product and its proposed uses are identical or substantially similar to a currently registered pesticide, or any differences between the product and currently registered pesticide do not significantly increase the risk of unreasonable adverse effects on the environment; and

(2) approving the registration or amendment would not significantly increase the risks of any unreasonable adverse effect on man or the environment.

New Uses of Currently Registered Products. Under FIFRA § 3(c)(7)(B), EPA may conditionally amend the registration of a product to permit new uses if:

(1) the applicant submits satisfactory data regarding the proposed new use; and

(2) approving the amendment would not significantly increase the risk of any unreasonable adverse effect on the environment.

New Active Ingredients. Under FIFRA § 3(c)(7)(C), EPA may grant a conditional registration for a new active ingredient only if:

(1) insufficient time has elapsed since EPA first imposed the new data requirements for the applicant to have developed the new data;

(2) EPA determines that the use of the pesticide during the period necessary to generate the required data will not cause any unreasonable adverse effect on man or the environment; and

(3) such use of the pesticide is in the public interest.

Under EPA regulations, the Agency will choose at its own discretion whether to review an application under the unconditional criteria of FIFRA § 3(c)(5) or the conditional registration criteria of FIFRA § 3(c)(7). The Agency's choice depends primarily on the extent to which EPA already may have reviewed the data base for other products containing the same active ingredient. Instead of reviewing data on a product-by-product or application-by-application basis, EPA conducts "generic" data reviews for each pesticide active ingredient, according to the Agency's own priorities, which presently are established primarily by the reregistration program. Thus, rather than commence a complete review of the data base on a given chemical in response to an application for registration to reach a decision for unconditional registration under FIFRA § 3(c)(5), EPA may choose to review the application under the conditional criteria set forth in FIFRA § 3(c)(7). For this reason, virtually all registrations issued by EPA since FIFRA § 3(c)(7) was enacted have been conditional.

Incremental Risk Assessment

EPA may grant conditional registrations if approval would not significantly increase the risk of unreasonable adverse effects on the environment. This so-called "incremental risk assessment" is critical to the Agency's conditional registration decision, and differs depending upon whether the registration sought is for a generic product, a new use of a previously registered product, or a new active ingredient.

Generic Products: EPA operates under the premise that the conditional registration of generic products (i.e., products that are identical or substantially similar to previously registered products) poses no incremental risk beyond that of the previously registered products. To be considered identical or substantially similar to a previously registered product, a generic product must: (1) contain an active ingredient obtained from an EPA-registered establishment; (2) have a formula identical to a previously registered product; and (3) have uses identical to the previously registered product. (The procedures for the "identicality" determination are described more fully in Registration Division Standard

Operating Procedure (SOP) No. 3068.2.) The registrant of a generic product is capable of determining whether the product uses are identical and whether the product is purchased from a registered establishment, but only EPA can determine whether the formulas of the proposed and the previously registered products are identical, because formulas are confidential. If EPA determines that a generic product is identical to a previously registered product, no incremental risk assessment will be required.

A generic product may fail to satisfy the identicality requirement due to differences in the inert ingredients and impurities contained in the products, variations in their formulations, or variations in their use patterns. If the products are not identical, EPA must determine the extent of any incremental risk based upon a comparison of the products. In most cases the incremental risk assessment does not require additional data. EPA may, however, request the applicant to submit additional toxicity studies or other pertinent data. In performing an incremental risk assessment, the Agency generally will review only those data necessary to make the incremental risk determination and will not review the entire data base. FIFRA § 3(c)(3)(B) establishes an expedited "fast track" procedure for processing and reviewing applications for generic, or me-too, registrations. These procedures are discussed in Chapter 6.

New Uses of Previously Registered Products: EPA uses the following procedure to evaluate incremental risks posed by new uses of previously registered products:

Step 1: The Agency first determines whether new or revised residue tolerances are needed. If so, it will proceed to Step 2. If not, it will proceed to Step 4.

Step 2: If the new use involves a food or feed crop, the Agency checks whether a notice has been issued that the pesticide meets or exceeds risk criteria associated with human dietary exposure. If so, conditional registration may not be granted under FIFRA § 3(c)(7)(B).

Step 3: If such a notice has not been issued, EPA will assess the incremental increase in dietary exposure that would result from the new use. If significant data requirements have not been satisfied for the previously registered uses (i.e., significant data gaps exist), and the Theoretical Maximum Residue Contribution (TMRC) increases more than 1 percent with application of the new use, EPA will not grant the conditional registration.

Step 4: EPA will assess the risk of toxicity posed by the new use and determine whether any risk criteria defined in § 162.11 are exceeded. For this assessment, the applicant must submit data addressing new toxic effects caused by the new use. These include only data pertinent to the new use.

Step 5: EPA will evaluate the incremental toxic effects resulting from increased exposures due to the new use. This assessment is different from the dietary exposure analysis performed in evaluating new or revised residue tolerances (steps 2 and 3). It includes exposures to applicators, field workers, and other nontarget species. The applicant may be asked to supply additional data for evaluation of these exposures.

Step 6: After performing the incremental risk analysis, the Agency will accept or deny the conditional registration for the new use. EPA also may require label restrictions that would reduce exposure, such as the use of protective clothing, reentry intervals, and geographical or timing restrictions.

New Active Ingredients: Issuance of a conditional registration for a new active ingredient, while permitted, is extremely rare. Such a registration may be granted only under the following circumstances and conditions: (a) insufficient time has elapsed since EPA first imposed new requirements for the applicant to have developed the new data; (b) the applicant must submit the missing data within a period of time reasonably sufficient for generation of the data; (c) the pesticide must not meet or exceed risk criteria set forth in the regulations; (d) the use of the product prior to the submission of such data must not cause unreasonable adverse effects on the environment; and (e) the use of the product must be in the public interest. FIFRA § 3(c)(7)(C). See also 40 C.F.R. § 152.114.

Other criteria that must be satisfied to obtain a conditional registration are the same as those which must be satisfied to obtain an unconditional registration under FIFRA § 3(c)(5). The application must be complete as to all procedural and administrative matters, and must include appropriate offers to pay compensation for the use of any data relied upon. In the case of antimicrobial disinfectant products, efficacy data are required. The Agency will review the packaging and labeling for compliance with the requirements of 40 C.F.R. part 157 and § 162.10 and to ensure that the product is not misbranded. Where applicable, requirements for tolerances and food additives under FFDCA §§ 408 and 409 must be satisfied. Finally, if the pesticide is also a drug under FFDCA § 201(g), then notice to FDA and compliance with requirements imposed by that Agency are necessary. §§ 152.113(a)(3), .114.

REGISTRATION PROCEDURES

Any person or company which submits a properly supported application for registration may obtain a pesticide registration. § 152.40. The registrant need not be the inventor of the product, or even the manufacturer. Nor are there any restrictions regarding residence or nationality, although applicants for registration residing outside the United States must appoint a U.S. agent authorized to act for the applicant on all registration matters. § 152.50(b).

After a registration has been issued, it may be amended from time to time to reflect, for example, changes in composition, labeling, and packaging. The registration may be amended only by the registrant. § 152.50.

In general, any application for registration of a new product or any significant change in the composition, uses or labeling of a product already registered, may be granted or permitted only upon the complete review and express approval of the Agency. There is a small category of amendments to a registration which may be effected merely upon notification of the Agency, as well as a category of changes which are expressly permitted without approval or notification. These amendments and changes are discussed more fully below.

Application Requirements

The application procedures for obtaining a registration and amending registration are the same. An application must be submitted on EPA form 8570-1. For reasons associated with EPA's application tracking system, an authentic original of the EPA application form, bearing a red serial number in the upper right hand corner, must be submitted; applications submitted on a photocopy of the form may be rejected.

The application must indicate the identity of the applicant, a current United States address, the applicant's company number (if any), and the name of any agents authorized to act on the applicant's behalf. If an applicant designates an agent, the applicant must also submit to EPA a separate letter stating the Agent's name and United States address. The applicant (or registrant) remains free, of course, to change or terminate an agent's authority at any time, notwithstanding the designation to EPA. The Agency should promptly be notified in writing, however, of any such changes.

Each application must include a list of data submitted in support of the application with a brief description of the study results. In addition, the following information is required to identify the product for which the application is being submitted: the product name; any trade names; and (in the case of an application for an amendment) the current product registration number.

The applicant also must submit five legible copies of draft labeling. EPA encourages applicants not to print labeling in final form before the draft labeling has been reviewed and approved by the Agency. In this regard, EPA regulations specifically permit the submission of typewritten labeling, or label mockups. Similarly, in the case of applications to amend labeling, the regulations permit submissions limited to the portion of the labeling that is being amended.

An application for registration must include a certification that the product meets all Agency requirements for child resistant packaging. If the applicant objects to the Agency's classification of the product (i.e. for restricted use only), the applicant must submit a request for reclassification, together with supporting documentation. If the proposed labeling indicates that the product is intended for use as a food or feed crop, or for uses which may result in residues in food, the applicant must submit a statement declaring that such residues are within applicable FFDCA tolerances, or a petition requesting that tolerances be established. (A complete discussion of procedures for establishing tolerances pursuant to the FFDCA appears in Chapter 5.)

Supporting Data

An application must be accompanied by test data demonstrating that the product will perform its intended function without unreasonable adverse effects on the environment. FIFRA § 3(c)(5)(C). In implementing this criterion, EPA has prescribed in considerable detail the studies which must be conducted, and prescribed protocols for each study. (The preparation and submissions are addressed separately in Chapter 4.)

Citation of Data

As an alternative to generating and submitting data, FIFRA § 3(c)(1)(D) permits an applicant to provide a "citation to data that appear in the public literature or that previously had been submitted to EPA." Such use of data submitted by other registrants is permitted only subject to the use-compensation provisions of FIFRA § 3(c)(1)(D)(i), (ii) and (iii). (The subject of data compensation is treated separately in Chapter 7. Procedural requirements for the citation of data are discussed below.)

EPA regulations specify two distinct methods by which applicants for registration may cite data already in EPA files. The "cite-all" system allows applicants to cite (and offer to pay for) all relevant data in EPA's files which are available for data licensing and for which submitters are entitled to compensation under § 3(c)(1)(D). The "selective" method allows an applicant to identify the particular data requirements that apply to its product and selectively cite its own data or another company's data to satisfy each data requirement.[6]

The cite-all method described in § 152.86 requires that the applicant cite all relevant data previously submitted to EPA by other companies. An applicant can establish its right to cite all relevant data by either obtaining permission from, or making an offer to pay to, each company on EPA's list of pesticide data submitters for a particular active ingredient. If the data are governed by FIFRA § 3(c)(1)(D)(ii), the applicant may simply make an offer to pay. If the data are subject to exclusive-use protection under FIFRA § 3(c)(1)(D)(i), an applicant must obtain permission from the data submitter in order to cite such data in support of an application. Absent permission, the applicant must submit the tests itself. Performing independent testing may not be feasible if the pesticide is under another company's patent. [7]

Under the cite-all method, an applicant does not have to determine which data requirements actually apply to its product. By securing the right to cite all relevant data in EPA files, the applicant obviates the need for identifying specific data requirements, specific studies, or data submitters for specific studies. All data in Agency files that are pertinent to its consideration of the requested registration are relied on by EPA. § 152.86.

---

6 Until 1983, EPA regulations *required* applicants to cite *all* pertinent data in the Agency's files when applying for registration. EPA's mandatory cite-all policy was declared unlawful in *National Agricultural Chemicals Association v. EPA*, 554 F. Supp. 1209 (D.D.C. 1983). The court found that under § 3(c)(1)(D) an applicant has the right to submit and rely on its own data, and cannot be compelled to cite, or pay for, other persons' data if it has sufficient data of its own.

7 In *Roche Products, Inc. v. Bolar Pharmaceutical Co.*, 733 F.2d 858 (Fed. Cir.), *cert. denied*, 469 U.S. 856 (1984), the court held that testing of a patented drug by a generic drug manufacturer prior to expiration of the patent, for purposes of preparing to submit a generic drug application, constituted patent infringement. Although the decision was legislatively overruled in the context of generic drug testing in § 202 of the Drug Price Competition and Patent Term Restoration Act of 1984, Pub. L. No. 98-417, 98 Stat. 1585, the decision still may serve as authority in the context of pesticide testing under FIFRA.

EPA regulations specify that the applicant must submit a written offer to all of the persons listed in the EPA list of "Pesticide Data Submitters by Chemical" for the active ingredient in the applicant's product, and must submit evidence to the Administrator of delivery of the offers. § 152.86(b). The offer must promise to pay compensation "to the extent required by FIFRA." § 152.86(b)(ii).

The advantage of the cite-all method is that it is uncomplicated. There is a substantial disadvantage, however. This method requires applicants to offer to pay compensation for *all* data on file with EPA, regardless of whether those data are identified, or whether the applicant has data of its own on file.

The selective method enables an applicant to obtain a registration by relying on only a select group of studies that satisfy the minimum registration requirements. The use of the selective procedure could reduce an applicant's compensation obligations under § 3(c)(1)(D). The selective approach is considerably more flexible than the cite-all approach since the applicant can satisfy each individual data requirement by any one of several methods.

Section 152.90 requires the applicant who chooses the selective method to identify and list the data requirements that apply to its products, its ingredients and uses. Most applicants will use 40 C.F.R. part 158 (Data Requirements For Registration) to determine their data requirements. Where the Agency has promulgated a Registration Standard for the active ingredient used in the applicant's pesticide, the applicant may rely on the Standard as presumptively establishing the data requirements applicable to its application. § 152.90(a)(1).

Once the data requirements have been identified, the applicant is permitted to satisfy each data requirement by any one of the following methods: generating its own new data, citing its own previously submitted data, citing data previously submitted by others, relying on public literature, seeking a waiver from a data requirement, or demonstrating the existence of a "data gap." A brief discussion of these methods follows.

Submit a New Study. An applicant has the right to submit its own valid study to satisfy a data requirement (as long as the test on the pesticide does not violate another registrant's patent). § 152.92. EPA requests that each study be accompanied by the following additional information:

(1) A title page containing certain identifying information about the study.

(2) A statement concerning its trade secret status under FIFRA § 10, and any claims of confidentiality made under that section.

(3) A certification concerning compliance with the Good Laboratory Practice standards of Part 160.

(4) An English translation, if the study report is written in a foreign language.

49 Fed. Reg. 30,895 (1984).

Cite a Previously Submitted Study. Under § 152.93, any valid study already in EPA's possession can be cited to demonstrate compliance with a data requirement. The applicant may submit another copy of such a study, or may simply reference it appropriately.

If the study is owned by another registrant, the applicant must determine whether the study falls into the category of exclusive-use data, or data subject to compensation under FIFRA § 3(c)(1)(D), in order to determine the procedures for proper citation. The procedures for obtaining permission and making an offer to pay are exactly the same as those under the cite-all method, except that the applicant must write to a specific data submitter concerning a specific study rather than to all data submitters on the Pesticide Data Submitters List. 40 C.F.R. § 152.93(b)(1), (2).

If data are not subject to § 3(c)(1)(D), the applicant may cite them without obtaining authorization from, or offering to pay, the original data submitter. § 152.93(b)(3).

Rely on Public Literature. FIFRA § 3(c)(1)(D) allows an applicant to cite data that "appear in the public literature" to satisfy a data requirement. Under the procedures established by § 152.94, an applicant is permitted to cite an article in a journal. However, journal articles and other sources of public literature rarely offer the detailed information (such as raw data results) required by the Agency to satisfy data requirements. 49 Fed. Reg.

30,896 (1984). Furthermore, compensation may be required for use of a published study which was generated and paid for by another registrant.

Obtain a Data Waiver. Any applicant may claim the existence of a waiver or request a waiver of a data requirement. A waiver will excuse the applicant from having to fulfill a data requirement that normally would apply to its product. A waiver can be obtained if a study would not be useful to the Agency in evaluating the risks and benefits of a particular product and use pattern. 49 Fed. Reg. at 30,895. To utilize a previously existing waiver, the applicant must provide a reference to the Agency record that describes the waiver and must explain its applicability to the product. § 152.91(a). EPA will make existing lists of waivers available if a request is filed under the Freedom of Information Act, 5 U.S.C. § 552 (1988).

To request a new waiver of a data requirement, the applicant must submit the information specified in § 158.45. If a request for a new waiver or extension of an existing waiver is denied, the applicant must choose another method to satisfy the data requirement.

Document a Data Gap. Under FIFRA § 3(c)(7), EPA is authorized to conditionally register some pesticide products based on the submission of less than the complete set of data that would currently be required for full registration under § 3(c)(5). An applicant may defer its obligation to satisfy an applicable data requirement if it can demonstrate a "data gap," i.e., that none of the other registrants for "identical or substantially similar" pesticides has submitted a study that satisfies the data requirement.

If an applicant is registering a pesticide whose data requirements are established by a Registration Standard, data gaps may be documented in the Registration Standard. § 152.96(b). If no Registration Standard exists, the applicant must write to each data submitter for the pesticide on the data submitters list and ask whether a study was submitted that fulfills the data requirement. Data submitters have sixty days to respond to data gap inquiries. If any person responds within the sixty-day period that he has submitted such a study, the applicant may not claim that a data gap exists. Failure to respond limits the data submitter's challenge rights after registration.

The data gap procedures may not be used in three situations:

1. To conditionally register a *new* active ingredient.

2. To satisfy data requirements for which each applicant must develop and submit data on his *own* product (e.g., efficacy data for anti-microbial products).

3. Data needed to make an incremental risk assessment under FIFRA § 3(c)(7)(B), typically for a new use.

49 Fed. Reg. 30,897 (1984).

Agency regulations establish procedures for resolving disputes between registrants concerning adherence to data compensation duties and to arbitration procedures and awards. § 152.99. The validity of a follow-on registration may be challenged if the registrant has, for example, failed to satisfy all applicable data requirements, failed to follow required procedures, improperly made a certification regarding an offer to pay or a data gap, or improperly cited an invalid study. § 152.99(a)(2). A challenge must be filed with EPA within one year after the Agency makes public its listing of recently approved applications. § 152.99(b).

Formulator's Exemption from Data Requirements

Section 3(c)(2)(D) of FIFRA establishes a significant exemption from the data requirements for pesticide "formulators" (i.e., companies which manufacture pesticide products from pesticides already manufactured by and purchased from other companies, rather than by manufacturing the pesticide ingredients themselves). Under this so-called formulator's exemption, "an applicant who proposes to purchase a registered pesticide" for such formulation is not required to "submit or cite data pertaining to such purchased product." FIFRA § 3(c)(2)(D).

As originally enacted in 1978, this provision applied only to "safety" data, and required that the registered pesticide be purchased for the purpose of manufacturing an end-use pesticide product. The provision was amended in 1988, however, to delete both of these restrictions. Thus, the exemption now applies to types of data additional to safety data (e.g., product chemistry and product performance data), and permits formulators to avail themselves of the exemption to manufacture active ingredients and manufacturing-use products, as well as end-use products.

It is important to note that the exemption, even as amended, applies only to data requirements relating to the purchased, registered material. Thus, formulators still are required to submit data that are unique to their own manufactured products. For example, formulators are required to submit a package of acute toxicity data in support of applications for their end-use products, even if they purchase the active ingredients from a registered source.

The rationale underlying this provision is that companies which purchase registered products for formulation are indirectly paying for use of the registration data in the purchase price, and should not be required to pay additional compensation. Thus the exemption does not apply to data concerning the safety or efficacy of the formulator's end-use product, for which data must be submitted or compensation must be offered in compliance with § 3(c)(1)(D).

A formulator that claims entitlement to the exemption must list in its confidential statement of formula a registered pesticide as the source of each active ingredient in the product. § 152.85(b). If a formulator changes the source of an active ingredient, an amended registration indicating the new source must be filed with EPA. § 152.85(c). A formulator can only maintain entitlement to the exemption if the new source is a registered pesticide.

Amendments Not Requiring EPA Approval

In order to streamline its registration processes, EPA established by regulation a category of registration amendments which are effective merely upon notice to the Agency. The registrant need not await EPA approval of such amendments, but may distribute or sell the modified product immediately upon providing notice to the Agency. Based upon such notice, EPA may subsequently require submission of an application for amended registration, if the Agency believes a formal registration amendment is appropriate. Failure to comply with such a requirement may result in cancellation proceedings under FIFRA § 6. 40 C.F.R. § 152.46.

Modifications which may be effected upon notification include: labeling revisions which involve no changes in the statement of ingredients, precautionary statements, or directions for use; addition or substitution of brand names; changes in sources of starting materials, provided that such changes do not result in increases in impurities of toxicological significance or the formation of new impurities; a change in the source of registered active ingredient; a change in the source of an inert ingredient whose source EPA has previously required to be specified; a change in the nominal concentration of an inert ingredient whose chemical identity or composition is known to the registrant; and changes in the production process of a non-integrated system, provided that the certified limits of active and inert ingredients do not change as a result. These modifications are addressed in greater detail at § 152.46(a).

Modifications Not Requiring Notice

Registrants may make the following changes in product composition, labeling, and packaging without providing notice to the Agency: corrections of typographical or printing errors in labeling; changes in package size and label net contents, provided that no change in requirements for use directions or child-resistant packaging result; revisions to non-mandatory label statements, including additions or corrections required by federal statutes or regulations; changes in the label regarding the name or address of the registrant, except for those changes resulting from a transfer in ownership; and revisions to label format not involving the text, provided that such changes are consistent with EPA regulations. § 152.46(b).

EPA Processing and Review

When the Registration Division receives an application for a new registration, it assigns the application to an EPA product manager and informs the applicant of the product manager's name and telephone number. An application involving a new active ingredient may be assigned to any of the product managers in the appropriate branch; an application for a follow-on product or a new use or formulation of a previously registered product will be assigned to the product manager responsible for the active ingredient contained in the product. Applicants that have never previously registered a product with EPA will be assigned a permanent company number. Notice of applications involving new active ingredients and changed use patterns is published in the *Federal Register* shortly after receipt.

The initial step in evaluating an application is to screen it for completeness. EPA first will determine whether all of the items required to be submitted with the application are included or have been correctly submitted. § 152.104.

EPA will not begin or continue reviewing any application the Agency determines to be incomplete. Instead, the Agency will notify the applicant of the deficiencies and allow the applicant seventy-five days to make corrections or complete the application. If the applicant has not cured the application within that time, EPA will terminate all action on the application and treat it as withdrawn. § 152.105.

The review of data submitted in support of an application is the most time consuming and resource intensive portion of EPA's registration process. In order to conserve resources, EPA normally will review only those data submitted to the Agency for the first time. EPA reviews other data submitted or cited by the applicant only: for purposes of reregistration; when reviewing an application for a product containing a new active ingredient; where such data have been "flagged" for potential adverse effects; or if such review otherwise is in the public interest.

By contrast, EPA reviews all draft labeling. Upon Agency approval of draft labeling submitted in support of an application for registration or amendment, the registrant is required to submit a copy of the final printed labeling, incorporating any changes or additions that the Agency requires. The Agency reviews final labeling selectively for compliance.

If an application is deemed complete and acceptable for purposes of review, EPA will notify the applicant by telephone, and will forward the application to appropriate branches of the Hazard Evaluation Division for scientific review. With the exception of certain deadlines for processing "fast-track" applications (addressed below), neither FIFRA nor the EPA regulations prescribe a specific time limit for the Agency to act on an application for registration or amendment. However, in a 1986 PR Notice, EPA did indicate that it would complete its substantive review of applications within the following time periods:

New chemical applications for food uses or non-food uses requiring chronic toxicity data	375 days
New chemical applications for non-food uses that do not require chronic toxicity data	250 days
Applications for the first days food use of a previously registered product	205-375 days

PR Notice 86-4 (April 15, 1986). Although these guidelines do not bind the Agency and are often observed in the breach, they do provide the registrant community a fairly accurate time table for planning purposes.

The exception arises in the case of so-called "fast track" registrations for generic products and minor amendments. The 1988 FIFRA amendments contain a provision requiring expedited review of (1) applications for generic end use products and (2) registration labeling amendments, provided that such applications require no review of scientific data. In expediting such reviews, EPA must notify the applicant within forty-five days after receiving the application whether it is complete. If the application is incomplete, the Agency may reject the application. Within ninety days after receiving a complete application, EPA must issue a decision granting or denying the application.

Approval of a Registration or Amendment

If upon completion of its substantive review EPA determines that the statutory criteria for registration are satisfied, and any technical deficiencies in the application are resolved, the EPA product manager will register the product. EPA will notify the applicant for a new registration by issuing a Notice of Registration, and notify the applicant for amendment by letter.

As noted above, EPA presently reviews virtually all applications under the conditional registration criteria of FIFRA § 3(c)(7). Thus the Notice of Registration will in most cases indicate certain conditions of registration which must be satisfied before the product may be distributed or sold. In the case of me-too registrations and registrations for new uses, such conditions may typically include the submission or citation of data required for *un*conditional registration under FIFRA § 3(c)(5), but which have not yet been submitted by registrants of previously registered products. In the case of new active ingredients, conditions may include submission of additional data in accordance with a schedule determined by the Agency. Other conditions, e.g. suggested additions to or deletions from proposed labeling, may be imposed on a case-by-case basis. Failure to satisfy any such condition of registration may result in cancellation under FIFRA § 6(e).

If a registration application is for a product that contains an active ingredient for which studies submitted by other registrants are entitled to "exclusive use" protection under FIFRA § 3(c)(1)(D)(i), then EPA must notify the original submitter of the protected study that an application for such a product is under consideration. Such notice must be provided at least thirty days before the new product is registered, and is intended to provide the original data submitter an opportunity to examine the new applicant's means of satisfying the data requirements for those new products. If the original data submitter believes that the application indeed relies upon the exclusive use study, it may (a) challenge the issuance of the new registration, or (b) certify that it has no objection to such reliance. § 152.116.

Public Notice

After a registration for a product that contains a new active ingredient or a new use is granted, EPA will publish a notice to that effect in the *Federal Register.* Within thirty days after the registration is issued, EPA must make available for public inspection the materials submitted by the registrant in support of the application. This specifically includes the test data on which the Agency based its registration decision; however, such data may not be released to or reviewed by any company that is a multinational pesticide producer within the meaning of FIFRA § 10(g).

Denial of an Application

EPA may deny an application for registration only if it does not satisfy the statutory criteria for registration under FIFRA § 3(c)(5) or § 3(c)(7). If EPA denies an application, it must do so by a certified letter setting forth the Agency's reasons why the product does not meet this standard. An applicant has thirty days from receipt of the letter to correct any specified defects in the application; the applicant also may choose to withdraw the application. If the applicant fails to correct the deficiency within thirty days and does not withdraw the application, EPA will publish a notice of denial in the *Federal Register.* Within thirty days, the applicant or any interested person with the written authorization of the applicant (e.g., affected user groups, trade associations, etc.) may request an administrative hearing regarding the denial.

Supplemental Registrations

FIFRA § 3(e) provides that:

> Products which have the same formulation, are manufactured by the same person, the labeling of which contains the same claims, and the

> labels of which bear a designation identifying the product as the same pesticide may be registered as a single pesticide; and *additional names and labels shall be added to the registration by supplemental statements.*

(Emphasis added.) Under this section, a registrant may authorize other companies (usually distributors) to market the registered product using the distributor's own brand names.

In accordance with this statutory authority, EPA regulations expressly authorize a registrant to market its product through a distributor, and to modify its label to bear the distributor's brand name and name and address. This distribution and sale is referred to as "supplemental distribution," and the product is referred to as a "distributor product." For practical purposes, the distributor is considered an agent of the registrant; both the registrant and the distributor, however, may be held liable for any violations pertaining to the distribution product.

To obtain a supplemental registration, a basic registrant need only submit EPA Form 8570-5 (notice of supplemental registration of distributor). This form requires identifying information, including the distributor's company number and the brand name used on the distributed product. Distribution under the supplemental registration may begin without EPA's approval as soon as the form is mailed, provided that the conditions described below are met.

The application must be signed by both the registrant and the distributor, and must include the names and addresses of both parties, the distributor's company number, the additional brand name(s) to be used, and the registration number of the registered product. § 152.132(a). In addition, the distributor product must be manufactured, packaged and labeled in a registered establishment operated by the same producer who manufactures, packages or produces the registrant's product, or by a party operating under contract in conformance with EPA's contract manufacturing policy. § 152.132(b). In addition, the distributor product must remain unopened in the producer's original container. § 152.132(c).

The label of the distributor product must be the same in all material respects as that of the registrants products. The distributor's label may bear a different product name, and the name and the address of the distributor may appear on the label instead of the name and address of the registrant. The registration number of the registered product must be followed by a dash and the distributor's company number. The establishment number must be that of the final establishment at which the product was produced. Finally, as a supplemental registrant, the distributor may label its product for all, or for any subset, of the uses for which the original product is labeled. § 152.132(d).

Registration Fees

EPA presently charges no fees for registration, except those imposed for review of tolerance petitions under FFDCA § 408. By contrast, substantial fees are imposed for *re*registration under FIFRA § 4. Moreover, annual maintenance fees must be paid for registrations after they are issued. See Chapter 6.

This fee system resulted from a provision in the 1988 FIFRA amendments, and should remain in place until the reregistration program is completed in 1997, absent further legislative changes. On May 26, 1988, just months before the 1988 amendments were enacted, EPA had established by regulation a "user fee" system imposing substantial fees for the review of applications for registration, registration amendments and EUPs. Congress suspended the program, however, when it established the reregistration and maintenance fee system, amending FIFRA § 4(i)(6) to provide that "EPA may not levy any other fees for the registration of pesticides until September 30, 1997."

Transfer of Registrations

FIFRA does not expressly provide for the transfer of registrations. Nevertheless, EPA has historically permitted the practice as an administrative convenience. Thus, in the case where a pesticide manufacturer has been acquired by another corporation or has sold a portion of its pesticide business, the manufacturer is permitted to transfer ownership of the registrations. The procedure also is frequently used by a prospective registrant as an alternative to applying for a registration in its own right. A prospective registrant may avoid the time and expense of obtaining a generic registration simply by purchasing a previously existing registration from another company.

EPA will approve only those transfers which are unconditional and irrevocable. A transferred registration may not serve as collateral or other security for a loan. Nor may a transfer be used to escape regulatory requirements. EPA deems the rights and duties of a transferee to be the same as those of the transferor at the time the transfer is approved.

EPA has promulgated by regulation strict procedures for effecting a transfer. § 152.135. Provided that the regulations are complied with, however, EPA approval of a transfer is a routine matter. Parties to a transfer must submit a document, signed by authorized representatives of the companies, acknowledging the conditions described above, and providing the following information: the names, addresses, and states of incorporation of the transferor and transferee; the names and registration numbers of the products being transferred; and a description of the underlying transaction. The transferor also must provide a notarized statement affirming that the person signing the transfer agreement is authorized to bind the transferor; that no court order prohibits the transfer; and that the transfer is authorized under federal, state, and local laws, and relevant corporation charters, bylaws, or partnership agreements. § 152.135.

Any false statement made in an application for registration is punishable under the False Statements Act. 18 U.S.C. § 1001 (1988). Rights to the use of data, or compensation for such use under FIFRA § 3(c)(1)(D), are separate from the pesticide registration itself and may be retained or transferred independently.

REGULATION OF INERT INGREDIENTS

An ingredient is considered "active" if it will prevent, destroy, repel or mitigate any pest, or function as a plant regulator, dessicant or defoliant. 40 C.F.R. § 152.3(b). Inert ingredients are substances that have no pesticidal activity, but are included in formulated pesticide products to serve as carriers, solvents, surfactants, fillers, propellants, etc. They include a wide variety of substances, ranging in complexity from clay, cellulose, and corn cob grits used as carriers, to highly sophisticated surfactants.

EPA and its predecessor agencies, the U.S. Department of Agriculture and the Food and Drug Administration, have regulated inert ingredients under both FIFRA and the FFDCA since 1969. Because FDA determined that inert ingredients are subject to the tolerance requirements under FFDCA § 408, no substance can be included in a food use pesticide (even as an inert ingredient) unless a tolerance is established or an exemption from tolerance is granted. In addition, EPA reviews the inert ingredients of pesticide products, as well as their active ingredients, to determine whether they may cause unreasonable adverse effects.

The permissible tolerances and exemptions for all pesticide ingredients (active or inert) are set forth at 40 C.F.R. part 180. An exemption from the tolerance requirement is granted only if EPA determines that a tolerance is "not necessary to protect the public health."

EPA's statement of regulatory policy regarding inert ingredients appears at 52 Fed. Reg. 13,305 (1987). As the policy statement indicates, EPA has reviewed data concerning approximately 1,200 inert ingredients presently used by pesticide manufacturers to determine their potential for carcinogenicity, reproductive or developmental toxicity, neurotoxicity, or other chronic hazards for such products. EPA has categorized the inerts according to their level of toxicity, and has taken regulatory action where the level of toxicity warranted action. 52 Fed. Reg. at 13,306-07.

Out of the 1,200 inert ingredients presently in use, EPA placed fifty-seven on List 1 — Inerts of Toxicological Concern. The Agency requires registrants of any products containing these substances as inert ingredients, irrespective of amount, to include on their labels the following legend:

"THIS PRODUCT CONTAINS THE TOXIC INERT INGREDIENT
____________________ (NAME OF INERT)."

No product containing a List 1 inert ingredient may be distributed in commerce without this warning.

In addition, EPA has actively encouraged registrants whose products contain List 1 inerts to substitute products of lesser toxicity. Registrants that declined to substitute less toxic substances for inert ingredients were subject to extensive FIFRA § 3(c)(2)(B) "data call-ins" for test data concerning the toxicological properties of the substances, as well as residue and environmental fate data indicating their persistence in food products or the environment. Following its review of this data, EPA may take further regulatory action either to require registrants to substitute other inert ingredients, or to further reduce the levels of inerts which pose a concern.

EPA has placed sixty-six chemicals on List 2 — Potentially Toxic Inerts. This determination was made on the basis of structural similarities to chemicals on List 1 or data suggesting a basis for concern. EPA is in the process of collecting information on List 2 inerts to determine whether further regulatory actions are necessary.

Approximately 300 inert ingredients are generally regarded as innocuous, and thus appear on List 4 — Inerts of Minimal Concern. These include such substances as cookie crumbs, corn cobs, clay, and synthetic substances "generally recognized as safe" under the FFDCA tolerance procedures.

The remaining inert ingredients appear on List 3 — Inerts of Unknown Toxicity. EPA is unaware of any serious toxicological concern regarding these substances, and has no data requiring their inclusion on Lists 1 or 2. In the absence of a positive finding that they pose no concern, however, the Agency has chosen to preserve its regulatory options by placing the substances on List 3.

EPA published revisions and modifications to the Inert Ingredients list on November 22, 1989. The revised lists appear at 54 Fed. Reg. 48,314 (1989).

LABELING AND PACKAGING REQUIREMENTS

An application for registration must include all proposed labeling. EPA requires applicants to submit all labeling in draft form; typewritten text is sufficient. Final printed labels need not be submitted until the product has been conditionally registered.

"Labeling" includes all labels and other written material that accompany a product (or to which reference is made on the label or in literature accompanying the product). The "label" itself constitutes only the written material that is placed securely on the product or its immediate container. FIFRA § 2(p).

All registered products must bear a label or labeling setting forth precautionary statements and directions for the use of the product. A product whose label or labeling does not contain the information required by EPA or which sets forth false or misleading information is misbranded. FIFRA §§ 2(q), 12(a)(1)(E).

EPA regulations at § 156.10(a)(1) require that all registered labels clearly and prominently bear the following information:

- name, brand, or trademark name and address of producer, registrant, or person for whom produced
- net contents
- EPA registration number
- EPA establishment number of the final production point
- ingredient statement
- precautionary statements
- classification statement (if restricted)
- directions for use (which may accompany the product on separate printed material if permitted by EPA)

40 C.F.R. § 156.10(a)(1). Wording must be legible, conspicuous, and easily understood. The label text may not be printed in smaller than six-point type. The label language must be in

English; signal words for Category I and II products must be in Spanish as well. § 156.10(a)(2), (3).

The ingredient statement must include the common name and the chemical name of each active ingredient in the product, its percentage by weight, and the total inert ingredient percentage. All percentages must be as precise as possible, reflecting good manufacturing practices. The terms "Active Ingredients" and "Inert Ingredients" must have the same prominence and size. § 156.10(g).

Precautionary statements on the label are based on EPA's review of toxicity, environmental, and product chemistry tests, and thus vary with each product. All products must include the statement "Keep Out of Reach of Children" on the front panel. Depending upon the product's toxicity category, an appropriate signal word also must be placed on the front panel: "Danger" is required for Category I; "Warning" for Category II; and "Caution" for Categories III and IV. A product assigned to Category I on the basis of its oral, dermal, or inhalation toxicity also must bear on the front panel the word "Poison" in red letters and the skull and crossbones symbol. The table on the following page illustrates the toxicity levels for each of the categories.

A statement of practical treatment (first aid), as well as a description of injuries that may result from oral, dermal, inhalation, and eye exposure must appear on the label of all products. Environmental and physical hazard statements are described in § 156.10(h) and Subdivision H (proposed) of the Pesticide Assessment Guidelines. All precautionary statements must appear together on the label, divided among the subheadings "Hazards to Humans and Domestic Animals," "Environmental Hazards," and "Physical or Chemical Hazards."

All directions for use must be approved by EPA, and, once approved, cannot be changed without further EPA review. Under FIFRA § 12(a)(2)(A), it is unlawful for any person to alter a label in any respect. Specific requirements for directions for use are set forth at § 156.10(i). Ordinarily, the directions for use must appear on the label. (Exceptions to this requirement are discussed at § 156.10(i)(1)(iii)). The directions must bear the heading Directions For Use. They must contain a statement of use classification (either "General Classification" or "Restricted-Use Pesticide") and the statement: "It is a violation of federal law to use this product in a manner inconsistent with its labeling." Directions for use must identify the site of application, rate, method, and timing of application, and the pests controlled. Other appropriate directions, e.g., those concerning reentry intervals, rotational crop restrictions, and specific warnings to prevent injury to crops, also must appear. Detailed directions can be omitted from the labels of manufacturing-use products and products used only by physicians or veterinarians.

The Label Improvement Program

EPA's Label Improvement Program, in addition to the statutory and regulatory provisions discussed above, is a third source through which labeling requirements are imposed. The program was established in 1980 as a means of directing registrants to update and improve their labeling. 45 Fed. Reg. 37,884 (1980). Numerous labels for pesticides registered many years ago are out of date and do not reflect new technologies or trends in pesticide application, or the latest knowledge regarding hazards to man or the environment. The Label Improvement Program permits EPA to address these inadequacies without awaiting reregistration under FIFRA § 3(g).

Specific procedures for modification of labeling in accordance with the Label Improvement Program are determined on a case-by-case basis. Because EPA confers with affected registrants and users in advance, a label improvement notice normally does not come as a surprise.

Label improvement directives usually are issued in the form of Pesticide Registration (PR) Notices from the Registration Division and are sent by certified mail to affected registrants. A notice identifies specific labeling changes to be made and establishes a schedule for compliance, taking into account the need to dispose of existing stocks. If an individual registrant fails to comply, EPA may issue a cancellation notice pursuant to FIFRA

§ 6(b). Theoretically, within thirty days after such a cancellation notice is issued, a registrant may file an objection and request a hearing to challenge the need for the required label changes.

Applicants for registration should be aware of the following label improvement notices when composing proposed labeling:

1. Deletion of salt water emesis statements (PR Notice 80-2 (Nov. 28, 1980)).

2. Upgrading directions for use of enclosed-area fumigants (Label Improvement Notice to Registrants dated Dec. 4, 1980).

3. Change in test methods for and categorization of eye irritants (PR Notice 81-3 (Sept. 29, 1981)).

4. Revisions to accommodate new Association of Official Analytical Chemists (AOAC) methods of chemical analysis (PR Notice 81-4 (Sept. 30, 1981)).

5. Upgrading of termiticide labeling (Label Improvement Notice to Registrants dated Nov. 9, 1981).

6. Upgrading of anti-fouling paint labeling (Label Improvement Notice to Registrants dated Mar. 9, 1982).

7. Upgrading of farmworker label precautions for hand-laborpesticides (PR Notice 83-2 (Mar. 29, 1983)).

8. Revision of storage and disposal statements (PR Notice 83-3 (Mar.29, 1983)).

9. Clarification of PR Notices 83-2 regarding farmworker safety and 83-3 regarding pesticide storage and disposal instructions (PR Notice 84-1 (Feb. 17, 1984)).

10. Revisions to labels for fumigant products (PR Notice 84-5 (Nov. 15, 1984)).

11. Revision of PR Notice 84-5 regarding fumigant products (PR Notice 85-6 (Aug. 30, 1985)).

12. Upgrading labeling for pesticides applied through irrigation systems (chemigation) (PR Notice 87-1 (Mar. 11, 1987)).

13. Deletion of rotational crop restrictions from labeling (PR Notice 87-3 (Mar. 11, 1987)).

14. Required labeling for products containing flammable propellants (PR Notice 87-8 (Jun. 25, 1987)).

15. Withdrawal of PR Notice 87-8, labeling requirements for products containing flammable propellants (PR Notice 87-11 (Dec. 10, 1987)).

In 1984, EPA proposed regulations that would have formalized the existence of the Label Improvement Program. The Agency withdrew the proposal in 1988, in part due to objections from industry. EPA agreed with many manufacturers that the program is still evolving and that regulations fixing an LIP procedure are premature. EPA indicated, however, that it may propose such regulations again.

Proposed Changes in Labeling Requirements

EPA issued proposed regulations in 1984 which would vastly revise and expand many of the labeling requirements described above. 49 Fed. Reg. 37,960 (1984). Although thc proposed regulations have not yet been finalized, they do remain pending and represent in substantial part the Agency's present thinking on labeling requirements.

The proposed regulations would affirmatively require pesticide registrants to add to their labels a good deal of information not presently called for, and will relax some present labeling prohibitions. Significant changes under the proposed regulations are summarized below.

Labeling Format. The proposed regulations establish a standardized label format to make label information readily identifiable. In addition, the regulations mpose requirements to ensure appropriate color contrast. Dark colored text on a light background is recommended; EPA reserves the right to reject combinations that do not contrast clearly. When a container is transparent, the color of the label must contrast with the contents. When the label is required to bear in red the word POISON, the surrounding text cannot be in red. § 156.10(b) (proposed). The Agency also is proposing to change its expression of label dimensions and print sizes to metric units and to change the type size requirements for many label headings. § 156.10(c) (proposed).

EPA further proposes to permit the use of graphic symbols, as an accompaniment to label text, to assist in communicating hazard information. § 156.10(a) (proposed). Specifically under consideration is a system of colored bands at the bottom of the label to identify the acute toxicity level of the product. 49 Fed. Reg. 37,962 (1984).

Finally, EPA is considering allowing registrants to provide certain information regarding hazards and warnings that is on the label and that is not required by the Agency or the regulations. This would permit registrants to voluntarily restrict a product's use. *Id.*

Product Identification. EPA proposes to eliminate the registrant's discretion under the present regulations to provide a name and address on the label for either the producer, registrant, or person for whom a pesticide is produced. Under the proposed regulations, the label must contain the name and address of the registrant (or distributor). Foreign registrants specifically will be required to comply with this provision, although they may include as well the main address of their U.S. agent. § 156.24 (proposed).

EPA also proposes to change the requirements for identification of a product's ingredients. With regard to active ingredients, EPA proposes that the label identify active ingredients (by percentage) and the amount of active ingredient found in a typical sample of the product (the "normal concentration"). § 156.34(a) (proposed). The Agency will continue to use the lower certified limit of active ingredients for enforcement purposes, but desires that the normal concentration be stated for purposes of calculations of dosages and dilutions.

EPA also proposes to require identification on the label of any inert ingredients which it believes contribute significantly to the acute overall hazard of the product, and disclosure of which is necessary to protect against an unreasonable risk of illness or injury to health or the environment. Required label statements would take the form of identification of the inert ingredient and precautionary statements to address its particular hazard.

The proposed regulations will eliminate the present requirement that a product bear an expiration date after which it cannot be sold or used, if its composition is known to change significantly after manufacture. All products nevertheless would be required to meet all composition claims while in channels of trade. The Agency thus will permit registrants to include an expiration date on the label to protect against potential enforcement action. The Agency also will require that use directions instruct the user how to compensate for the lower level of active ingredient present after such composition changes. § 156.72(i) (proposed).

Toxicity Category Revisions. Proposed revisions to the Toxicity Category criteria are summarized in the following chart:

Hazard Indicators	Category I	Category II	Category III	Category IV
Oral LD_{50}	Up to and including 50 mg/kg	50 thru 500 thru mg/kg	500 thru 5000 mg/kg	5000 mg/kg
Dermal LD_{50}	Up to and including 200 mg/kg	200 thru 2000 mg/kg	2000 thru 5000 mg/kg	5000 mg/kg
Inhalation C *(actual chamber concentration measured for a 4-hour exposure)*	Up to and including 0.06 mg/liter	0.05 thru 0.5 mg/liter	0.6 thru 5 mg/liter	5 mg/liter
Eye effects	Corrosive *(irreversible destruction of ocular tissue)* corneal involvement or irritation persisting for more than 21 days	Corneal involvement or irritation clearing in 8-21 days	Corneal involvement or irritation clearing in 7 days or less	Minimal effects clearing in less than 24 hours
Skin effects	Corrosive *(tissue destruction into the dermis)*	Severe irritation at 72 hours *(severe erythema or edema)*	Moderate irritation 72 hours *(moderate erythema)*	Mild or slight irritation *(no irritation or slight erythema)*

Flammability Statements. EPA desires to change its flammability criteria to make them consistent with regulations under the Hazardous Materials Transportation Act and proposed regulations under the Federal Hazardous Substances Act. Thus the Agency proposes to require flammability statements for all products with a flash point between 20°F and 100°F. 40 C.F.R. § 156.58(a) (proposed). This change will extend the upper limit from 80°F, as established under existing regulations. EPA also proposes to permit, for the first time, a declaration of non-flammability, under certain specified criteria. § 156.58(b) (proposed).

Classification Statements. Section 3(d) of FIFRA requires classification of pesticides for restricted or general use. (EPA's pesticide classification system is discussed below). Because the terms "restricted use" and "general use" may be confusing to the public, EPA proposes to change the required labeling statements regarding classification. Under the proposed regulations, pesticides classified for general use will not be permitted to bear a statement of classification. Restricted use pesticides will continue to bear the designation RESTRICTED USE PESTICIDE, and must bear the additional statement: "For use only by a certified applicator for uses authorized by his certification, or by persons under his direct supervision." § 156.62 (proposed).

Disposal Statements. EPA's present requirements for disposal statements are derived from the disposal guidelines promulgated by the Agency at 40 C.F.R. part 165, subpart C. These guidelines have been affected by the enactment of the Resource Conservation and Recovery Act (RCRA) and a series of regulations implemented under its authority. EPA now has proposed labeling requirements that would be integrated with the RCRA disposal requirements. § 156.70 (proposed).

Child-Resistant Packaging

Child-resistant packaging (CRP) is packaging constructed to be "significantly difficult for children under five years of age to open or obtain a toxic or harmful amount of the substance contained therein within a reasonable time." § 157.21(b). Such packaging also must meet the effectiveness standards prescribed by the Consumer Product Safety Commission (CPSC) at 16 C.F.R. § 1700.15(b). See discussion in Chapter 17.

Products for residential use that exceed certain toxicity triggers and that are not classified for restricted-use must be packaged in child-resistant containers. The toxicity characteristics for this requirement are as follows:

acute oral LD_{50}	1500 mg/kg or less
acute dermal LD_{50}	2000 mg/kg or less
inhalation LC_{50}	2 mg/liter or less
eye irritation	corrosive (irreversible destruction of ocular tissue) or corneal involvement or irritation persisting 21 days
skin irritation	corrosive (tissue destruction into the dermis and/or scarring) or "severe" skin irritation at seventy-two hours (severe erythema or edema)
other effects	human data, use history, accident data, or other information causing EPA to determine that a serious hazard exists which child-resistant packaging could reduce

EPA has granted general exemptions for products that are classified for restricted use and for products packaged in large sizes.

The packaging sizes which trigger a general exemption are as follows:

Solid pesticide products	50 pounds or greater (regardless of type)
Swimming pool products	Greater than 7.5 gallons by volume
Liquid products (for use other than swimming pools)	5 gallons or greater by volume
Aerosol products (regardless of type)	2 pounds or greater

Notwithstanding the general exemptions, child-resistant packaging under some circumstances may be required. EPA may require child-resistant packaging for products in large sizes upon a written determination that the product is intended to be sold to homeowners or members of the general public, and for restricted use pesticides upon a written determination that a product poses a serious risk of illness or injury. A registrant receiving such notice is entitled to a hearing on the Agency's determination. § 157.24(a)(1), (2).

EPA also may grant exemptions from the child-resistant packaging requirement upon a showing that the hazards indicated by the toxicity criteria above are not indicative of hazards to man, or that such packaging is not technically feasible. § 157.24(b). If an exemption is not sought, the applicant must certify in the application (EPA Form 8570-1) that child-resistant packaging will be used that will satisfy the standards in § 157.32 for effectiveness, compatibility, and duration. Registrants must retain a description of the package (dimensions, composition, and description of the child-resistant mechanism) and records of the scientific evidence demonstrating that their product complies with these standards. § 157.36.

CLASSIFICATION FOR RESTRICTED OR GENERAL USE

FIFRA § 3(d) requires EPA to classify all pesticides for general use or restricted use, or both, as part of their registration. Classification is based principally upon the acute toxicity of the product and the site at which it is to be used, i.e., domestic, non-domestic, or outdoors. Domestic applications include those in areas associated with the household, patient care areas of health-related institutions, and areas where children spend time; and applications made directly to humans or pets.

If the Agency determines that a pesticide, when applied in accordance with its warnings, cautions, and directions for use, *or* in accordance with widespread and commonly recognized practice (which may involve misuse of the product), will not generally cause unreasonable adverse effects on the environment, it will classify the product for general use.

FIFRA § 3(d)(1)(B). If it determines that, under such conditions and without additional regulatory restrictions, the product may generally cause unreasonable adverse effects on the environment, it must classify the product (or the particular uses that pose the risk of adverse effects) for restricted use. FIFRA § 3(d)(1)(C).

Classification Criteria

EPA's classification criteria are set forth in regulations at § 152.160. This regulation was adopted in 1988 and represents a substantial change from the previous classification system.

The thrust of the present system is to identify products that should be classified for restricted use only. EPA will not normally classify products for general use; rather, products which do not fit the criteria for restricted use will simply remain unclassified. EPA may restrict a product, or certain uses of a product, to use by or under the supervision of a certified applicator. The Agency also may impose restrictions regarding product composition, labeling, packaging uses, distribution and sale, or on the status or qualifications of the user.

EPA's criteria for classification of a product for use by certified applicators are set forth at § 152.170, and appear on the following page:

RESIDENTIAL/INSTITUTIONAL USE PRODUCTS

Criterion	Product is candidate for restricted use if:
Oral toxicity	1.5 g/kg on product as diluted for use
Dermal toxicity	2000 mg/kg on formulated product
Inhalation toxicity	0.5 mg/l on product as formulated
Eye irritation	Toxicity Categories I and II
Skin irritation	Toxicity Categories I and II
Oral toxicity	50 mg/kg or less on product as formulated
Dermal toxicity	(1) 200 mg/kg on formulated product (2) 16 g/kg on product as diluted for use
Inhalation toxicity	0.05 mg/l on product as formulated
Eye irritation	Current Toxicity Category I
Skin irritation	Current Toxicity Category I

PRODUCTS INTENDED FOR OUTDOOR USE

Criterion	Product is candidate for restricted use if:
Mammalian toxicity	Occurs as residue of 1/5 acute LD_{50}
Bird subacute toxicity	Occurs as residue of 1/5 subacute dietary LC_{50}
Aquatic organism toxicity	Occurs in water at 1/10 acute LC_{50}
Bird acute toxicity	For granular products, acute mammalian or avian LD_{50} of formulated product 50 mg/kg

In addition to these specific hazard criteria, EPA may determine on the basis of such evidence as field studies, use history and accident data that a product in use poses a serious hazard that may be mitigated by restricting its use to certified applicators. Even if EPA makes such a finding, it still may not impose use restrictions unless: (i) the product's labeling is inadequate to mitigate the hazards identified; (ii) restriction of the product to use by certified applicators would decrease the risk of adverse effects; and (iii) the decrease in risks as a result of the restriction would exceed the decrease in benefits.

Once a product is classified for restricted use, it is subject to a number of regulatory requirements not imposed on products that remain unclassified. Labeling requirements for such products are set forth at § 156.10. Restrictions on the distribution and sale of such products appear at § 152.167. In addition, any advertising for a restricted use product must contain a statement of its restricted use classification. § 152.168.

Changes in Classification

Although EPA may conduct a classification review of a registered pesticide at any time, such reviews typically are conducted only at the time of a new product's initial registration. § 152.164(b). Products normally are reclassified only as part of the review of an application for a new use of a previously registered product, as part of the process of developing a registration standard, or as part of a Special Review. Id.

Several statutory and regulatory procedures have a bearing on EPA decisions to reclassify a product. FIFRA provides that EPA may change a classification for general use to restricted use if the change is necessary to prevent unreasonable adverse effects in the environment. FIFRA § 3(d)(2). The Agency must provide written notice to the registrant forty-five days in advance, and must publish the change in the Federal Register. A registrant who opposes such a change is entitled to a hearing under § 6(b).

EPA also may change the classification of a pesticide from restricted to general use upon a petition by the registrant demonstrating that classification for restricted use is not necessary, and that reclassification to general use will not cause unreasonable adverse effects.

FIFRA § 3(d)(3). An unsuccessful petitioner is entitled to judicial review of the petition in a United States district court.

EPA regulations establish two reclassification procedures. The first is by regulation, which requires notice and comment rulemaking procedures. Within sixty days after the effective date of a final rule, the registrant may be directed to submit to EPA either: (1) a copy of the amended labeling reflecting the reclassification; (2) a statement that the registrant will comply with the reclassification decision within time frames required under the final rules; or (3) an application for amended registration to delete the restricted uses for the label, or to "split" the label into two registrations, one for restricted uses and the other for unclassified ones. § 152.164(c)(1).

EPA regulations also provide for an alternative reclassification procedure whereby EPA is required merely to notify the registrant of its decision to reclassify a product. The registrant subsequently is required to submit amended labeling reflecting the reclassification upon threat of cancellation proceedings if the registrant does not comply. § 152.164(c)(2). The validity of this procedure, however, is subject to question.

In a similar setting, EPA announced in a Registration Standard a unilateral decision to reclassify products containing the active ingredient simazine for restricted use only, and sent all simazine registrants a mailgram directing them to submit amended labeling reflecting the reclassification. EPA threatened to initiate misbranding proceedings against any company that failed to comply.

One of the registrants challenged this order on the ground that EPA was obligated to follow the reclassification process under FIFRA § 6(b) before initiating misbranding proceedings. EPA withdrew the order before the legal issue was decided. *Ciba-Geigy Corp. v. EPA,* 801 F.2d 430 (D.C. Cir. 1986).

EMERGENCY EXEMPTIONS UNDER § 18

FIFRA § 18 permits EPA to exempt state governments or federal agencies from the requirements of FIFRA where necessary, for example, to control unexpected pest outbreaks, the loss of crops in unusual conditions, or to prevent the spread of disease. EPA may issue such an exemption if it determines that an emergency exists and that the emergency use of the pesticide will not cause unreasonable adverse effects on the environment. An exemption may not be granted to allow the use of a pesticide whose registration was cancelled pursuant to FIFRA § 6, unless EPA specifically modifies the terms of the suspension or cancellation after a hearing.

This provision of FIFRA saw limited use in the early years of its implementation. In 1978, for example, only 199 applications were submitted for emergency exceptions, and seventeen applications were submitted for crisis exceptions. An EPA audit over the next five years showed such a substantial increase (to 724 applications for emergency exemptions and 145 crisis exceptions in 1982) that the Agency expressed its concerns that emergency exceptions (as well as special local needs exemptions under FIFRA § 24(c)) were being used to circumvent the § 3 registration requirements. PR Notice 84-2 (Apr. 20, 1984). EPA thus established a program to monitor the issuance of exemptions and special local need registrations. The Agency also established a policy restricting the issuance of emergency exceptions for multiple pesticides for use on the same pest on one site. Further, where the justification for an emergency exemption was a lack of registered pesticides, the Agency would issue an exemption only where the pest was new to a site or, though previously present, had not been a critical problem; the introduction of a new crop to an area would not be considered a valid basis for exemption. PR Notice 84-2.

These and other concerns are reflected in the Agency's emergency exemption regulations, codified at 40 C.F.R. part 166. The four types of emergency exemptions permitted under § 18 are the specific, quarantine, public health, and crisis exemptions.

Specific Exemptions

A specific exemption permits the use of an unregistered pesticide to control a particular pest. It may be authorized in an emergency condition to avert a significant economic loss or a significant risk to an endangered species, a threatened species, beneficial organisms, or the environment. To obtain a specific exemption, the governor of a state or the head of a federal agency must submit a written request to the Registration Support and Emergency Response Branch of EPA's Registration Division. The request must describe: (a) the emergency (including the specific part to be controlled, the events which brought about the emergency, and the anticipated risks or economic losses that would be averted by the proposed use of the pesticides); (b) the pesticides proposed for use under the exemption; and (c) the nature of the proposed use (including the sites and total acreage to be treated, the method of and rates of application, the amount of the pesticide product and active ingredient to be used and all applicable restrictions and requirements concerning the proposed use, and the qualifications of the applicators using the pesticide). The application must contain a detailed explanation of the reasons that currently registered pesticides are not available or effective to control the emergency and the reasons that alternative practices, if available, are not adequate. In addition, the application must contain data to show that the pesticide will be effective for the proposed use, a listing of all foods likely to contain residues and the amounts of such residues, and a discussion of risk information with references to test data. The applicant must notify any other federal or state agencies likely to be concerned with the application, notify the registrant or basic manufacturer, and provide a proposed enforcement program. § 166.20.

Specific exemptions are granted only for as long a period as expected to be necessary and are never granted for longer than one year. Any unexpected adverse effects must be reported immediately to the Agency. Following the expiration of the exemption, the user must submit a report of the total acreage and amount of commodity treated and quantity of pesticide used, the effectiveness of the pesticide, a description of any adverse effects, results of any required monitoring, any enforcement actions taken, and methods of disposition. Records of such information must be maintained for two years.

Quarantine Exemptions

Quarantine exemptions are used to control the introduction or spread of foreign pests in the United States. They are sought by states or federal agencies and generally are processed by EPA in the same manner as specific exemptions. In addition to the information required for a specific exemption, the application must indicate the origin of the pest and the means of introduction or spread, and the anticipated impact of a failure to control the pest. A quarantine exemption may last as long as the product is deemed necessary, but in no case longer than three years. The reporting and recordkeeping requirements for quarantine exemptions are the same as those for specific exemptions.

Public Health Exemption

A public health exemption may be authorized in an emergency condition to control a pest that will cause a significant risk to human health. In addition to the general information required above, an application must contain the following specific information: (a) the name of the pest to be controlled and, if the pest is a vector, a description of the disease it is expected to transmit; (b) the magnitude of the health problems expected to occur without the pesticide use; and (c) the availability of medical treatment of the health problem. The reporting and recordkeeping requirements for a public health exemption are the same as those for a specific exemption. An exemption granted for this purpose may last as long as reasonably expected to be necessary, but may not exceed one year.

Crisis Exemptions

Crisis exemptions may be used by states and federal agencies without prior EPA approval in emergency situations where there is not sufficient time to request use of the other exemptions discussed above. This authority may not be invoked to permit the use of new chemicals, pesticides that have been cancelled or suspended, or where the proposed use would be the first food use of the pesticide. Nor may the pesticide be used if the EPA has instructed the governor of the state or the head of the federal agency not to issue the exemption.

A federal agency or state which issues a crisis exemption must notify EPA and the registrants of the pesticide at least thirty-six hours before the exemption is utilized and not less than twenty-four hours after making the decision to utilize the exception. Any adverse effects noted must be reported to the EPA immediately; a detailed report of the treatment under the exemption must be filed within three months of the last application. If the crisis will result in residues in or on food, then EPA will notify the FDA and other appropriate agencies. EPA also is required to announce the issuance of crisis exemptions at least quarterly in the *Federal Register*, and to announce annually the number of crisis exemptions issued and revoked.

Crisis exemptions may be authorized only for so long as necessary to control the emergency condition, up to a maximum of fifteen days, unless an application for a specific quarantine or public health exemption has been submitted within that time.

Chapter 4
PREPARATION AND SUBMISSION OF DATA

As discussed in Chapter 3, an application for an EPA pesticide registration or registration amendment must be accompanied by test data that demonstrates that the product will perform its intended function without unreasonable adverse effects on the environment within the meaning of FIFRA § 3(c)(5) or 3(c)(7). To ensure uniformity in the application of the statutory criteria, EPA regulations establish requirements for specific studies which must be performed, the manner in which they must be conducted, and the form in which they must be submitted. These requirements are discussed below.

DATA REQUIREMENTS

EPA data requirements are set forth in regulations at 40 C.F.R. part 158. The regulations contain tables which indicate, by use pattern, the types of studies which must be submitted in support of an application. The tables indicate which tests are required (R), which tests are not required (--), and which are conditionally required (CR). Data designated as conditionally required must be submitted where a product meets criteria specified in notes accompanying the data requirements tables.

In some cases, "tiered" testing is permitted. The first tier study may be all that is required to support a registration, unless the results of this first tier study show that toxicity levels exceed certain prescribed thresholds. In such cases, second- and third-tier studies may be required in order to quantify or further define the toxic effects.

The tables also specify the test substance that must be used for each test. Required test substances differ depending upon whether the applicant is seeking a registration for an end-use product (EP) or a manufacturing-use product (MP).

An application for an unconditional registration under FIFRA § 3(c)(5) must be accompanied by all of the data indicated on the tables, unless EPA has approved an application for a waiver. Applications for conditional registrations under § 3(c)(7) require fewer data. An applicant for a conditional registration of a new chemical under FIFRA § 3(c)(7)(A) need not submit data for which the data requirements were imposed so recently that there has been insufficient time to produce the data. Applicants for conditional registration under § 3(c)(7)(B) for new uses of products already registered must submit product chemistry data, product performance data (in the case of antimicrobials) and data from the tables which apply solely to the new use. Applicants for any form of conditional registration need not submit data for which the requirements have been waived.

The data required under part 158 can be divided into basic categories which identify the composition of the pesticide and reflect the effect of the pesticide on man and the environment: product chemistry, residue chemistry, environmental fate, toxicology, reentry protection, wildlife and aquatic, plant protection, and product performance. Several of the more significant categories of data requirements are discussed briefly below. (For a discussion of data requirements for biochemical and microbial pesticides, see Chapter 12.)

Product Chemistry (§§ 158.150-.190)

Product chemistry data provide a profile of the product's chemical and physical characteristics and address any aspect of the manufacturing process which could result in the formation of impurities. Data include the starting ingredients, processing equipment, conditions such as temperature and pressure, relevant chemical reactions, and the duration of each reaction step. In addition, the applicant must provide qualitative and quantitative data on impurities in the technical and end-use products.

The registrant also must certify the upper and lower limits of each active ingredient and provide an analytical method that can be used to verify active ingredient levels in the product as it is distributed in commerce. The analytical method must be releasable to the public and capable of sufficient accuracy to serve the purpose of enforcement.

Toxicology (§ 158.340)

Toxicology data are among the most time-consuming and costly to generate. They are designed to indicate toxicological properties of the pesticide according to the route of exposure (oral, dermal, inhalation, or ocular), the duration of exposure (acute, subchronic, or chronic), and the target effects (neurotoxicity, teratology, oncogenicity, reproduction, mutagenicity, etc.).

The objectives of acute toxicity testing are to measure the dose which produces lethality, to make gross observations of other effects, and to allow relative comparisons with other compounds. These studies involve a single dose and exposure of a small number of animals, followed by observation for two weeks. The results from acute studies are generally expressed as a median lethal dose (LD_{50}) — the dose which kills 50 percent of the animals— or median lethal concentration (LC_{50}) when measuring inhalation exposure.

The main objectives of subchronic toxicity testing are to obtain more information about the test compound by evaluating cumulative effects, if any, and to identify target organs. The results from the subchronic studies in turn are used to determine exposure levels for chronic testing. Subchronic studies also are used to define other toxicological end points of concern, such as reproductive effects.

Chronic toxicity data are necessary to determine the long term effects of a pesticide. Effects of concern include carcinogenicity, teratogenicity, mutagenicity and metabolic effects. The primary animals used for carcinogenicity testing are rats and mice. In testing for other toxicological effects, the test species generally are rodents and dogs. Chronic studies in rodents normally last for two years.

The cost of toxicology testing is enormous. The cost to conduct all of the rat and mouse feeding studies necessary to satisfy registration requirements presently ranges from $1-3 million. The additional costs of conducting subchronic studies, such as a reproduction study, may be as much as one million dollars. The full battery of toxicology studies usually requires at least three years to complete.

Wildlife and Aquatic (§ 158.490)

The primary objective of wildlife testing is to determine toxicity to non-target species. Tests are normally conducted on avian species such as mallards and bobwhite quail. The oral median lethal dose and dietary median lethal concentration can be determined in the laboratory. Requirements for reproduction and field studies can be triggered by data suggesting persistence in the field, bioaccumulation, or reproductive effects in mammalian studies. These studies can require large areas and, in the case of field studies, elaborate protocols and control fields.

The primary concerns of aquatic testing are acute and chronic toxicity and bioaccumulation. Acute toxicity is determined for fish and aquatic invertebrates, and, if applicable, marine and estuarine organisms. Chronic aquatic studies in fish and/or aquatic invertebrates are required whenever exposure to aquatic environments will be continuous, the acute LC_{50} or EC_{50} (the median lethal environmental concentration) is less than 1 milligram per liter, the estimated environmental concentration is equal to or greater than 1 percent of the acute EC_{50} or LC_{50}, bioaccumulation is likely, or the pesticide has a half-life in water of four days or more.

On a case-by-case basis, mesocosm or actual field tests can be required. The current trend is toward mesocosm studies because the results, based on multiple treatment levels and replicated sites, can be applied to a broader range of use sites. The results of actual field studies may only be representative of the environment where the test was run.

As with chronic toxicity testing, mesocosm and actual field studies are very time-consuming and expensive. Extensive contact with EPA in the development of test protocols, site selection, and review of interim progress reports is necessary to produce acceptable studies and to protect the registrant's investment.

Residue Chemistry (§ 158.240)

Pesticides registered for use on a food or feed crop require either the establishment of a tolerance or an exemption from the tolerance requirements for each raw agricultural commodity (RAC) derived from the crop and for any processed food derived from the RAC.[8] EPA uses residue chemistry data submitted with an application in order to set and enforce pesticide tolerances. (For a discussion of the requirements and procedures for establishing pesticide tolerances, see Chapter 5.)

As the first step in this process, studies must be performed to identify the nature of the residues that accumulate in plants and livestock as a result of treatment with the pesticide. These studies are generally performed with radiolabeled pesticide active ingredients.

Field studies later are performed on crops to determine the magnitude of the residues that accumulate in each crop. Consideration is given to the geographical distribution of the crop, use rate and typical harvesting practices. Samples of both food and feed items should be taken. For crops that are processed into dried food, oils, pastes, etc., residue levels may be determined by actually processing treated samples of the RAC.

Animal studies must be performed to determine residues in meat, various organs, milk and eggs. Dose levels should be high enough to represent the maximum residue in treated feed. In a similar fashion, residue levels can be determined for fish, potable water, and food exposed in a food handling establishment.

The ultimate purpose of residue testing is to establish an acceptable residue level for each food and feed item intentionally exposed to the pesticide. Food or feed items for which a tolerance is not established, or in which residues exceed the tolerance, are considered adulterated and cannot lawfully be distributed or sold in commerce.

---

8 EPA cannot require a registrant to submit "field residue data from a geographic area where the pesticide will not be registered for such use." FIFRA § 3(c)(2)(A), *as amended by* the Food, Agriculture, Conservation, and Trade Act of 1990, Pub. L. No. 101-624, sec. 1492, 104 Stat. 3359, (Nov. 28, 1990).

Environmental Fate (§ 158.290)

As concerns have grown in recent years concerning the accumulation of pesticides and pesticide residues in the environment, additional environmental fate data requirements have been introduced, especially regarding ground water. For pesticides used outdoors, studies are required to determine their rates of hydrolysis, photodegradation, leaching and dissipation. Data also may be required to demonstrate the metabolism of the pesticide in water or soil. Depending on the nature of the pesticide and the proposed uses, data also may be required to show the product's volatility, and its rate of accumulation in crops from residues in soil and accumulation in aquatic organisms. In certain cases, EPA may require prospective and retrospective ground water monitoring.

Product Performance (§ 158.640)

Product performance (or efficacy) data generated by field testing a product for the purpose of labeling the product or refining the instructions for its use historically have been the most voluminous data generated by pesticide manufacturers. EPA's early experience in pesticide registration indicated that reviewing efficacy data for registration purposes placed too great a burden on the Agency's resources. Thus, in an interim rule published in May 1979, EPA waived the automatic submission of efficacy data for most pesticide products. That conditional waiver subsequently has been incorporated into the Agency's list of data requirements. 40 C.F.R. § 158.640 n.1. Pursuant to the waiver, EPA generally does not require submission of efficacy data for agricultural pesticides, disinfectant products whose failure would not cause human health consequences (e.g., antibacterial products used to control deterioration or odor in paint, industrial fluids or leather), algicides, and slimacides.

EPA continues to require automatic submission of performance data for all antimicrobial pesticides intended for use on human pathogens and for vertebrate control products. This requirement applies to each end-use application. Thus an applicant may not rely on data previously submitted in support of similar formulations, unless "bridging" data are submitted to demonstrate that any differences in the product chemistry or manufacturing processes do not change the product's effectiveness. With respect to fungicides, EPA requires performance data for products intended to control organisms that produce mycotoxins. EPA also may require performance data for products whose efficacy cannot readily be determined by the end user, such as agents to control subterranean termites. Finally, EPA requires a full complement of performance data for any product that has been suspended or cancelled for reasons of human safety or is the subject of a Special Review, if the risks that brought about those regulatory actions are applicable to the proposed new or added uses.

Applicants for registration should not interpret EPA's waiver of the submission of efficacy data as excusing them from conducting product performance studies necessary to provide adequate directions for use. Notwithstanding the waiver, a product which is ineffective does not meet the standards for registration under FIFRA § 3(c)(5)(A). EPA thus retains the right to require the submission of efficacy data for "any product registered or proposed for registration." 40 C.F.R. § 158.640(b)(1). EPA may cancel the registration of any product shown to be ineffective, or treat the product as misbranded.

Waivers of Data Requirements

EPA's data requirements were devised by the Agency as a regulatory tool for determining whether a pesticide satisfies the standard for registration under FIFRA § 3(c)(5). Thus they are not express statutory obligations and the Administrator may waive them under appropriate circumstances. EPA's registration regulations contain an express provision permitting the waiver of any data requirement. Waivers can be granted where it is established that the composition, degradability, proposed patterns of use, and other chemical or physical properties relating to an evaluation of the pesticide's effects on man or the environment are fundamentally different from the factors considered by the Agency in establishing the data requirement. 40 C.F.R. § 158.45.

An applicant that believes that a data requirement should be waived may submit a written request based on the standard above. EPA is required to consider the request together with other available information, and make a written finding whether the properties of the pesticide are fundamentally different from the factors considered by the Agency in establishing the data requirement. If the Agency reaches such a determination, then it has discretion to waive the data requirement, provided that the Agency further determines that the data are not necessary for deciding whether the pesticide will cause unreasonable adverse effects on man or the environment. Where an application for a new registration is granted on the basis of a waiver of a data requirement, the Agency's *Federal Register* notice indicating the approval must note and state the basis for the waiver.

Pesticide Assessment Guidelines

In addition to the data requirements at 40 C.F.R. part 158, EPA publishes Pesticide Assessment Guidelines that set forth standards and protocols for testing that are acceptable to the Agency. Separate guidelines have been written to address each of the major categories of the part 158 requirements, as well as particularized requirements such as Experimental Use Permits, labeling, and biorational pesticides.

To assist in the reporting of studies carried out to support regulatory actions, EPA has also published two sets of study-specific guidance, Data Reporting Guidelines (DRGs), and Standard Evaluation Procedures (SEPs). The DRGs provide guidance for writing final reports for each of the studies and setting out the specific data EPA expects to find. Use of the DRGs ensures that EPA reviewers will be able to locate the data necessary to validate the study. The SEPs explain how EPA reviews each study and how the data are treated in validating and interpreting the results.

These guidelines are available from:

National Technical Information Service
5285 Port Royal Road
Springfield, Virginia 22161
Tel. (703) 487-4650

GOOD LABORATORY PRACTICE REGULATION

EPA has prescribed by regulation Good Laboratory Practices (GLPs) which must be adhered to in conducting any study submitted to satisfy FIFRA data requirements. 40C.F.R. pt. 160. EPA has also promulgated separate GLP standards under the Toxic Substances Control Act (TSCA). Although EPA has attempted to harmonize the GLPs under TSCA and FIFRA, minor differences do exist.

EPA first published GLP guidelines for both FIFRA and TSCA on November 29, 1983. 48 Fed. Reg. 53,922, 53,946 (1983). The intent of these regulations was to establish standards for the operation of establishments and the conduct of studies that generated data on health and safety for submission to EPA.

On August 17, 1989, EPA published expanded GLP regulations for both FIFRA and TSCA. The new regulations enlarge the scope of studies falling under the regulation to include ecological effects, chemical fate, residue chemistry and efficacy testing. 54 Fed. Reg. 34,034, 34,052 (1989). In the case of physical and chemical characterization studies (Guideline 63), all provisions of the GLP regulations apply to studies designed to determine stability, solubility, octanol-water partition coefficient, volatility and persistence. All other Guideline 63 studies must conform to most of the requirements governing the facility and its management, but are exempt from part of the requirements governing the Quality Assurance Unit. 40 C.F.R. § 160.35.

The GLP regulations set forth requirements regarding facilities, laboratory management, personnel and equipment, and establish standards for operating procedures, use of test and control substances, handling of test organisms, storage of data, and design of and adherence to test protocols. The regulations also establish strict recordkeeping and reporting requirements and provide for routine auditing of facilities that perform GLP studies, as well as the study reports themselves.

EPA's GLP standards are consistent in all material respects with those promulgated by the Food and Drug Administration. The minor differences between the two are explained in the preamble to the EPA standards found at 54 Fed. Reg. 34,052. EPA standards also are consistent in nearly all respects with those adopted by the Organization for Economic Cooperation and Development.

Under the GLP regulations, EPA may refuse to accept as reliable, for purposes of supporting a FIFRA application or permit, any study that is not conducted in accordance with the standards. 40 C.F.R. § 160.17(a). The Agency requires any such application to be accompanied by a statement signed by the applicant, the sponsor and the study director indicating that the study was conducted in accordance with the GLPs, or that describes in detail all deviations from the GLP standards. Alternatively, if the submitter was not the sponsor or the study director and does not know whether the study was conducted in compliance with the GLPs, the submitter may so state. Submission of a statement that is false may result in cancellation, suspension, modification, or denial of the application or permit, and may serve as the basis for civil or criminal penalties under FIFRA § 14 or the False Statements Act, 18 U.S.C. § 1001 (1988). EPA also requires the operator of a testing facility to permit audit inspections of the facility, its records, and all laboratory specimens at any reasonable time. EPA will not consider in support of any FIFRA application or permit any test data generated by a laboratory or test sponsor that refuses such inspections.

In addition to prescribing procedures for conducting studies, the GLP standards also set forth detailed requirements for storage and retention of records. All raw data, documentation, records, protocols and final reports, as well as all correspondence and other documents relating to the interpretation and evaluation of data, must be retained. Such material must be archived so as to be readily retrievable, and must be indexed by test substance, test system, nature of study and study date.

Most records required under the GLPs must be retained for the period that the sponsor (e.g., a FIFRA registrant) holds any research or marketing permit to which the study is pertinent, or for five years after the data are submitted, whichever is longer. 40 C.F.R. § 160.195. In cases where the test data are not submitted to the Agency, GLP records must be maintained for at least two years after the study is terminated. These retention requirements apply to all test documentation, raw data, specimens, master schedule sheets, protocols, inspection records, training and personnel records, and all equipment maintenance and calibration records required under the standards. Specimens which are fragile and degrade during storage must be retained only so long as their quality will permit evaluation. If a test facility or archive facility goes out of business, all required data must be transferred to the test sponsor and written notice provided to EPA. Following a GLP audit, EPA may inform the sponsor or testing facility in writing that specimens, samples, and other non-documentary materials for those studies determined to have been conducted in compliance with the GLPs no longer need be stored.

DATA SUBMISSIONS

EPA has issued strict requirements regarding submission of data. Data generally should be submitted with each study in a separately bound volume. However, all of the required product chemistry studies for end-use products may be bound in a single volume. Product chemistry data for technical products, manufacturing-use products, Experimental Use Permits, food tolerances and end-use products produced from unregistered source ingredients should be bound in one volume for each series. Batteries of acute toxicity tests performed on the same end-use product and submitted under the same cover page may be bound together.

EPA also has issued guidelines under which studies must be formatted. See Pesticide Registration (PR) Notice 86-5. The title page of the submission (page 1) must contain the study title, including the name of the test article, the data requirement for which the study is being submitted, the study author(s), the study date, the name and address of the performing laboratory, the laboratory's study identification number, a reference to any previously submitted study to which the study is related, and publication information (if the study is from public literature).

The title page of the study also should contain the registration number of the test article (if available), the name and address of the registrant, the name and address of the submitter (if different from that of the registrant), and the total number of pages in the report.

The second page of each study must include a statement of confidentiality claims, through which the submitter must either waive confidentiality of the data or state that confidential material has been removed and submitted separately in a confidential attachment. Any claim of confidentiality should be made under FIFRA §§ 10(d)(1)(A), (B) or (C). Attempts to claim confidentiality for data not falling within these provisions will result in significant delays of data reviews while confidentiality disputes are resolved. Where confidential information has been removed and submitted separately, the submitter must clearly cross-reference each excerpted passage, to facilitate the review.

In order to claim confidentiality for data on a basis other than FIFRA §§ 10(d)(1)(A), (B) or (C), the submitter must supply a full substantiation. The information claimed confidential must be clearly marked in the body of the study and identified as subject to a claim of confidentiality. A Supplemental Statement of Data Confidentiality Claims must be submitted and appended to the report behind any other confidential appendix. The statement must identify the specific page and line number(s) of the study claimed confidential; explain why the information qualifies for confidential treatment; indicate how long confidentiality is to be maintained; document measures that have been taken by the submitter to guard against disclosure; describe the extent to which any disclosure has already occurred and the precautions taken to prevent those disclosures; enclose copies of determinations of confidentiality made by EPA, other federal agencies, or courts; explain why disclosure would result in substantial harm to the submitter; and, if the information is voluntarily disclosed, indicate whether (and how) disclosure of this information might lessen the future availability of such information to EPA.

Page three of each study must contain a statement of compliance with Good Laboratory Practices. This statement must be signed by the study director, sponsor, and data submitter.

Subsequent pages in the study volume must be sequentially numbered and also indicate the total number of pages (e.g., 4 of 10, 5 of 10, etc.). The numbering should continue sequentially through all appendices and attachments, including confidential attachments.

Each submission should be accompanied by a transmittal document that includes the name and address of each submitter; the regulatory action for which the submission is made; the name and registration number of the product; the transmittal date; a list of the volumes as they appear in the submission, with their titles and data guideline numbers; the name, signature and company name of the official making the submission; and the contact person and appropriate phone number. Frequently, this information is included in a letter which serves as the transmittal document. In such cases, a copy of the letter should be attached to the top of the submission for use in the preliminary screening of the submission. A copy of the transmittal document will be returned by EPA with the Master Record Identification Numbers (MRID) for each study when the submission has been determined to be in compliance with the formatting guidelines.

All studies submitted in support of applications for registration, registration amendments, or Experimental Use Permits must be submitted in triplicate. EPA uses one copy to prepare microfiche for storage. For this reason, all documents must be submitted on eight- by eleven-inch paper. The Agency catalogues each study submitted. Complete listings of these studies are available to the public.

Chapter 5
PESTICIDE TOLERANCES AND FOOD ADDITIVE REGULATIONS

Before a pesticide can be approved for use on food or animal feed, EPA must establish a tolerance, or an exemption from the requirements for a tolerance, for each of the product's active and inert ingredients. The Agency will not approve an application for registration under FIFRA until the applicant obtains the necessary tolerances and exemptions. 40 C.F.R. § 152.112(g). A tolerance is the maximum residue level of a pesticide (usually measured in parts per million, or ppm) that legally can be present in or on raw agricultural commodities, food, or feed transported in interstate commerce. A tolerance must be set for each raw agricultural commodity (RAC) and all processed commodities, rotational crops, and livestock in which residues of the pesticide may occur.

EPA establishes pesticide tolerances for RACs (including fresh fruits and vegetables, grains, nuts, eggs, raw milk, and meats) under § 408 of the Federal Food, Drug, and Cosmetic Act (FFDCA), 21 U.S.C. §§ 301-393 (1988), and for processed commodities under FFDCA § 409 (a § 409 tolerance technically is referred to as a "food additive regulation"). A tolerance must be set at a level that is sufficiently low to protect public health, yet high enough to allow for residues likely to remain when the pesticide is used in accordance with its EPA-approved labeling.

In establishing a tolerance, the EPA Administrator must consider several factors, including the contribution of the pesticide to the production of an adequate, wholesome, and economical food supply; the ways in which the consumer might be affected by the pesticide; and the usefulness of the pesticide. Should the tolerance record fail to demonstrate proper consideration of these factors by EPA, the tolerance is unlawful and may be set aside on judicial review. *National Coalition Against the Misuse of Pesticides v. Thomas*, 809 F.2d 875, 881-83 (D.C. Cir. 1987).

EPA publishes in the *Federal Register* notices of the receipt and withdrawal of permanent tolerance petitions, of their approval, and of the reasons for their approval. Notices of the establishment of temporary tolerances are also published, but the Agency does not publish notices of the receipt or withdrawal of temporary tolerance petitions. The Agency sets forth established tolerances and exemptions for RACs in 40 C.F.R. part 180. A short list of pesticides that are widely recognized as "safe" are identified in 40 C.F.R. § 180.2. Tolerances established for pesticides in or on raw agricultural commodities are set forth in 40 C.F.R. §§ 180.101-180.449. EPA has identified exemptions from the tolerance requirements for certain compounds when applied to growing crops in 40 C.F.R. § 180.1001(b) and for inert ingredients in 40 C.F.R. §§ 180.1001(c)-(e). EPA also has provided a list of chemical-, commodity-, and use-specific exemptions for active ingredients in 40 C.F.R. §§ 180.1002-.1092. A list of tolerances established for pesticides in processed foods is published at 40 C.F.R. part 185; a list of animal feed pesticide tolerances is published at 40 C.F.R. part 186.

FFDCA § 408 — TOLERANCES FOR RAW AGRICULTURAL COMMODITIES

The Requirement for a Tolerance

FFDCA § 301 prohibits shipment in interstate commerce of "adulterated" food. Under FFDCA § 402(a)(2)(B), a food is adulterated "if it is a raw agricultural commodity and it bears or contains a pesticide chemical which is unsafe within the meaning of section 408(a)." A raw agricultural commodity is defined in FFDCA § 201(r) as "any food in its raw or natural state, including all fruits that are washed, colored, or otherwise treated in their unpeeled natural form prior to marketing." Any pesticide is "unsafe" under FFDCA § 408(a), and a commodity containing the pesticide is adulterated under FFDCA § 402(a), unless: (1) a tolerance for the pesticide is established in or on the raw agricultural commodity and the quantity of the pesticide is within the limits of the prescribed tolerance, or (2) the pesticide has been exempted from tolerance requirements.

Tolerance Procedures

FFDCA § 408 and EPA's implementing regulations at 40 C.F.R. §§ 180.7-.35 provide detailed procedures for the establishment of pesticide tolerances for raw agricultural commodities. Effective January 5, 1991, EPA issued rules to clarify the procedures regarding objections, requests for hearings, and formal evidentiary hearings under § 408. These rules amend 40 C.F.R. part 180 and add a new part 178 (regarding objections to regulations establishing, modifying or revoking a tolerance and requests for hearings) and a new part 179 (regarding formal evidentiary public hearings).

Statutory Procedures. Section 408 of FFDCA establishes procedures for setting pesticide tolerances. Any person who holds or has applied for a pesticide registration may file a tolerance petition. FFDCA § 408(d)(1). In addition, EPA has the discretionary authority to establish tolerances and exemptions on its own initiative or in response to a request of any interested person. FFDCA § 408(e).

Once a petition for the establishment of a tolerance or exemption has been filed with EPA, the Agency is required to provide public notice in the *Federal Register* of the filing within thirty days. After reviewing the petition, EPA must either establish a tolerance for the purposes for which the product is certified as useful, or grant an exemption from the tolerance requirements. FFDCA § 408(d)(2). Alternatively (in theory at least), the petition may be referred, either by EPA or at the request of the petitioner, to an advisory committee. FFDCA § 408(d)(3). The advisory committee procedure is rarely if ever employed by the Agency.

EPA also has the authority to prescribe tolerances or exemptions by regulation, either on its own initiative or upon the request of any interested person. Thirty days after publication of such a proposed rule, EPA may publish a final regulation establishing the tolerance or exemption. However, any person who holds or has applied for a FIFRA registration for a product containing the pesticide affected by the proposal may, during the thirty-day publication period, request that the proposal be referred to an advisory committee for independent study. The advisory committee is not required to hold a hearing. Within sixty days, the committee must certify a report and make recommendations to EPA. The Administrator may, within thirty days of this certification, publish a regulation establishing a tolerance or an exemption. FFDCA § 408(e).

Any person adversely affected by EPA's decision to establish a tolerance or issue an exemption may file written objections with the Agency within thirty days after EPA's publication of the regulation. If objections are filed by a person other than the petitioner, the petitioner is provided an opportunity to review the objections and make a written reply. A public hearing is held to receive evidence relevant to the objections; at least one member of the advisory committee must be made available to testify. Following the hearing, EPA will rule on the objections and, by order, issue a regulation. FFDCA § 408(d)(5).

An order by EPA setting a tolerance or issuing an exemption is subject to judicial review in a U.S. Court of Appeals. FFDCA § 408(i). However, an Agency refusal to exercise its discretionary authority under § 408(e) to propose or amend a tolerance in response to a request is not an appealable order under § 408(i). *Nader v. EPA*, 859 F.2d 747, 750 (9th Cir. 1988).

Regulatory Procedures. EPA adopted regulatory procedures effective 1991 which set forth rules governing objections, requests for hearings, and formal evidentiary hearings under both §§ 408 and 409. 55 Fed. Reg. 50,282 (1990).

To foster similarity among the governing agencies' regulations, the part 178 procedures regarding objections and requests for hearings are derived primarily from FDA's regulations at 21 C.F.R. part 12. The new rules provide that an objection to a proposed tolerance must be in writing, signed by the objector, and must specify the basis for the objection and the relief sought. The objection must be received by the hearing clerk no later than thirty days after publication in the *Federal Register* of the order to which the objection is being taken. 40 C.F.R. § 178.25. The objector also may submit a written request for an evidentiary hearing. The mere filing of an objection will not automatically trigger a hearing. Hearings will be held only when questions of fact exist, not to settle disputes over the Agency's conclusions of law or policy. 40 C.F.R. § 178.32(b)(1).

The part 179 rules governing formal evidentiary public hearings also are derived primarily from FDA's regulations at 21 C.F.R. part 12. The Administrator must publish a notice of the hearing in the *Federal Register.* 40 C.F.R. § 179.20(a). The rules establish discovery procedures and provide that the evidence-taking portion of the hearing will be conducted before an administrative law judge. Preliminary conferences, to be held at the discretion of the presiding judge, may be employed to simplify the issues and order the presentation of evidence prior to the taking of testimony. 40 C.F.R. §§ 179.86, 179.87. Under the discovery provisions, the Office of Pesticides and Toxic Substances must produce all documents of fact and opinion within sixty days of publication of the notice of hearing and other parties must provide such information within seventy days of the publication of the notice of hearing. 40 C.F.R. § 179.83. The burden of persuasion that a tolerance regulation satisfies the statutory criteria rests with the party so arguing. 40 C.F.R. § 179.91.

The part 180 rules establish that the Administrator has discretion to provide additional procedures governing the Agency's review of a request from an interested person (under FFDCA § 408(e)) to propose or revise a tolerance. 40 C.F.R. §§ 180.29, 180.32. These additional procedures may include the Agency's publishing a notice in the *Federal Register* inviting public comment on the request. 55 Fed. Reg. at 50,287.

The rules also govern judicial review of § 408 tolerance decisions. Specifically, § 180.30 provides that judicial review of an EPA regulation establishing, modifying, or revoking a tolerance or exemption cannot be obtained prior to objections having been filed under part 178 and finally acted upon by the Agency under part 178 or part 179. A denial of a request under § 408(e) for EPA to initiate rulemaking for a tolerance is to be considered a final Agency action for which no judicial review is available under FFDCA, although review may be available under other federal statutes such as 5 U.S.C. §§ 701-706 (1988) (the Administrative Procedure Act) or 28 U.S.C. § 1331 (1988) (general federal question jurisdiction). 40 C.F.R. § 180.30(b).

The Tolerance Petition

The petition for a tolerance should follow the outline prescribed by EPA regulations set forth at 40 C.F.R. § 180.7. The petition should be assembled in accordance with the formatting requirements applicable to a pesticide registration application set forth in 40 C.F.R. § 152.50 and should follow the specific requirements of Pesticide Registration (PR) Notice 86-5 (e.g., the requirements for confidentiality claims, Good Laboratory Practice, and flagging statements) for the submission of individual study reports. The following paragraphs outline the kinds of data that normally must be submitted with the tolerance petition:

Section A must contain product chemistry data, including the chemical identity and composition of the pesticide. 40 C.F.R. § 158.155.

Section B must include the directions for use, specifying the amount, frequency, and time of application, preferably expressed in terms of the active ingredient instead of the formulated product.

Section C must contain full reports of toxicological investigations made with respect to the pesticide chemical, including detailed data from appropriate animal or other biological experiments in which the methods used and results obtained are set forth. 40 C.F.R. § 158.340.

Section D must contain data on the amount of residue remaining, including data on the analytical methods used. 40 C.F.R. §§ 158.180, 158.240. Along with the petition and data, a petitioner usually must submit an analytical grade standard sample of the pesticide so that the adequacy of the residue detection method can be evaluated.

Section E must contain data concerning the practicable methods used for removing residues that exceed the proposed tolerance. If a registrant demonstrates in Section D that residues will not exceed the proposed tolerance, however, a method for extracting residues need not be included in Section E.

Section F must include a statement of proposed tolerance levels for the pesticide and all components of the pesticide which are subject to regulation, such as its degradation products and metabolites.

Section G must demonstrate reasonable grounds to support the issuance of the proposed tolerance, including an explanation of how the submitted data corroborate the relative safety and appropriateness of the tolerance. In this section, the petitioner should compare the no-effect level shown in the most sensitive species tested in toxicity studies to the theoretical or anticipated amount of residues expected to occur in man's daily diet. This comparison is discussed in greater detail below under the heading "Safety Assessment."

Petition Fees

Section 408(o) of FFDCA requires EPA to collect fees sufficient to cover the Agency's costs in processing tolerance petitions. EPA's first fee schedule, set in 1972, went unchanged until February 1986. Under current regulations, the schedule is required to be changed annually by the same percentage as any change in the federal pay scale. In addition, the Agency is required to review processing costs "periodically" and adjust the fee schedule accordingly. 40 C.F.R. § 180.33(o). The present fee schedule is summarized in the chart below:

FEE CATEGORY	FEE
Each petition for a new tolerance or for a higher tolerance than already established up to and including nine raw agricultural commodities	$48,225, plus $1,200 for each commodity more than nine
Each petition for a lower tolerance or tolerance on additional commodities of the same level	$11,025, plus $775 for each raw agricultural commodity including the first

FEE CATEGORY	FEE
Each petition for exemption from a tolerance or repeal of an exemption	$8,875
Each petition for temporary tolerance or temporary exemption $19,250	
Renew or extend temporary tolerance	$2,725
Each petition for temporary tolerance at the same or higher level for pesticides having tolerances for other uses	$9,625, plus $775 for each raw agricultural commodity, including the first
Each petition for repeal of a tolerance	$6,025
Each petition for a crop group tolerance	100 percent of the analogous category for a single tolerance that is not a crop group petition, regardless of the number of raw agricultural commodities
Filing of objections	$2,425
Advance deposit to cover costs of advisory committee	$24,050
Further advance deposits	$24,050
Petition for judicial review of an order under FFDCA §§ 408(d)(5) or (e)	Cost of preparing the record on which the order is based but automatically waive this requirement for those who have no financial interest in the petition for judicial review
Fee waivers for the program	Fees will be automatically waived for the IR-4 program and may be waived for small businesses

The Administrator has the discretionary authority to waive or refund part or all of any tolerance petition fee if he determines that such a waiver or refund will promote the public interest or that payment of the fee would work an unreasonable hardship. 40 C.F.R. § 180.33(m). Ironically, EPA requires a fee of $1,200 for each request of a waiver or refund. However, if EPA grants a requested waiver, this fee will be refunded as well. If a petition or request is not accepted for processing because it is technically incomplete, EPA will return the application and the fee, less $1,200 for handling and initial review. 40 C.F.R. § 180.33(g). The fee (but not the petition) is sent to EPA's Headquarters Accounting Branch in Pittsburgh. In order for payments to be properly credited and acknowledged, the check must be accompanied by a copy of the letter or petition requesting the tolerance.

EPA Review

Once a tolerance petition is submitted to the Agency, the product manager responsible for the subject chemical will forward the accompanying data to the residue chemistry and toxicology branches for review. These EPA branches will then determine whether the data support the establishment of the tolerance. Although FFDCA § 408 requires EPA to review petitions within ninety days, reviews typically take a year or longer to complete.

Safety Assessment

The EPA reviewers determine whether the pesticide can be used with reasonable certainty that injury to man will not result even after a lifetime of exposure. A risk assessment based on the toxicology and residue data is performed in the following manner:

Non-Carcinogenic Endpoints.

1. The no-observable-effect level (NOEL) for the pesticide is determined. This is the dosage level at which the pesticide shows no observable or measurable effects in dosed animals when compared to control animals. The NOEL is expressed in terms of the amount of the pesticide consumed (in milligrams) by the test animal per day for every kilogram (kg) of body weight.
2. The NOEL is divided by a safety factor to determine the Reference Dose (RfD) (formerly the Acceptable Daily Intake) which is the level of daily exposure which is not expected to create appreciable risks over a human lifetime, for the pesticide residues. A safety factor is used by EPA to compensate for the uncertainty in extrapolating laboratory animal data to

man. This safety factor will vary depending upon the type of data and chemical evaluated and may range from 100 to 2,000. EPA may use a different safety factor if the pesticide is a cholinesterase inhibitor. The Agency's policy on the regulation of cholinesterase inhibition as an adverse effect, however, is in the process of being formulated and presently is unclear.

3. The maximum permissible intake (MPI) of residues for an average human adult is calculated by multiplying the RfD by 60 kg, the assumed average weight of an adult.

4. The reviewers determine the theoretical maximum residue contribution (TMRC), which is the total theoretical amount of the pesticide's residues that may be expected to occur in a person's daily diet, assuming that all commodities are treated and contain residues at the tolerance levels. Alternatively, EPA may evaluate the anticipated residue contribution (ARC) where sufficient acceptable residue data have been submitted to warrant the Agency's diversion from the more conservative TMRC method of analysis. The TMRC or ARC is calculated by multiplying the tolerance level or anticipated residue level, respectively, by the daily quantity of each commodity consumed and totalling the values for all commodities containing residues. In performing this dietary exposure analysis, EPA will use its Dietary Risk and Exposure System (DRES), which incorporates average consumption data for the general population as well as specific, potentially sensitive population subgroups such as pregnant women, infants, and adolescents.

5. The RfD is compared with the TMRC or ARC. If the TMRC or ARC is less than the RfD, the tolerance may be granted; if higher, the petition ordinarily will be subject to further review. In some instances, EPA will establish a tolerance even though the TMRC exceeds the RfD. The Agency recognizes that the TMRC calculation provides an unrealistic, worst-case assessment since residues ordinarily would not be expected to occur at the tolerance levels in

all commodities, the pesticide rarely is applied to the total production of a commodity, and the processing of a commodity before it is eaten usually reduces residue levels.

Carcinogenic Endpoints. If the pesticide is found to be a carcinogen, the Agency usually performs a quantitative risk assessment for the cancer risk in addition to the above-mentioned calculations. See Chapter 9.

Temporary Tolerances

A registrant may wish to obtain a temporary tolerance for experimental products or uses that have not been registered. Under FFDCA § 408(j), EPA may establish a temporary tolerance on its own initiative or at the request of any person who holds an Experimental Use Permit (EUP). A temporary tolerance allows participants in the EUP program to sell or use commodities harvested from the EUP test plots.

The types of data needed to establish a temporary tolerance vary with the particular use and toxicity of the subject pesticide. The data requirements are similar to those for permanent tolerances, but generally do not include chronic feeding, oncogenicity, or reproduction study requirements. 40 C.F.R. pt. 158.

Import Tolerances

Residues of a pesticide not registered in the United States may be present in raw agricultural commodities and processed foods that are produced abroad and imported into the United States. Under FFDCA §§ 408 and 409, these imported commodities are considered adulterated unless EPA has established a tolerance for the pesticide.

Generally, a petition requesting a tolerance must demonstrate that the petitioner "has registered or has submitted an application for the registration of" the pesticide under FIFRA. 40 C.F.R. § 180.7(b). In some instances, however, a particular pesticide (or a particular use of that pesticide) may not be intended for use in the United States and, therefore, has never been the subject of an application for United States registration under FIFRA. Under such circumstances, the petitioner must request EPA to exercise its discretion to propose an "import tolerance" under the authority of FFDCA § 408(e).

The concept of an "import tolerance" is ill-defined. EPA regulations are silent as to situations where a tolerance should be sought for a pesticide not registered in the United States and contain no mention of import tolerances. The Agency has not issued any regulatory guidance as to when an import tolerance should be sought, or established any procedures or data requirements for obtaining an import tolerance. Import tolerance petitions in practice are processed in the same general manner as petitions for domestically registered products. EPA has asserted, however, that data submitted to the Agency in support of an import tolerance by a person other than a registrant or an applicant for registration are not protected by the rights to data confidentiality and compensation provided by FIFRA §§ 3 and 10, although petitioners may otherwise claim that the submitted data are confidential. 55 Fed. Reg. 50,282, 50,284 (1990); 40 C.F.R. § 177.81(f).

Exemptions from Tolerance Requirements

EPA will grant an exemption from the tolerance requirement when it appears that no hazard to public health will result from residues of a pesticide. 40 C.F.R. § 180.1001(a). The procedures for exempting a pesticide from the tolerance requirement are the same as those for establishing a permanent tolerance.

Exemptions from the tolerance requirement must be obtained for all inert ingredients in formulations applied to raw agricultural commodities (RACs). Requirements for inert ingredient exemptions are not well defined but are described in general terms in an EPA policy statement set forth at 52 Fed. Reg. 13,305 (1987). Requests for such exemptions are handled on a case-by-case basis. The primary types of data needed to support these exemptions are residue chemistry, product chemistry, and toxicology data, including subchronic toxicity, teratology, and mutagenicity studies. In some instances, ecotoxicity and environmental fate data also may be required, depending upon the intended uses of the inert ingredient. Before compiling data to support a tolerance exemption for an inert ingredient, an applicant should discuss with the Registration Division the types of data required to support the exemption. Exemptions for certain inert ingredients — for example, chemicals approved by FDA as food additives — may be obtained with minimal data.

Action Levels, Regulatory Limits, and Revocation of Tolerances

FDA's basic enforcement authority provides the Agency with the discretion to decide when an enforcement action is warranted. In its efforts to exercise this discretion in an orderly manner, FDA has supplemented tolerance setting and enforcement procedures with "action levels" and "regulatory limits." Like tolerances, action levels and regulatory limits are set to limit the quantity of a pesticide in or on food commodities to the extent necessary to protect the public health. Action levels were first used as an informal enforcement method to deal with low levels of unavoidable residues in situations where a pesticide tolerance either had not been issued or had been revoked, but the pesticide did not degrade quickly and persisted in the environment.

Action levels, to the extent they were originally used by FDA to specify acceptable levels of chemical substances in foods, were found to constitute substantive rules rather than general policy statements. *Community Nutrition Institute v. Young*, 818 F.2d 943 (D.C. Cir. 1987) (*CNI v. Young*). Because they had not been promulgated in accordance with the notice and comment procedures of the Administrative Procedure Act, FDA's then-existing action levels were found to be invalid. *CNI v. Young*, 818 F.2d at 950.

Following *CNI v. Young*, FDA amended its procedures for action levels for residues of certain cancelled pesticides in food and feed under FFDCA § 408 and for added poisonous or deleterious substances in food under FFDCA § 406. FDA continues to use action levels to deal with unavoidable pesticide residues in food and feed (including residues of pesticides whose tolerances have been revoked), but has stated that action levels merely provide informal guidance and are not binding on the courts, the public, or FDA. 55 Fed. Reg. 14,359 (1990). Action levels do not have the force of law of substantive rules, nor do they create immunity from enforcement for the food or feed industries. 55 Fed. Reg. at 14,360.

In 1990, FDA developed "regulatory limits," an additional mechanism for enforcement of pesticide and other residues under FFDCA. 55 Fed. Reg. 20,782 (1990). Regulatory limits, unlike action levels, are substantive rules created by FDA through informal notice and comment rulemaking, and thus have the force of law and are binding on FDA, the courts, and the public. FDA adopted the concept of regulatory limits in response to public comments that non-binding action levels by themselves would be insufficient to regulate unavoidable illegal residues in food and feed. According to FDA's policy, a regulatory limit for a pesticide residue may be established when: (1) the substance cannot be avoided by current good manufacturing practices; and (2) there is no tolerance established for the substance in the particular food under FFDCA sections 408 or 409. 55 Fed. Reg. at 20,785.

FFDCA § 409 — FOOD ADDITIVE REGULATIONS FOR PROCESSED FOODS

The Requirement for a Food Additive Regulation

Under FFDCA § 402(a)(2)(C), a food is "adulterated" if it contains a food additive which is "unsafe" within the meaning of FFDCA § 409. FFDCA § 201(s) broadly defines a food additive as any substance added directly or indirectly to food which is not generally regarded as safe, but excludes "a pesticide chemical to the extent that it is intended for use or is used in the production, storage, or transportation of any raw agricultural commodity." Significantly, the definition of food additive does not exclude pesticide residues which appear in processed foods. Thus, pesticides that are present in processed foods, whether added directly (e.g., during processing), or indirectly (e.g., through use on the raw agricultural commodities being processed or on processing equipment), intentionally or unintentionally, are regulated as "food additives" under FFDCA § 409.

Pesticidal food additives include residues from fumigants, insecticides, or other pesticides used in processing and packaging foods. (For a list of pesticides permitted in food for human consumption, see 21 C.F.R. part 193, subpart A; for pesticides which become food additives through contact with containers or equipment, see subpart B.) Since the definition of "food" also includes animal food under FFDCA § 201(f), EPA has established tolerances under § 409 for pesticides used in animal feed. These tolerances have been codified at 21 C.F.R. part 564.

FFDCA § 402(a)(2) adds an important proviso governing processed foods resulting from the canning, cooking, freezing, dehydrating, or milling of raw agricultural commodities that contain residues of pesticides used on the RACs in accordance with an exemption or tolerance under FFDCA § 408. Under this proviso, a pesticide residue in a processed food is not unsafe (and thus the food is not adulterated) if the residue "has been removed to the extent possible in good manufacturing practice and the concentration of such residue in the processed food when ready to eat is not greater than the tolerance prescribed for the raw agricultural commodity." FFDCA § 402(a)(2)(C). Consequently, processed foods containing pesticide residues at or below the FFDCA § 408 tolerance level for the subject raw agricultural commodities are covered by the RAC tolerance and do not require an FFDCA § 409 tolerance (i.e., food additive regulation). If, on the other hand, processing concentrates a pesticide residue so that it exceeds the established FFDCA § 408 tolerance for the raw commodity, the pesticide is considered an "unsafe" food additive and renders the food adulterated, unless a § 409 tolerance is established.

The issue of whether pesticide residues may be considered a "food additive" within the meaning of FFDCA § 201(s) was addressed in *United States v. Ewig Bros. Co.*, 502 F.2d 715 (7th Cir. 1974). The case arose from the discovery of DDT and dieldrin in smoked chubs. The court concluded that, before processing, DDT and dieldrin were "pesticide chemicals" in a raw commodity, and that, after processing, the chemicals were "food additives." Since, at that time, no tolerances had been established for DDT or dieldrin in fish, the presence of these pesticides in the raw fish rendered the fish adulterated due to the lack of a tolerance under FFDCA § 408. Similarly, the processed fish were adulterated because they contained unsafe pesticidal food additives as a result of the absence of a tolerance under FFDCA § 409.

Food Additive Regulation Procedures

FFDCA § 409 provides detailed procedures and criteria regarding the establishment of food additive regulations, or tolerances, for pesticide residues in or on processed foods. Section 409, however, does not spell out the procedures to be used to modify or revoke a processed food tolerance. Unlike EPA's approach to RAC tolerances under § 408, the Agency historically had not established procedures governing § 409 tolerances. In the past, this lack of regulatory procedures was not problematic— EPA generally followed the existing rules established under § 408. In order to provide for consistent rules applicable to tolerance procedures under both sections 408 and 409, EPA adopted rules effective in 1991 which establish procedures for tolerance petitions, modifications, and revocations under § 409 and make applicable to § 409 the same rules regarding objections, requests for hearings, and formal evidentiary hearings that apply to § 408. 55 Fed. Reg. 50,282 (1990) (codified at 40 C.F.R. pts. 177-180).

Statutory Procedures. Section 409 provides detailed procedures regarding the establishment of processed food tolerances. Any person may file a petition proposing the issuance of a § 409 tolerance. FFDCA § 409(b)(1). In addition, under § 409(d), EPA has the authority to establish tolerances on its own initiative. Although § 409 vests authority to make tolerance decisions in the Secretary of the Department of Health and Human Services (and not the EPA Administrator), the Secretary's authority under § 409 with regard to pesticide tolerances has been delegated to the Administrator of EPA.

Under § 409(b)(5), EPA is required to publish in the *Federal Register* notice of the filing of a petition for the establishment of a processed food tolerance within thirty days. After reviewing the petition, the Agency must determine the conditions under which the pesticide chemical may be safely present in or on the subject processed food and must either deny the petition, or by order, issue a regulation establishing the requested tolerance. FFDCA § 409(c)(1). Although the Administrator must take such action within ninety days (extendable to 180 days) according to § 409(c)(2), EPA rarely is able to meet this statutory response period.

EPA also has the authority to propose a § 409 tolerance regulation on its own initiative. After thirty days following the publication of such a proposed rule, the Agency may, by order, establish a final regulation setting the subject processed food tolerance. FFDCA § 409(d).

Any person adversely affected by an EPA order establishing or denying a tolerance under § 409 may file written objections with the Agency within thirty days of EPA's publication of its order. Following the submission of such objections, EPA is to promptly hold a public hearing for the purpose of receiving evidence relevant to the objections. As soon as practicable after the hearing, EPA will rule on the objections by order. FFDCA § 409(f).

An EPA order setting a processed food tolerance is subject to judicial review in a U.S. Court of Appeals. FFDCA § 409(g). In *Nader v. EPA*, 859 F.2d 747 (9th Cir. 1988), *cert. denied*, 109 S. Ct. 1931 (1989), the U.S. Court of Appeals for the Ninth Circuit ruled that judicial review under § 409(g) cannot be invoked by an affected party unless that party has filed written objections with the Agency under § 409(f) and such objections have been ruled upon by EPA.

Regulatory Procedures. Regulatory procedures under § 409 include part 177 (regarding the establishment, modification, and revocation of food additive regulations) and parts 178 and 179 (discussed above with regard to FFDCA § 408) which apply equally to processed food tolerance procedures for objections and hearings.

There are two important points that should be noted with regard to the rules governing § 409 tolerances. First, the § 409 rules do not limit the time EPA may have to take action on a petition. FFDCA § 409 requires the Agency to issue a decision on a tolerance petition within ninety days (extendable to 180 days) after it is filed. The rules do not include such a "clock," because the Agency views such a timetable as unrealistic. Furthermore, if any petitioner insists on having a decision on its petition within the statutory time limit, EPA expects to issue a denial of the petition if its evaluation is not complete on the ground that the Agency has not been able to conclude that the statutory criteria for approval have been met. 55 Fed. Reg. at 50,283-84. Second, the rules require a § 409 petition to include both a full statement of the factual and legal grounds as to why approval would be reasonable and any other information known to the petitioner that is unfavorable to the petition. 40 C.F.R. § 177.102(h), (i). The petitioner must also include a summary of the application suitable for publication in the *Federal Register.* 40 C.F.R. § 177.102(j). EPA believes that, because FFDCA § 409(b) does not require publication of a proposed rule establishing a processed food tolerance, the requirements for inclusion of the factual and legal grounds for the petition and a publishable summary of the petition should allow for more informed public comment in response to an Agency notice regarding the filing of a petition. EPA further has noted that, instead of merely approving or denying a petition, it may issue a proposed rule and invite public comment in cases where the complexity or controversiality of the issues raised by the petition warrant the Agency's seeking public comment. 53 Fed. Reg. 41,126, 41,128 (1988).

The Delaney Clause

The Delaney Clause creates a prohibition on the use of any "food additive" which has been found, "after tests which are appropriate for the evaluation of the safety of food additives, to induce cancer in man or animal." FFDCA § 409(c)(3)(A). See also Chapter 9. The Delaney Clause is found not only in the food additive provision, but also at FFDCA § 512, concerning new animal drugs, and at FFDCA § 706, concerning color additives. Each version of the Delaney Clause also contains the so-called "DES proviso," which provides that the Delaney Clause does not apply if the use of certain oncogenic substances in animal feed leave no residue in the animal after slaughter, or in any food product obtained from the animal, and which therefore would be safe for human consumption.[9]

The Delaney Clause does not automatically apply to pesticides used on food crops or in animal feed unless a § 409 tolerance is required. In *Environmental Defense Fund, Inc. v. HEW*, 428 F.2d 1083, 1091 (D.C. Cir. 1970), the court held that the Delaney Clause does not apply to pesticide chemicals because FFDCA § 201(s) expressly excludes from the term "food additive" pesticides present in raw agricultural commodities and pesticides used in the production, storage, or transportation of such commodity.

However, the Delaney Clause *does* apply automatically to pesticides used in food processing and packaging, because they require food additive tolerances under FFDCA § 409. *United States v. Ewig Bros. Co.*, 502 F.2d 715, 723 (7th Cir. 1974) ("the food additive definition [is] broad enough to encompass pesticide chemical residues in processed food"). The Delaney Clause also applies when the processing of a raw agricultural commodity causes the pesticide residue to concentrate so as to exceed the RAC tolerance, because in that case a food additive regulation must be established. Reorg. Plan No. 3 of 1970, 40 C.F.R. pt. 1 (1970), *reprinted in* 5 U.S.C. app. at 1343-47 (1988), *and in* 84 Stat. 2086 (1970).

There is widespread scientific agreement that the Delaney Clause now reflects dated judgment concerning chemical carcinogenesis, at odds with advances in scientific understanding. Despite this greater understanding, the Delaney Clause makes no distinction regarding dose levels, potency, risk, or correlation between cancers in particular animals and in man and, therefore, displaces a rational approach to regulating cancer risk.

---

9 The Delaney Clause initially did not distinguish between compounds added directly to human food and compounds that might indirectly enter human food by virtue of having been administered to food-producing animals. FDA interpreted the Delaney Clause as prohibiting the approval of a carcinogenic substance for use in animals. Congress thereafter amended the three versions of the Delaney Clause by making them inapplicable to chemicals when used for food-producing animals under certain conditions.

In an extended written discussion of the Delaney Clause published in 1982, FDA noted that scientists' capability to detect minor constituents in chemical substances has grown tremendously and that many such minor chemical constituents are carcinogenic. Indeed, FDA acknowledged its "frank expectation that a growing number of additives will be found to contain a carcinogenic chemical in future years," and that, interpreted without qualification, the Delaney Clause would force FDA to "refuse to approve or to terminate approval of the use of each of these additives, even though they themselves may be safe." 47 Fed. Reg. 14,464, 14,466 (1982). In order to obviate this result, FDA and EPA have advanced three interpretations of the Delaney Clause: the *de minimis* concept, the constituents doctrine, and the sensitivity-of-method (SOM) approach.

De Minimis Concept

The *de minimis* approach originates from case law which states that an administrative agency generally has authority to avoid applying terms of a statute literally when to do so would yield pointless results. 53 Fed. Reg. 41,104, 41,107 (1988); *Alabama Power Co. v. Costle*, 636 F.2d 323 (D.C. Cir. 1979). EPA has pointed out that there are two conditions necessary to permit such an interpretation:

> First, the problem that would be addressed by regulation must be trivial in fact, such that no real benefit would result from regulation. Second, the legislative design must allow the Agency not to apply the statute literally in such a case.

53 Fed. Reg. at 41,107.

The *de minimis* exception to the Delaney Clause was first employed by FDA. In *Monsanto Co. v. Kennedy*, 613 F.2d 947 (D.C. Cir. 1979), the U.S. Court of Appeals for the D.C. Circuit held that "there is latitude inherent in the statutory scheme to avoid literal application of the statutory definition of 'food additive' in those *de minimis* situations that, in the informed judgment of the Commissioner, clearly present no public health or safety concerns." 613 F.2d at 954. While *Monsanto* overtly dealt with the question of physical migration of components of a putative food additive into food, FDA construed it to stand for the broader principle that the FFDCA permits regulatory flexibility with respect to chemical substances that present no real safety concern.

However, a more recent case struck down FDA's *de minimis* approach to the Delaney Clause with regard to color additives. *Public Citizen v. Young*, 831 F.2d 1108 (D.C. Cir. 1987), *cert. denied*, 485 U.S. 1006 (1988). The court held that the legislative history of the FFDCA color additive provisions and the language of the provision itself were rigid and left no room for a *de minimis* interpretation. 831 F. 2d at 1113. The court deliberately left open, however, the question of whether a *de minimis* interpretation would be acceptable for the food additive provisions. 831 F.2d at 1119-20.

Since the court's decision in *Public Citizen*, and in response to the 1987 report by the National Academy of Sciences (NAS), commissioned by the Agency in 1985, EPA has announced that, "for pesticides that pose at most a negligible risk of cancer and whose use requires section 409 clearances, EPA will change its current practice to the extent that, in the future, EPA will propose to issue food additive regulations on the basis of the *de minimis* doctrine." 53 Fed. Reg. at 41,105.

The NAS study, "Regulating Pesticides in Food: The Delaney Paradox," suggested, in pertinent part, that pesticide residues in food, whether raw or processed or governed by new or old tolerances, should be based on consistent standards. Furthermore, the study recommended that a negligible risk standard for carcinogens in food could dramatically reduce total dietary exposure to oncogenic pesticides with modest reduction of benefits. 53 Fed. Reg. at 41,104. The "paradox" arises because a literal reading of the Delaney Clause interferes with EPA's ability to discriminate between significant and insignificant risks. EPA endorsed the study's suggestions and noted that "[n]ew pesticides that pose lower cancer risks than pesticides currently on the market have been denied registration while older, more hazardous pesticides remained in use." 53 Fed. Reg. at 41,108.

EPA's initial response in October 1988, announcing it would apply a negligible risk standard to the Delaney Clause in future rulemaking proceedings for the adoption of tolerances under § 409, was followed in April 1990 with EPA's announcement that it would apply the same standard to currently-registered pesticides. This policy was made clear in the Agency's notice of response to a petition to revoke food additive regulations for certain uses of benomyl, chlordimeform, dichlorvos, dicofol, mancozeb, phosmet, and trifluralin. 55 Fed. Reg. 17,560 (1990). By consistently applying the *de minimis* exception, the Agency believes it can give practical construction to the Delaney Clause and apply FIFRA's basic risk/benefit approach to pesticide regulation. "The *de minimis* exception simply recognizes that the Delaney Clause should not be a bar to the use of a pesticide that satisfies the statutory standard of FIFRA for a carcinogenic pesticide which presents a trivial risk." 55 Fed. Reg. at 17,565-66.

More recently, EPA reiterated its commitment to a *de minimis* standard for both new and previously registered pesticides in its action of August 29, 1990. In that action, EPA refused a petition to revoke the food additive tolerance for ethylene oxide (EtO), a fumigant used on spices and seasonings, on grounds that the current data base was insufficient to estimate the risk from consumption of EtO and, therefore, revocation was not warranted. 55 Fed. Reg. 35,347, 35,351 (1990).

A case challenging EPA's interpretation of the Delaney Clause has been filed by the State of California and others. *Van de Kamp v. Reilly*, No. 89-0752 (E.D. Cal. filed May 2, 1989). Plaintiffs in *Van de Kamp* asserted that EPA was not properly applying the Delaney Clause by failing to require and evaluate data and establish § 409 tolerances for pesticides that concentrate in processed foods.

Legislative Initiatives

As a result of requests by EPA, as well as increased attention from the media and the public, Congress and the Administration have proposed measures to modify the existing food safety laws. The Administration's plan, unveiled in October 1989, included a proposal to replace the Delaney Clause with a negligible risk standard. Other proposed legislation in Congress also has addressed the Delaney Clause, but differences between the congressional initiatives and the Administration's plan have not been resolved.

Constituents Doctrine

The constituents doctrine holds that the Delaney Clause applies only where additives are carcinogenic as a whole. Relying on the language of the clause itself, constituent elements of food additives, even if carcinogenic, are not held to be covered by the Clause. Under the constituents doctrine, FDA approved the color additive D&C Green No. 6, even though its p-toluidine constituent caused cancer in mice. 47 Fed. Reg. 14,138, 14,139 (1982).[10]

Following FDA's lead, EPA also has used the constituents doctrine, as when EPA decided to establish a tolerance for residues of the pesticide dicamba in or on sugarcane molasses, even though dicamba contains a carcinogenic constituent. EPA explained its decision as follows:

> The Delaney Clause requires the disapproval of any food or feed additive that has been shown to be a carcinogen in animal studies *but does not say that an additive shall be disapproved if any of its impurities are found to induce cancer.* . . . The impurity should be judged under the general safety provisions of the FFDCA, using risk assessment as one of the decision making tools.

48 Fed. Reg. 34,024, 34,025 (1983) (emphasis added). EPA added that its constituents policy applies to impurities resulting from manufacture of a pesticide and not to intentionally added active or inert ingredients. 48 Fed. Reg. at 34,024.

Sensitivity-of-Method Approach

In 1982, FDA announced its decision to adopt the SOM approach to implement the DES proviso of the Delaney Clause. 47 Fed. Reg. 14,464 (1982). Under the SOM

---

10 Under the constituents doctrine, FDA has approved several other food and color additives: C.I. Vat Orange No. 1., 50 Fed. Reg. 20,405 (1985); a mixture containing 5-chloro-2-methyl-4-isothiazolin-3-one, 2-methyl-4-isothiazolin-3 one, and magnesium nitrate, 50 Fed. Reg. 4643 (1985); and dimethylamine-epichlorohydrincopolymer as a flocculent and/or decolorizer in the clarification of refinery sugar liquors and juices. 49 Fed. Reg. 13,016 (1984).

approach, FDA declines to classify as a food additive per se any carcinogenic constituent (of a non-carcinogenic food or color additive) that is undetectable by an FDA-approved method, as long as the cancer risk posed by any amount of the constituent present is below the lifetime cancer risk of one in one million. 52 Fed. Reg. 49,572 (1987).

EPA has used the SOM approach in two actions establishing food additive regulations, one concerning cyromazine and the other concerning thiodicarb. For cyromazine, EPA had to decide whether to issue a "feed additive" regulation to permit residues of the insecticide cyromazine in or on poultry feed. 49 Fed. Reg. 18,120 (1984); 50 Fed. Reg. 20,370 (1985). For thiodicarb, EPA had to make the same decision with respect to residues of thiodicarb in or on cottonseed and soybeans. 50 Fed. Reg. 27,452 (1985); 50 Fed. Reg. 41,341 (1985). Cyromazine was intended for use as an additive to the feed of poultry or other animals which are raised for production of eggs, meat, or milk for human consumption; thiodicarb was intended for use as an additive for cottonseed and soybeans, the hulls of which are used as animal feed for animals which may be consumed by humans. EPA decided to follow FDA's lead in interpreting the DES proviso. 49 Fed. Reg. at 18,123.

As a result of the foregoing approaches employed by FDA and EPA, the Delaney Clause has been diminished as an obstacle to regulatory approval of the use of putative carcinogenic food or color additives, including pesticide chemicals.

MONITORING AND ENFORCEMENT

The regulatory, monitoring, and enforcement responsibilities of the federal government with regard to pesticide residues in food are divided among three authorities — EPA, FDA, and USDA — under six federal statutes — FIFRA, 7 U.S.C. §§ 136-136y (1988); FFDCA, 21 U.S.C. §§ 310-393 (1988); the Pesticide Monitoring Improvements Act, 21 U.S.C. §§ 1401-1403 (1988); the Federal Meat Inspection Act, 21 U.S.C. §§ 601-695 (1988); the Poultry Products Inspection Act, 21 U.S.C. §§ 451-470 (1988); and the Egg Products Inspection Act, 21 U.S.C. §§ 1031-1056 (1988).

As discussed above, EPA establishes pesticide tolerances for raw agricultural commodities, processed foods, and feed for animals used in food production. EPA also validates the analytical methods used for enforcement purposes. FDA and USDA share the responsibility for monitoring pesticide residues and, through seizure and injunctive actions, for enforcing the pesticide tolerances set by EPA. USDA monitors meat, poultry, and egg products (except whole, raw eggs) for pesticide residues and reports this information to FDA. FDA is responsible for monitoring pesticide residue incidences and levels in all other pertinent commodities.

FDA

Under FFDCA §§ 402(a)(2)(B) and 408(a), a raw agricultural commodity that contains an illegal pesticide residue — i.e., a residue which exceeds the tolerance established by EPA or for which there is no applicable tolerance or exemption — is an "adulterated" food. Likewise, §§ 402(a)(2)(C) and 409(a) of FFDCA provide that a processed food that contains an illegal pesticide residue also is considered an "adulterated" food. FDA has authority under FFDCA § 304 to seize, condemn, and destroy any food that is adulterated within the meaning of the Act.

Under its monitoring program, FDA collects and analyzes samples of individual lots of domestically produced and imported foods for pesticide residues. FDA has the ability to test for residues of approximately 250 different pesticide chemicals, and the analytical methods used to do so generally are capable of detecting and quantifying residue levels well below the EPA-established tolerance limits.

In 1988, for example, FDA analyzed over 18,000 samples of food from the United States and eighty-nine foreign countries for pesticide residues. Food and Drug Administration Pesticide Program: Residues in Foods - 1988, 72 J.A. Official Analytical Chemists (Sept./Oct. 1989). FDA anticipates that sampling will increase, and has announced a plan to analyze about 20,000 samples of food commodities from the United States and over 100 foreign countries for residues of 600 to 700 pesticide active ingredients.

In general, FDA considers that the presence of pesticide residues in the U.S. food supply does not pose a significant risk to human health. Of the over 18,000 samples of domestic and imported food analyzed by FDA in 1988, over 96 percent did not contain violative residues. No residues were detected in 61 percent of the analyzed samples, and less than 1 percent contained residues in excess of tolerance levels established by EPA. The remainder of the violative residues — i.e., those that were violative because of reasons other than exceeding an EPA-set tolerance — arose where no applicable tolerances existed for either the detected pesticide or the particular pesticide/commodity combination.

In 1988, Congress enacted the Pesticide Monitoring Improvements Act (PMIA), 21 U.S.C. §§ 1401-1403 (1988). FDA is implementing the PMIA in connection with its pesticide tolerance monitoring and enforcement duties under FFDCA §§ 408 and 409. The PMIA has three basic mandates: (1) the implementation of computerized management systems to allow FDA to record and evaluate the results of its pesticide residue monitoring program for food products, and assist EPA in carrying out its responsibilities under FIFRA; (2) the pursuit of cooperative agreements with foreign countries that are major sources of foods imported into the United States for the purpose of obtaining information regarding pesticides used on food products intended for export to the United States; and (3) the improvement of FDA's monitoring program through the implementation of an extensive analytical methodology research program, including the development of new and improved methods for rapid residue detection and for simultaneous detection of multiple pesticide residues. In the future, EPA and FDA may require pesticide registrants to validate multiple residue analytical methods in order to obtain registration of compounds for use on food or feed crops.

USDA

The United States Department of Agriculture is responsible for inspecting meat, poultry, and egg products (except whole, raw eggs, which are subject to inspection and monitoring by FDA) to ensure that they are not unwholesome, adulterated, or misbranded. USDA's authority to conduct such inspections is established under the Federal Meat Inspection Act (FMIA), 21 U.S.C. §§ 601-695 (1988), the Poultry Products Inspection Act (PPIA), 21 U.S.C. §§ 451-470 (1988), and the Egg Products Inspection Act (EPIA), 21 U.S.C. §§ 1031-1056 (1988). Meat, poultry, and egg products are deemed "adulterated" if they are found to contain pesticide residues at levels prohibited by the FMIA, PPIA, and EPIA implementing regulations, or in excess of the levels permitted under FFDCA. 9 C.F.R. §§ 301.2(c)(2)(ii), (iii), 318.16 (pesticide chemicals and other residues in meat products); §§ 381.1(b)(4)(ii)(b), (c), 381.74 (biological residues in poultry products); 7 C.F.R. §§ 59.5, 59.435(a)-(c) (chemical additives and other substances used in the manufacture or preparation of egg products). USDA has authority to detain, seize, and condemn any meat, poultry, or egg product that is adulterated within the meaning of the pertinent acts. PPIA §§ 467a, 467b, 467f; 9 C.F.R. §§ 381.210, .217; FMIA §§ 672, 673, 679; 9 C.F.R. §§ 329.1, .6; EPIA §§ 1033(a), 1034(c), 1048, 1049; 7 C.F.R. §§ 59.5, .240, .422.

Inspections of meat and poultry are conducted by USDA's Food Safety and Inspection Service (FSIS). Egg products (other than whole, raw eggs) are inspected by USDA's Agricultural Marketing Service.

Under its National Residue Program, the FSIS conducts regular sampling and analysis of domestic and imported meat and poultry products for illegal residues of pesticides, animal drugs, and other potentially hazardous chemicals.

Meat and poultry samples are not analyzed for all pesticides, only those which are most likely to present the greatest potential risk. The list of compounds for which the FSIS conducts testing varies from year to year, with compounds being deleted or added depending on the likelihood of residues and the potential risks to humans. The FSIS currently intends to analyze over 70,000 domestic samples and over 22,000 samples of imported meat and poultry products for residues of 113 compounds (including pesticides, animal drugs, and other chemicals).

The FSIS generally considers that pesticide residues in meat and poultry do not pose a threat to the public health in the United States, and, in agreement with the positions taken by both EPA and FDA, believes that the United States food supply is safe. Of the over 41,000 domestic samples analyzed by the FSIS for pesticides, animal drugs, and environmental contaminants during its 1989 National Residue Program, only 0.3 percent of the samples indicated illegal residue levels.

Chapter 6
REREGISTRATION, DATA CALL-INS AND JOINT TESTING VENTURES

Reregistration is the process by which EPA reassesses older pesticides under current regulatory and scientific standards. In an effort to accelerate reregistration, Congress in 1988 amended FIFRA and added a new five-phase reregistration program. EPA is required to reregister all pesticides in nine years. As part of or independent of reregistration, FIFRA authorizes the Administrator to require any registrant to submit additional data to the Agency. FIFRA § 3(c)(2)(B). This chapter reviews EPA's reregistration program as amended by the FIFRA '88 amendments, the Agency's data call-in authority, and the joint venture approach to developing data.

Before FIFRA was amended in 1988, FIFRA § 3(g) required the Administrator to reregister all pesticides "in the most expeditious manner practicable." Despite this command, EPA's reregistration of the approximately 600 existing active ingredients has been slow. In response to this lack of progress Congress amended FIFRA in 1988, adding a new § 4, which requires EPA to reregister within nine years all pesticides first registered before November 1, 1984, in a five-phase program.

Section 4, as amended, requires registrants of active ingredients to identify data gaps. EPA will then publish a data call-in notice under FIFRA § 3(c)(2)(B) requiring all registrants of the active ingredient to submit this additional data within a specified period. After data on the active ingredient and end-products are submitted, EPA will determine if the pesticide is eligible for reregistration. In addition to requiring additional data as part of reregistration, FIFRA § 3(c)(2)(B) authorizes the Administrator to require registrants to submit additional data at any time. Registrants may develop data required for reregistration or in response to a data call-in notice either on their own or together with other registrants.

THE REREGISTRATION PROGRAM

Congress and Reregistration

Since 1972 Congress has urged EPA to examine the safety of pesticides containing previously registered pesticides using current regulatory and scientific standards. The 1972 amendments to FIFRA required EPA to reregister all pesticides within four years. Congress extended this deadline in 1975 to October 21, 1977.[11] This deadline passed, and Congress in 1978 added the term "reregistration" to FIFRA for the first time.

> The Administrator shall accomplish the reregistration of all pesticides in the most expeditious manner practicable: *Provided*, That, to the extent appropriate, any pesticide that results in a postharvest residue in or on food or feed crops shall be given priority in the reregistration process.

7 U.S.C. § 136a note, subsec. (g) (1988).

EPA's inability to reregister pesticides was examined in an April 1986 General Accounting Office (GAO) report entitled "Pesticides, EPA's Formidable Task To Assess and Regulate Their Risks." GAO concluded that despite the fact that EPA had not formally reregistered a single active ingredient, "the alternatives available for significantly accelerating this time-consuming and resource-intensive effort appear limited because of the tasks involved in technically reviewing a large volume of data and making complex regulatory decisions for a large number of pesticides." GAO recommended that Congress consider an approach that would set reasonable deadlines and interim milestones, an approach ultimately adopted by Congress in 1988 when it added a new § 4 to FIFRA entitled "Reregistration of Registered Pesticides."

---

11 Federal Insecticide, Fungicide, and Rodenticide Act, Pub. L. No. 94-140, 4(ii), 89 Stat. 751, 752 (1975).

EPA's Reregistration Program Before 1988

The Registration Standard. EPA's regulations define reregistration as "the systematic review of pesticides."[12] 40 C.F.R. § 152.144. Before Congress added § 4 to FIFRA in 1988, EPA's reregistration program centered on the Agency's development of pesticide Registration Standards. In developing a Registration Standard for an active ingredient, EPA would review all previously submitted studies on the active ingredient that might satisfy a pesticide data requirement, determine whether the studies were adequate, identify resulting data gaps, and arrive at an interim regulatory position based on the previously submitted studies. The Agency would then publish the results of this review in a Registration Standard for the active ingredient. "Based on the Registration Standard, the Agency may require a registrant to change a product's composition, labeling, packaging, or uses in order to be reregistered and to maintain his registration in compliance with FIFRA." 40 C.F.R. § 152.144.

A Registration Standard typically includes a FIFRA § 3(c)(2)(B) Data Call-In Notice issued by the Administrator requiring technical registrants of the active ingredients to develop additional data to satisfy the data gaps identified in the Registration Standard. A standard might also require the registrant to amend its product labels to conform with the Registration Standard's labeling requirements, including perhaps making all or certain uses restricted uses.

The Legal and Regulatory Status of Registration Standards. EPA characterizes Registration Standards as "Agency position documents, which do not constitute a final agency determination." 49 Fed. Reg. 37,916, 37,920 (1984). In the Agency's view, therefore, the issuance of a Registration Standard does not create a right to request a FIFRA § 6(b) hearing. "[S]uch requests are proper only after a registration action is taken under the Standard." 49 Fed. Reg. at 37,920. Most Registration Standards provide that if a registrant

12 General Accounting Office, Pesticides, EPA's Formidable Task to Assess and Regulate Their Risks 47 (GAO/RCED-86-125, 1986).

does not take action required by the Registration Standard (such as relabeling or reclassifying products), the Agency may seek to cancel the registration under FIFRA § 6.

The Ciba-Geigy Decision. EPA's Simazine Registration Standard required registrants to develop additional ground water data and imposed a deadline of December 31, 1984, for relabeling simazine products for restricted use and for including a warning against using simazine products in hydrologically vulnerable areas. In July 1984, the Agency notified registrants that EPA would initiate cancellation proceedings for all registrations not in compliance after the December deadline. Then, in December 1984, EPA advised registrants of a new deadline (January 30, 1985), and cautioned registrants that "[a]ffected stocks shipped after that date without the required changes [could] be considered misbranded." After the January 1985 deadline expired, EPA again advised registrants that products not relabeled were subject to a misbranding action.

Ciba-Geigy filed an action for declaratory judgment in district court seeking relief against EPA for imposing labeling changes and use restrictions without affording the company a FIFRA § 6 cancellation hearing. *Ciba-Geigy Corp. v. EPA*, 801 F.2d 430, 433 (D.C. Cir. 1986). The district court dismissed the action for lack of ripeness, but the court of appeals reversed, holding the action was ripe. 801 F.2d at 439. The action was settled and the district court never ruled on whether EPA may declare a product misbranded for failure to comply with labeling and restricted use requirements imposed in a Registration Standard without first holding a cancellation hearing. It should be noted that after Ciba-Geigy filed the declaratory action, the Agency advised the company that although it would continue to require immediate compliance with ground water label warnings, it had decided to seek restricted use classification of simazine through a cancellation proceeding. 801 F.2d at 433 n.4. The labeling issues were settled and there was no cancellation hearing.

FIFRA '88 AND § 4

FIFRA § 4 now requires EPA to complete the reregistration of all pesticides by 1997 in five phases. 54 Fed. Reg. 18,076 (1989); 54 Fed. Reg. 22,706 (1989). Each product containing an active ingredient that was first registered before November 1, 1984 must be reregistered. This deadline applies as well to pesticides containing active ingredients for which Registration Standards had already been published.

Phase 1 — Listing of Active Ingredients Subject to Reregistration

In Phase 1 EPA published four lists of pesticides — A, B, C and D. List A, published on February 22, 1989, contained 194 active ingredients for which EPA already had issued Registration Standards. 54 Fed. Reg. 7740. These active ingredients are exempt from Phases 2, 3 and 4, but the Administrator must reregister these products within the nine-year period. Lists B and C each contain 150 active ingredients for which Reregistration Standards have not been issued. List D contains all remaining active ingredients. List B was published on May 25, 1989, 54 Fed. Reg. 22,706, and List C on July 24, 1989, 54 Fed. Reg. 30,846. List D was published on October 24, 1989, 54 Fed. Reg. 43,388. The publication date is important because it starts the clock running for the registrant and the Agency for completing activities under Phases 2, 3 and 4. The listing of a pesticide is not subject to judicial review. FIFRA § 4(c)(3).

Phase 2 — Registrants' Intent to Seek Reregistration

In Phase 2 registrants were required to advise the Agency within three months of a product's listing whether they intend to seek reregistration; identify studies required by 40 C.F.R. part 158 and any missing or inadequate data; and commit to support reregistration by submitting replacement studies for those studies identified by the registrant as missing or inadequate. For the purpose of a registrant's Phase 2 response, studies were considered inadequate in two circumstances. First, a study was presumed to be inadequate unless the registrant could certify that he possesses or has access to the raw data used in or generated by the study. Second, a study was considered inadequate if it was submitted to EPA before January 1, 1970, unless the registrant could demonstrate that the data should be considered in support of a pesticide's reregistration. FIFRA § 4(d)(3)(B).

EPA previously defined "raw data" in its Good Laboratory Practice Standards:

> [A]ny laboratory worksheets, records, memoranda, notes, or exact copies thereof, that are the result of original observations and activities of a study and *are necessary for the reconstruction and evaluation of the report of that study*. . . . "Raw data" may include photographs, microfilm or microfiche copies, computer printouts, magnetic media, including dictated observations, and recorded data from automated instruments.

40 C.F.R. § 160.3(j) (emphasis added).

EPA's Phase 2 Reregistration Response Worksheet did not require that a registrant certify that it has access to or possesses raw data as part of its Phase 2 response, but the registrant was required to certify that it had access to raw data in Phase 3. The Agency's December 21, 1989 Phase 3 Reregistration Guidance document states that for studies initiated or conducted after December 23, 1983 the Agency will use the GLP definition as a guide to determine the adequacy of raw data for studies initiated prior to December 29, 1983. The Phase 3 guidance document, Appendix H, lists data and information the Agency considers to be "raw data" for studies in different subdivisions.

Any data gap identified in a registrant's Phase 2 response must be satisfied within a reasonable period of time, but no more than forty-eight months after its commitment is submitted. FIFRA § 4(d)(4)(B). For each required study, the Agency's Phase 2 worksheet, as revised, specified a period of time in which the study must be submitted.

The Administrator may issue a notice of intent to suspend if he determines that inadequate testing progress is being made or if the registrant fails to submit a study on time. FIFRA § 4(d)(6). He may extend the testing deadlines for an additional two years for "extraordinary circumstances beyond the control of the registrant." FIFRA § 4(d)(4)(B). EPA's Phase 2 Response Worksheet required registrants to file a status report on all reregistration testing within ninety days of the response due date and, thereafter, every twelve months after the response due date.

If a registrant fails to submit the required Phase 2 response within the prescribed time, the Agency is required to issue a notice of intent to cancel the registration. After a sixty-day comment period, the Agency may cancel the registration or take other appropriate action without a hearing.

Phase 3 — Registrants' Submission of Information

Under Phase 3, registrants must:

1. summarize each study previously submitted and considered by the registrant to be adequate;
2. summarize each study previously submitted that may not comply with EPA's requirements but which the registrant believes is adequate;
3. reformat data from each summarized study submitted to the Agency prior to Januaury 1, 1982, which concerns chronic dosing, oncogenicity, reproductive effects, mutagenicity, neurotoxicity, teratogenicity or residue chemistry;
4. reformat all adequate data from acute and subchronic studies submitted before January 1, 1982 if the data in 3 are not required;
5. submit adverse effects information (FIFRA § 6(a)(2));
6. submit any other information the registrant believes supports a registration;
7. certify that the registrant or the Administrator possesses or has access to the raw data used in or generated by the studies that the registrant summarized;
8. either commit to submit outstanding data or submit an offer to share in the development of the data; and
9. submit evidence of compliance with FIFRA § 3(c)(1)(D)(ii) for all previously submitted data.

The registrant must submit Phase 3 information within twelve months of an active ingredient's listing.

On December 21, 1989, EPA's Office of Pesticides and Toxic Substances issued FIFRA Accelerated Reregistration — Phase 3 Technical Guidance (Guidance Document). The Guidance Document contains criteria for determining the adequacy of studies, how studies must be summarized and reformatted and how to identify adverse effects information. The Guidance Document also explains how to determine whether to generate data on anticipated residues on food and feed. Using the Guidance Document, a registrant should first determine what studies may support registration and then determine whether those studies meet EPA's acceptance criteria. If a study does not meet EPA's acceptance criteria then the registrant must inform EPA in its Phase 3 response that it will conduct a new study or explain why in its opinion EPA should accept the study.

If a toxicology or residue study satisfies EPA's acceptance criteria, then the registrant must summarize the study in accordance with the Guidance Document. In addition, all studies conducted before 1982 must be reformatted in accordance with Pesticide Registration (PR) Notice 89-3, which is included as part of the Guidance Document. Two categories of studies do not require reformatting: (1) oncogenicity studies in the mouse that have been through the Office of Pesticide Program's peer review and have been classified by EPA, and (2) acute toxicity studies. For these studies the summary information registrants are required to supply in Phase 3 is sufficient to satisfy the reformatting requirement. Finally, the Guidance Document contains information on identifying unreasonable adverse effects information. These guidelines generally follow the Agency's prior guidance on § 6(a)(2) (see Chapter 15).

If the Administrator determines that the registrant did not make a good faith attempt to conform his Phase 3 submission to the guidelines, the Administrator may issue a notice of intent to cancel the registration. § 4(e)(3)(B)(iii)(I). While the registrant may request a hearing, the only issue for consideration at the hearing is whether the registrant made a good faith attempt to conform his submission to the guidelines.

Phase 4

In Phase 4 EPA will independently review the registrant's Phase 3 submissions. If the Agency disagrees with a registrant's position that a study should be accepted even if it does not meet EPA's acceptance criteria, it will issue a data call-in for the requirement. During this review, the Administrator may require the registrant to submit any studies in his possession. The Administrator must publish all data gaps in the *Federal Register* and issue a § 3(c)(2)(B) data call-in notice. The Administrator must complete his review within eighteen, twenty-four and thirty-three months of the publication of Lists B, C and D, respectively. The registrant is required to submit missing or inadequate studies in accordance with the schedule set forth on the Phase 3 Chemical Response Worksheet. If any required studies are not initiated within one year or the Administrator determines that the registrant cannot complete the studies on time, or the studies are not submitted on time, the Administrator may suspend the registration.

Phase 5

Phase 5 concludes the reregistration process. The Administrator is required to determine whether pesticides containing an active ingredient are eligible for reregistration within one year after *all* data on the active ingredient have been submitted to EPA. The Administrator can extend the review period by one year "for extraordinary circumstances." FIFRA § 4(g)(2)(A). After making a determination that products containing an active ingredient are eligible for reregistration, the Administrator may require product-specific data, but he must review the data within ninety days of its submission. The registrant must submit any product-specific data to EPA within eight months of the Administrator's determination that a product is eligible for reregistration. FIFRA § 4(g)(2)(B).

After completing his review of the active ingredient and any product-specific data, the Administrator must determine whether "when used in accordance with widespread and commonly recognized practice [the pesticide] will not generally cause unreasonable adverse effects on the environment." FIFRA § 3(c)(5)(D). If a pesticide does not meet this criteria the Administrator must take appropriate regulatory action, including data call-in, cancellation or Special Review. Criteria for initiating Special Review are described in 40 C.F.R. part 154 (see Chapter 8). At the conclusion of its Phase 5 review, EPA will develop a Reregistration Document. 54 Fed. Reg. 53,368 (1989). EPA intends that this document will be similar to the Registration Standard.

Reregistration for Minor Uses and Low Volume Pesticides

The magnitude of the expense of reregistering an active ingredient is determined by the use patterns of its end-use products. Thus, the cost of reregistering an indoor use product is increased significantly if it is used also on terrestrial or aquatic sites. FIFRA § 3(c)(2)(A) requires the Administrator to consider minor uses when "establishing standards for data requirements." Historically, EPA has considered relieving a registrant from reregistration data requirements for minor or low volume uses only if all uses of the ingredient are low volume. EPA's Guidance for Phase 2 of Reregistration for Pesticide Minor Uses and Low Volume Pesticides[13] continues this policy. Under the 1990 amendments to FIFRA, the Administrator has discretion to waive or reduce the fee if he "determines that the fee would significantly reduce the availability of the pesticide." FIFRA § 4(i)(5)(A), *as amended by* the Food, Agriculture, Conservation, and Trade Act of 1990, Pub. L. No. 101-624, sec. 1493, 104 Stat. 3359, (Nov. 28, 1990).

Reregistration Docketing and Public Participation Procedure

In response to the criticism (and a lawsuit) from public interest groups that EPA in the early 1980's met too frequently with pesticide registrants behind closed doors, the Agency in November 1985 established procedures that provide specific opportunities for public participation in the development of pesticide Registration Standards. 50 Fed. Reg. 48,998 (1985) (codified at 40 C.F.R. pt. 155). These procedures require EPA to maintain a docket

---

3 This guidance document was included in the Agency's response package.

for each Registration Standard once the Agency begins a review of data or includes the compound in its annual listing of Registration Standards that will be developed during the next year. The public docket must be indexed and made available to the public.

EPA's public participation rules contain several important features. First, any meeting request must be in writing. Second, government personnel must prepare for inclusion in the docket memoranda of all meetings with outside persons. Third, EPA must include in the public docket all information submitted in connection with the development of a Registration Standard. Finally, the public participation procedures prohibit the Agency from committing to take any particular action during discussions with parties outside the government. 40 C.F.R. § 155.27.

For active ingredients undergoing accelerated reregistration under FIFRA '88, EPA will establish and maintain a public docket for each active ingredient like the docket formerly maintained for list A active ingredients. The docket will be established at the time the Agency issues a notice in the *Federal Register* of initiation of scientific review under FIFRA § 4(g)(1). If a docket was previously established for an active ingredient under 40 C.F.R. § 155.32(a), the Agency will maintain the same docket. 54 Fed. Reg. 53,368 (1989).

DATA CALL-IN PROGRAM

Under FIFRA § 3(c)(2)(B)(i), the Administrator has authority to require a registrant to submit additional data "to maintain in effect an existing registration." Few pesticide registrants have avoided receiving such notices. Registration Standards, issued prior to the FIFRA '88 amendments, and Phase 4 notices require registrants to submit additional data.

A data call-in notice advises the registrant that it has the following options for responding to the notice:

1. Submit to EPA any existing data it believes are responsive to EPA's requirements.

2. Within ninety days of receipt of the notice submit to EPA a promise to develop the data alone or together with other registrants. In the latter case, the registrant must submit evidence of an agreement with other registrants.
3. Claim an exemption because it purchases and uses a registered pesticide ("the formulator's exemption"). To support this exemption under FIFRA § 3(c)(2)(D), a registrant must submit a current Confidential Statement of Formula listing its source(s). If the manufacturer of the active ingredient does not develop the data, the formulator must develop the data.
4. Request a minor use exemption. EPA's application of the minor use exemption under FIFRA § 3(c)(2)(A) is very narrow: the total production volume by all manufacturers must be small. 51 Fed. Reg. 11,341 (1986).[14]
5. Request that its registration not be suspended on the grounds that it made a firm and binding offer to participate in joint testing with another registrant who has committed to develop the required data and has refused its offer.
6. Request a voluntary cancellation or request EPA to issue a notice of intent to suspend.

Suspension

The registrant receiving a data call-in notice must act promptly. It must "provide evidence within ninety days after receipt of notification that it is taking appropriate steps to secure the additional data that are required." FIFRA § 3(c)(2)(B)(ii). This "evidence" is in the form of a completed data summary sheet which typically accompanies the DCI notice. The data summary sheet must be returned to EPA within ninety days, not the requested data themselves. The notice will specify a schedule for submitting the data.

---

[14] EPA cannot require a registrant to submit "field residue data from a geographic area where the pesticide will not be registered for such use" under the 1990 amendments to FIFRA. FIFRA § 3(c)(2)(A), *as amended by* the Food, Agriculture, Conservation, and Trade Act of 1990, Pub. L. No. 101-624, sec. 1492, 104 Stat. 3359, (Nov. 28, 1990).

If a registrant fails to respond to a call-in notice within ninety days or if it fails to submit data on time according to EPA's schedule, the Agency will issue a notice of intent to suspend (NOIS) — not to cancel — the registrant's registrations. FIFRA § 3(c)(2)(B) provides the authority for the NOIS. The NOIS becomes effective within thirty days of its receipt by the registrant unless (1) the registrant requests a hearing, or (2) satisfies the Administrator it has complied fully with the requirements of the Notice. FIFRA limits any hearing to two issues: "(1) whether the registrant has failed to take the action which served as the basis for the [NOIS] and (2) whether the Administrator's determination with respect to the disposition of existing stocks is consistent with this subchapter." FIFRA § 3 (c)(2)(B)(iv). Registrations suspended for failure to comply with a data call-in "shall be reinstated by the Administrator if the Administrator determines that the registrant has complied fully with the requirements that served as the basis for the suspension." § 3(c)(2)(B)(iv).

DATA DEVELOPMENT

Individual Data Development

The registrant facing the requirement to develop additional data in response to a Registration Standard, accelerated reregistration under FIFRA '88 or a data call-in notice, must decide whether to develop the data individually or together with other registrants who face the same data requirement. While FIFRA § 3(c)(2)(B) authorizes joint data development, and while joint development may cost less, a company may prefer to test alone. For example, a company may not wish to share testing know-how with other registrants or may believe that it has better control over the quality of the data call-in testing if it works alone.

Joint Data Development

FIFRA authorizes the formation of joint testing ventures. Section 3(c)(2)(B)(i), (ii) of FIFRA provides:

> If the Administrator determines that additional data are required to maintain in effect an existing registration of a pesticide, the Administrator shall notify all existing registrants of the pesticide . . . . Two or more registrants may agree to develop jointly, or to share in the cost of developing, such data if they agree and advise the Administrator of their intent within ninety days after notification.

Joint testing has the advantage of saving resources and lessening the expense. The most frequent form of organization used to conduct cooperative testing is the joint venture, commonly referred to as a task force. The joint venture is an unincorporated entity that operates much like a partnership except that, unlike a partnership, the joint venture's life is limited to accomplishing the objectives of the group.

Section 3(c)(2)(B) authorizes joint ventures to develop jointly or to agree to cost share. Under the former each participant normally has full data ownership rights. Under the latter, FIFRA affords a cost sharer only the right examine and to rely upon the data to support its EPA registration. If the registrants cannot agree on the terms of an agreement, then either party may initiate binding arbitration. For a full discussion of data arbitration, see Chapter 7.

Administration of the Organization

Typically, the registrant with a substantial interest in the product and in developing the data will invite other technical registrants to an organizational meeting. At the organizational meeting prospective members discuss the terms of the data development agreement, including the tests to be conducted, the rights and responsibilities of the members of the venture, and how costs will be shared. Based on these discussions, a draft memorandum of understanding is circulated and eventually executed by each company that wishes to participate in the program. Most joint ventures have a business group with a business chairman and a technical group with a technical chairman. The latter develops protocols, monitors the studies, and reviews the results. The former decides when assessments will be made for expenses, and decides if and when the scope of the testing program should be expanded beyond the original tests.

Cost-sharing

FIFRA does not specify how data development costs should be allocated or shared. Companies developing pesticide data together can choose from many types of possible arrangements. Participants may wish to share costs but not ownership of the data developed. They may wish to share costs and ownership equally. Costs of the testing program may be allocated pro rata or upon relative market share or a formula that combines both equal and market share allocation. Disputes can be resolved through private binding arbitration (see discussion of arbitration under § 3(c)(2)(B) in Chapter 7).

The Organizational Meeting

Two tasks should be accomplished at the organizational meeting of the joint venture. First, a draft of a suggestive agreement should be circulated and a procedure and schedule set for reaching agreement on the final terms of the agreement. Second, a preliminary consensus should be reached on what testing must be completed. Obviously, at the first meeting this will be an approximate estimate. If the organizational meeting follows the submission of a Phase II accelerated reregistration response, the necessary testing already will have been identified. Minutes of this and all other task force meetings should be maintained.

If a draft agreement is distributed for the first time at the organizational meeting, the group should focus on the anticipated testing program rather than the agreement. A second meeting should be held at which time comments are received on the draft agreement and revisions to the agreement decided upon. Inside or outside counsel should review the draft agreement as soon as possible.

THE JOINT TESTING VENTURE AGREEMENT

Membership — Who Can Join?

The agreement should clearly define who is eligible to join the task force. Typically, the agreement provides that any holder of a technical registration can join. Even though formulators that purchase registered technical pesticides are exempt from developing data on the active ingredients, some formulators for commercial reasons may wish to participate in a joint testing venture and therefore should probably consider participating unless the number of formulators would promise to make the group unwieldy.

Committees and Officers

Most task forces consist of a business committee and a technical committee and create the positions of Chairman of the Business Committee, Chairman of the Technical Committee, Treasurer and Secretary. The business committee is initially very active in organizing the task force and reaching agreement on the terms of the agreement but thereafter meets on a less frequent basis, perhaps twice a year, to approve major expenditures and decide any policy matters. The technical committee is responsible for the day-to-day running of the technical program, including the preparation of protocols, conducting analyses, receiving and evaluating protocols from testing laboratories, monitoring studies and submitting final reports to EPA.

The Test Substance

A task force can prepare and test a composite blend of the members' products or it can select and test a "representative" product: that is, a product whose purity and contaminants are representative of all of the members' products. The agreement should state how the test substance will be selected.

Members That Join Late

The agreement should specify what additional amount, if any, a late-joining member is required to pay and when any premiums become effective. Additional fees might include interest, a fee representing the reimbursement of charter members for their time and expenses devoted to organizing the task force and a testing risk assessment, which compensates charter members for the greater risk they assume than those joining the task force subsequent to disclosure of test results.

Defaulting and Withdrawing Members

The life of a testing venture is typically four to six years, depending on tests conducted. During this period, it is entirely possible that a charter member may wish to withdraw from the testing program for commercial reasons. While a member might simply transfer his membership as part of the sale of its business, in which the purchaser assumes the rights and responsibilities of the member, the member may be unable to find a buyer and wish therefore to withdraw from the task force. The agreement should specify under what circumstances a member can withdraw and its rights in existing and future data once it withdraws. Some of the options include:

1. A party may withdraw anytime before an assessment and, as long as it has paid prior assessments, may retain possession and use of any data it has previously received;
2. A party may withdraw at any time but it looses all rights to the possession or use of data that it has previously received;
3. Until the ceiling amount is reached, a party cannot withdraw from the testing program.

Assigning and Consolidating Membership

The agreement should specify under what circumstances a member can assign his membership to another company. The typical agreement will provide that a company can assign a membership as part of the sale of its business. Also, during the life of a testing program one member may acquire another member, either by acquiring the member's stock or by acquiring its assets, including its membership in the task force. The agreement should specify whether the purchaser holds one or two members. If the costs are allocated based on a market share formula, it will make no difference. If, however, costs are shared equally, the purchaser may have to pay assessments twice as a consequence of the membership. Typically agreements provide that the acquiring member will hold only one membership.

Chapter 7
EXCLUSIVE USE OF DATA, DATA COMPENSATION DISPUTES, AND PROTECTION OF TRADE SECRETS

Pesticide registration data are costly and time consuming to develop. Data submitters want to protect the substantial investments they have made in generating research data to support their products. Generic producers want to rely on the same data to obtain or renew "me-too" registrations. FIFRA strikes a balance between these competing interests:

1. Under the "exclusive use" provision, some data are temporarily protected from use by a data submitter's competitors. FIFRA § 3(c)(1)(D)(i).
2. Under the "data compensation" provision, most data can be used by any company willing to pay compensation to the data submitter. FIFRA § 3(c)(1)(D)(ii).
3. Under the "joint data development" provision, two or more registrants can agree to develop jointly, or to share the cost of, new data needed for reregistration or to respond to data call-ins. FIFRA § 3(c)(2)(B)(ii).
4. Under FIFRA's binding arbitration provisions, data compensation or cost sharing disputes can be resolved by a neutral arbitrator. FIFRA § 3(c)(1)(D)(ii), 3(c)(2)(B)(iii).

This scheme has worked well. Incentives for pesticide research and development have been maintained, while generic competition has increased. As a result, Congress has repeatedly rejected attempts to eliminate or restrict data submitters' rights.

HISTORY OF FIFRA § 3(c)(1)(D)

Enactment of § 3(c)(1)(D)

The original provisions for exclusive use of data and data compensation were enacted as part of the 1972 amendments to FIFRA. Federal Environmental Pesticide Control Act of 1972, Pub. L. No. 92-516, § 3(c)(1)(D), 86 Stat. 973, 979-80. Prior to 1972, FIFRA neither specifically authorized nor prohibited one company from using another company's test data to obtain pesticide registrations.

The 1972 version of FIFRA prohibited EPA from using a company's "trade secret" or "confidential" test data to support another applicant's registration. Data which were not trade secret or confidential could be used by other applicants if they offered to pay "reasonable" compensation to the data owner. The period during which data were eligible for compensation was unlimited. Compensation was determined by EPA through a hearing before an administrative law judge, and an award could be appealed to the federal courts. In 1975, § 3(c)(1)(D) was modified to provide that the duty to pay compensation applied to data submitted on or after January 1, 1970. Pub. L. No. 94-140, sec. 12, 89 Stat. 751, 755.

During the six-year period (1972-78) that the original § 3(c)(1)(D) remained in effect, much litigation was spawned over the question of which test data constituted trade secrets and, thus, were eligible for exclusive use protection. See, for example, *Chevron Chemical Co. v. Costle*, 443 F. Supp. 1024 (N.D. Cal. 1978); *Mobay Chemical Corp. v. Costle*, 447 F. Supp. 811 (W.D. Mo. 1978), *appeal dismissed*, 439 U.S. 320 (1979), *modified*, 517 F. Supp. 252 (W.D. Pa. 1981), *aff'd sub nom. Mobay Chemical Corp. v. Gorsuch*, 682 F.2d 419 (3d Cir.), *cert. denied*, 459 U.S. 988 (1982). However, only two compensation awards were made by EPA under the 1972 FIFRA. *Ciba-Geigy Corp. v. Farmland Industries, Inc.*, FIFRA Comp. Docket Nos. 33, 34, 41 (EPA 1981); *Union Carbide Corp. v. Thompson-Hayward Chemical Co.*, FIFRA Comp. Docket No. 27 (EPA 1982).

1978 Amendments to § 3(c)(1)(D)

In 1978 Congress enacted significant amendments to § 3(c)(1)(D). Federal Pesticide Act of 1978, Pub. L. No. 95-396, sec. 2(a)(1), 92 Stat. 819, 820-22. The 1978 amendments to § 3(c)(1)(D) became effective on September 30, 1978, and remain in effect today.

Although the 1978 amendments withdrew exclusive use rights for trade secret data, registrants are granted a ten-year period of exclusive use for data on active ingredients first registered after September 30, 1978. FIFRA § 3(c)(1)(D)(i). Compensation continues to be required whenever data submitted after December 31, 1969 are considered by EPA in support of another company's registration. § 3(c)(1)(D)(ii). The duty to pay compensation, however, ends fifteen years after the data are submitted, after which no further payment is required for use of the data. § 3(c)(1)(D)(iii). In the case of an active ingredient subject to exclusive use protection, the data are subject to compensation for five years after the ten-year period of exclusivity expires.

The 1978 amendments also made it clear that virtually all types of pesticide research data submitted to EPA are subject to compensation from me-too applicants:

1. Data submitted to support an experimental use permit.
2. Data submitted to support an application for registration.
3. Data submitted to support an amendment adding a new use to an existing registration.
4. Data submitted to support or maintain in effect an existing registration.
5. Data submitted for reregistration.

FIFRA § 3(c)(1)(D)(ii).

In 1978, Congress also removed from § 3(c)(1)(D) any suggestion of a compensation standard, and transferred the responsibility for resolving data compensation disputes from EPA to private arbitrators. This was due in part to EPA's admission that it lacked the requisite expertise to determine equitable compensation in each case. *PPG Industries, Inc.*

v. Stauffer Chemical Co., 637 F. Supp. 85 (D.D.C. 1986), *appeal dismissed*, No. 86-5502 (D.C. Cir. Nov. 4, 1987).

Another change made by the 1978 amendments was enactment of the "formulator exemption" from data compensation requirements. Under § 3(c)(2)(D), formulators are exempt from the requirement to submit and pay compensation for data concerning a pesticide active ingredient, as long as the formulator purchases a registered pesticide to formulate its product. 40 C.F.R. § 152.85. This exemption was somewhat broadened by the 1988 FIFRA amendments. Federal Insecticide, Fungicide, and Rodenticide Act Amendments of 1988, Pub. L. No. 100-532, sec. 102(b)(2), § 3(c)(2)(D), 102 Stat. 2654, 2667. See Chapter 3.

In 1988 Congress enacted a so-called "fast track" procedure for processing certain me-too applications. FIFRA § 3(c)(3)(B). See Chapter 3. This procedure does not alter the obligation of me-too registrants to pay compensation to data submitters.

Constitutionality of § 3(c)(1)(D)

The Supreme Court has twice upheld the constitutionality of the mandatory data licensing and compensation scheme under § 3(c)(1)(D). In *Ruckelshaus v. Monsanto Co.*, 467 U.S. 986 (1984), the Supreme Court considered whether the data compensation provisions of FIFRA § 3(c)(1)(D) and the data disclosure provisions of FIFRA § 10 effected a governmental "taking" of private property without just compensation, in violation of the Fifth Amendment. The Court recognized that some health and safety data submitted by data owners are trade secret property, but concluded that the compulsory licensing of such data to others under § 3(c)(1)(D) (and the disclosure of the data to others under § 10) is not unconstitutional — although EPA's consideration or disclosure of certain data could give rise to a government obligation to pay for them.

The Supreme Court held that EPA's consideration or disclosure of data submitted before October 22, 1972, or after September 30, 1978, does not create a government obligation to pay. However, the government *is* obligated to pay "just compensation" to a data owner if (1) data were submitted to EPA between October 22, 1972 and September 30, 1978; (2) the data were trade secrets; (3) the submitter designated the data as trade secrets at the time of submission; (4) the use or disclosure of the data conflicts with the confidentiality or exclusive use provisions contained in the statute during that six-year period; and (5) the arbitration provision of § 3(c)(1)(D) does not adequately compensate the data owner for the loss in market value of the data. 467 U.S. at 1013-14. Where these conditions are met, the data submitter can sue the United States for compensation in the U.S. Claims Court. 467 U.S. at 1019.

Only one year after the *Monsanto* decision was announced, the Supreme Court again confronted the constitutionality of FIFRA § 3(c)(1)(D). *Thomas v. Union Carbide Agricultural Products Co.*, 473 U.S. 568 (1985). In that case, several pesticide manufacturers challenged the FIFRA arbitration system on the ground that it delegated too much power to an arbitrator to determine compensation, without sufficient guidance from Congress on how compensation should be calculated, and without review of the soundness of arbitration awards by the federal courts. The Supreme Court concluded that this delegation of adjudicatory power to arbitrators, rather than the courts, does not violate the "separation of powers" required by the Constitution. 473 U.S. at 592-93. A federal district court subsequently held that the lack of a standard in FIFRA for measuring compensation is not unconstitutional. *PPG Industries v. Stauffer Chemical Co.*, 637 F. Supp. 85 (D.D.C. 1986), *appeal dismissed*, No. 86-5502 (D.C. Cir. Nov. 4, 1987).

EXCLUSIVE USE OF DATA

Under § 3(c)(1)(D)(i) of FIFRA, data owners are entitled to ten years of "exclusive use" for data submitted to support the registration of products containing new active ingredients first registered after the enactment of the Federal Pesticide Act of 1978 (September 30, 1978). During that period, no other applicant may use the data to support an application for registration. To be eligible for exclusive use, data must pertain to a "new" active ingredient, i.e., an active ingredient registered for the first time after September 30, 1978. See 40 C.F.R. pt. 152.

Section 3(c)(1)(D)(i) specifies only two types of data subject to exclusive use protection:

1. Data submitted to support the application for the original registration of the product containing the new active ingredient.
2. Data submitted to support an amendment adding a new use to the registration, and that pertain solely to the new use. 40 C.F.R. § 152.83(c) (definition of "exclusive use study").

Under the statute, "defensive" data are not eligible for exclusive use treatment. Defensive data are data submitted in response to a data call-in or a Special Review. Data submitted to satisfy the terms of a conditional registration are not "defensive" and can receive exclusive use protection.

Data submitted to support an Experimental Use Permit are eligible for exclusive use protection. However, the courts have held that such exclusive use rights cannot bar a subsequent applicant from using the data to obtain its own Experimental Use Permit under § 5. The right to exclusivity only bars subsequent applicants from using the exclusive data to obtain registrations under § 3. *Rohm & Haas Co. v. EPA*, 525 F. Supp. 921 (E.D. Pa.), *aff'd*, 651 F.2d 176 (3d Cir. 1981).

Data are entitled to exclusive use only during the ten-year period following the date of first registration of the pesticide. Data submitted to support a new use are not entitled to an independent ten-year exclusive use period; they receive exclusive use only for the unexpired period of the original ten-year term.

Data that concern the unique characteristics of several pesticide ingredients in combination are entitled to exclusive use when submitted in support of the initial registration of the combination product.

When an original registrant has exclusive use protection for a new active ingredient, a second registrant can obtain a registration for the same product by submitting its own complete data package. The second registrant, however, is not necessarily required to duplicate exactly the original submitter's data package. In *Rohm & Haas Co. v. EPA*, No. 86-0980 (D.D.C. Apr. 15, 1986), the court upheld EPA's decision to allow Rhone-Poulenc to obtain a registration for an exclusive use chemical without submitting four conditionally required residue studies that Rohm and Haas had been required to submit, because Rhone-Poulenc's product had a different residue profile. *Rohm & Haas*, slip op. at 5.

DATA COMPENSATION AND COST SHARING DISPUTES

Under FIFRA § 3(c)(1)(D)(ii), an applicant for registration must offer to pay compensation if it is relying in whole or in part on another company's data. 40 C.F.R. pt. 152, subpt. E; see also Chapter 3. In addition, every applicant for reregistration of a pesticide must provide evidence of its compliance with the compensation requirements of § 3(c)(1)(D). FIFRA § 4(e)(1)(I). Registrants may agree to develop jointly new data needed by EPA for reregistration. FIFRA § 3(c)(2)(B)(ii). Applicants for reregistration which are not developing reregistration data either alone or jointly must offer to share in the cost of the data being developed by other registrants. See FIFRA §§ 4(d)(3)(B)(ii), 4(e)(1)(H)(ii). See discussion of reregistration Phase 3 in Chapter 6.

FIFRA does not establish a formula or standard for determining the amount of compensation under § 3(c)(1)(D) or the manner in which costs should be shared under § 3(c)(2)(B). Instead, the statute leaves it to the parties themselves to work out compensation or cost sharing arrangements, or to agree upon a dispute resolution procedure. Any party, however, has the right to initiate binding arbitration proceedings in order to resolve a data compensation or cost sharing dispute. FIFRA § 3(c)(1)(D)(ii), 3(c)(2)(B)(iii).

Prior to 1978, EPA resolved data compensation disputes through an administrative hearing procedure. In 1978, however, Congress removed EPA from any substantive or even procedural role in determining compensation. Instead, Congress created a binding arbitration procedure to be administered by the Federal Mediation and Conciliation Service (FMCS). Because FMCS is primarily involved in resolving labor-management disputes, it was ill-equipped to administer a pesticide data dispute arbitration system. As a result, FMCS delegated that function to the American Arbitration Association (AAA), the nation's largest private, non-profit dispute resolution organization. See 29 C.F.R. part 1440 (Arbitration of Pesticide Data Disputes). FMCS, however, has retained the authority to formally appoint an arbitrator in each dispute (even though AAA actually selects the arbitrator), and to resolve any issues regarding AAA's interpretation of the FIFRA Arbitration Rules.

The statutory waiting periods for initiating binding arbitration under §§ 3(c)(1)(D) or 3(c)(2)(B), and the sanctions for failure to participate in an arbitration or comply with the arbitrator's decision, are summarized in figure 7-1. See 40 C.F.R. § 152.99. It should be noted that the arbitrator's award or determination is not subject to judicial review except in instances of fraud, misrepresentation or misconduct by the arbitrator or one of the parties. FIFRA §§ 3(c)(1)(D)(ii), 3(c)(2)(B)(iii). Moreover, an arbitrator's choice of a particular method of awarding compensation cannot constitute misconduct and is not reviewable by a court. *PPG Industries, Inc. v. Stauffer Chemical Co.*, 637 F. Supp. 85 (D.D.C. 1986), *appeal dismissed*, No. 86-5502 (D.C. Cir. Nov. 4, 1987).

FIFRA Arbitration Procedures

Arbitration is a trial-type procedure, but is supposed to be more informal, faster, and less expensive. Arbitrators are typically private attorneys or other professionals whose expertise or background makes them particularly desirable as the decision maker in a dispute.

FIFRA arbitrations involving disputes under §§ 3(c)(1)(D), 3(c)(2)(B), or both are governed by the FIFRA Arbitration Rules, 29 C.F.R. part 1440 appendix. The FIFRA Rules have been adapted from AAA's Commercial Arbitration Rules. 44 Fed. Reg. 43,292 (1979); 45 Fed. Reg. 28,105 (1980); 45 Fed. Reg. 55,395 (1980).

Arbitration is initiated by filing a written Demand for Arbitration with a Regional Office of AAA. FIFRA Arb. R. 3(a). The demand must identify the parties, the nature of the dispute, and the compensation or other remedy sought, and must demonstrate that the statutory waiting period for initiation of binding arbitration has passed. See fig. 7-1. Filing of an answer to the demand is permitted, but not required. FIFRA Arb. R. 3. The party requesting arbitration also may designate a locale for the hearing. If the other party seeks a different locale, AAA decides. FIFRA Arb. R. 4.

The Demand for Arbitration also must be accompanied by an administrative fee, as set forth in the AAA Administrative Fee Schedule, following the Rules. This fee, which is for AAA's services and not the arbitrator's, is advanced by the party requesting arbitration, and is allocated between the parties by the arbitrator at the close of the proceeding. FIFRA Arb. R. 38.

The dispute is heard by one neutral arbitrator, unless the parties agree to have a panel of three arbitrators. Unless the parties agree on whom the arbitrator(s) should be, the AAA Tribunal Administrator (an AAA staff person) provides lists of names and biographical sketches to the parties, who can cross off unacceptable names and rank the others in order of preference. AAA then selects the arbitrator(s). FIFRA Arb. R. 6 & 7. This procedure sometimes can take several months, especially if more than one list is necessary.

The parties pay the arbitrator's fees and expenses (as well as the AAA administrative fee). FIFRA provides that the parties must share the arbitrator's fee and expenses equally. FIFRA §§ 3(c)(1)(D)(ii), 3(c)(2)(B)(iii).

Once the arbitrator is formally appointed by FMCS, the claimant has sixty days to file a Statement of Claim, which is essentially an opening brief. FIFRA Arb. R. 13(a). The respondent then has sixty days to file a responsive Statement of Position. FIFRA Arb. R. 13(b).

As in other types of arbitration proceedings, prehearing discovery is quite limited. It only can be pursued in FIFRA arbitrations with prior approval of the arbitrator. FIFRA Arb. R. 23(a). There is, however, a mandatory prehearing exchange of witness lists, testimony summaries, and exhibits. FIFRA Arb. R. 23(d). A prehearing conference with the arbitrator is optional, but is typically held prior to the prehearing exchange. FIFRA Arb. R. 24.

The arbitration hearing involves direct and cross examination of witnesses, and presentation of documentary evidence, similar to a trial. The hearing can require from one day to scores of days to complete, depending on the complexity of the dispute.

Rules of evidence in arbitration proceedings are considerably more liberal than in judicial proceedings. See FIFRA Arb. R. 25. In general, any information which is helpful to the arbitrator can be presented by a party at the hearing. The arbitrator has the authority to enter a protective order for confidential information. FIFRA Arb. R. 20. The claimant has the burden of going forward with its case. See FIFRA Arb. R. 26, 28. Opening and closing arguments are at the discretion of the arbitrator.

Following the hearing, the parties submit closing briefs. FIFRA Arb. R. 30. The arbitrator then renders a decision, normally in writing. FIFRA Arb. R. 32. As discussed above, FIFRA expressly provides that the arbitrator's decision is final and binding and not subject to judicial review except in the extraordinary circumstance of fraud, misrepresentation, or other misconduct. FIFRA §§ 3(c)(1)(D)(ii), 3(c)(2)(B)(iii).

FIFRA Arbitration Decisions

Although the FIFRA arbitration procedure has been in place since 1978, pesticide registrants have used it sparingly. This may be due in part to the fact that most data submitters and me-too registrants would prefer to settle their disputes privately through negotiation (as many have done), rather than incur the burden and expense of arbitration proceedings, which in some cases have taken years to complete.

AAA treats individual arbitration proceedings as confidential. Four FIFRA arbitrations, however, have been widely publicized throughout the pesticide industry.

Stauffer Chemical Co. v. PPG Industries, Inc., No. 16199 077 82 FIFRA (Am. Arb. Ass'n June 28, 1983) (Birch, Smolka, Vassil, Arbs.). This was the first arbitration decision under FIFRA § 3(c)(1)(D) for use of data to obtain me-too registrations. Following a lengthy hearing, the panel of three arbitrators concluded that compensation should take into account the value received by the me-too registrant from use of the original registrant's data, as well as the full range of costs incurred by the original registrant in generating the data. The arbitrators determined that the value received encompasses virtually all data generation costs avoided by the me-too registrant, and the value to the me-too registrant of entering the market early by avoiding years of testing time.

The compensation awarded in the *Stauffer* arbitration reflected this "value" approach. The arbitrators awarded a lump sum of approximately $1.5 million, representing one-half (i.e., an equal share) of the testing costs avoided by the me-too registrant. They also awarded a running royalty, representing one-half of the projected profits to be earned by the me-too registrant during its first five years of sales. The arbitrators explained that the royalty was intended to provide compensation to the data submitter for the value received by the me-too registrant from entering the market early through use of the data, and also for reliance on the original registrant's product performance (i.e., efficacy) data when the me-too registrant copied the original registrant's product labeling.

Following the *Stauffer* award, PPG filed a suit in federal district court seeking to overturn the award on the theory that the arbitrators had engaged in "misconduct" by not utilizing a strict cost-sharing formula. The district court dismissed the case, holding that there could not have been such misconduct, since FIFRA § 3(c)(1)(D) does not establish a standard or formula for compensation. PPG filed a notice of appeal, but then voluntarily dismissed it after settling its dispute with Stauffer. *PPG Industries, Inc. v. Stauffer Chemical Co.*, 637 F. Supp. 85 (D.D.C. 1986), *appeal dismissed*, No. 86-5502 (D.C. Cir. Nov. 4, 1987).

FMC Corp. v. Tricon International, No. 16 199 0033 84G (Am. Arb. Ass'n Jan. 10, 1985) (Fay, Krister, Morris, Arbs.). This was the first significant FIFRA arbitration ruling under the cost-sharing provision of FIFRA § 3(c)(2)(B)(ii). The claimant contended that § 3(c)(2)(B) required the me-too registrant to pay a per capita share (i.e., 50 percent) of its testing costs for generating new data required by EPA. The me-too registrant argued that § 3(c)(2)(B) required a market-share allocation. The arbitrators ruled prior to hearing that § 3(c)(2)(B) neither requires a per capita nor a market share allocation of testing costs for generating new data. Following this ruling, the parties settled their dispute.

E.I. Du Pont de Nemours & Co. v. Griffin Corp., No. 16 171 0080 86M (Am. Arb. Ass'n Dec. 21, 1988) (Birch, Juten, Foy, Arbs.). This FIFRA arbitration involved separate claims against two me-too registrants, Griffin Corporation and Drexel Chemical Corporation, for compensation under § 3(c)(1)(D) and cost sharing payments under § 3(c)(2)(B). Following a lengthy hearing, the arbitrators awarded a combined total of approximately $1.5 million, using a current market-share method of allocation. Unlike the *Stauffer* case, the arbitration panel did not award compensation for efficacy, pre-1970 and certain other types of data, such as voluntarily submitted data. Nor did the panel award a royalty, based on their finding that there was no "early entry" benefit received by the me-too registrants under the particular facts of the case. It should be noted that one of the arbitrators on the *Du Pont* panel also sat on the *Stauffer* panel. For two differing viewpoints of the *Du Pont* decision, see 12 Chem. Reg. Rep. (BNA) 1681-88 (Feb. 17, 1989).

American Cyanamid Co. v. Aceto Corp., No. 13 171 0080 85 (Am. Arb. Ass'n Feb. 7, 1989) (Juten, Foy, Mathis, Arbs.). This § 3(c)(2)(B) cost sharing dispute was decided shortly after the *Du Pont* case by two of the same three arbitrators. The principal issue in *American Cyanamid* was whether costs for new data should be shared on a per capita (i.e., 50 percent) or a market share basis. Based on the facts of the dispute, the arbitrators found that the me-too registrant should pay a 35 percent share of the costs of providing EPA with the required data. This was basically a compromise between the per capita share approach the claimant sought and the market share approach advocated by the me-too registrant.

Abbott Laboratories v. Agtrol Chemical Products, No. 16 171 00536 89G (Am. Arb. Ass'n Sept. 6, 1990) (Birch, Boyd, Charnoff, Arbs.). In a one paragraph prehearing order, the arbitrators (one of whom also sat on the *Stauffer* and *Du Pont* panels) held as a matter of law that Abbott is not entitled to compensation for the cost of generating efficacy data, but also ruled that evidence on Abbott's efficacy data costs would be accepted to support the royalty portion of its claim, which the arbitrators refused to dismiss. This approach appears to be consistent with the landmark *Stauffer* decision, in which the royalty portion of the award encompassed the value of the me-too registrant's reliance on the data submitter's efficacy studies.

The varying results and interpretations in these arbitrations, in some instances by the same arbitrators, demonstrate that each arbitration is to be decided on the basis of its own facts, rather than pursuant to a pre-ordained compensation or cost sharing standard or formula. This is the very reason why Congress refrained from establishing a statutory standard or formula, and instead, gave each arbitrator or arbitration panel in each case the responsibility for reaching an equitable result under the particular facts presented. Thus, to reach an equitable result, each arbitration panel is permitted to interpret FIFRA and EPA regulatory requirements as it sees fit, even if other arbitration panels in other cases have interpreted the same provisions differently.

PROTECTION OF TRADE SECRETS

Confidential information submitted to EPA may be protected under FIFRA and EPA regulations. Test data on registered pesticides, however, are available for disclosure only to segments of the public not affiliated with international commerce in pesticides. Test data cannot be disclosed before a pesticide is registered or a tolerance is issued.

Trade Secrets and Confidential Business Information

FIFRA § 10(b) provides in general that trade secrets, and commercial or financial information that is privileged or confidential, shall not be disclosed by EPA. FIFRA incorporates the definition of a "trade secret" contained in the Restatement of Torts § 757 comment b, at 5 (1939). The factors which are considered in determining whether information is a trade secret include:

1. The extent to which the data are independently known to outsiders or used by outsiders.
2. The extent to which the data are known within the company.
3. The measures taken by the owner to guard the secrecy of the information.
4. The value of the data to the owner and to competitors, and the extent to which the data confer a competitive advantage on the owner.
5. The amount of effort or money expended in developing data.
6. The ease or difficulty with which the data could be acquired by others.

Commercial or financial information which is privileged or confidential also is protected under § 10 of FIFRA. This is the same standard contained in exemption 4 of the Freedom of Information Act, which protects confidential information if its disclosure would be likely to provide a competitive advantage to other firms. 5 U.S.C. § 552(b)(4) (1988). The protection of trade secrets and confidential information is non-discretionary: EPA is required not to disclose such information to the public. However, the determination of which information is a trade secret or confidential is committed to the Administrator.

Formulas of products, even if confidential, may be revealed to federal agencies, in public hearings, or in findings of fact issued by EPA, when necessary to carry out other provisions of FIFRA (e.g., in cancellation or suspension hearings). FIFRA § 10(b).

To identify confidential information, § 10(a) provides that an applicant *may* mark the portions which are claimed to be confidential and may submit the marked materials separately from other materials. EPA regulations *require* confidential information to be marked with an appropriate legend (e.g., "trade secret," "proprietary," or "company confidential") at the time the information is submitted to EPA. If information is not marked, EPA will feel free to disclose the information without further notice to the submitter. 40 C.F.R. §§ 2.203(b), (c).

Public Access to Test Data

Section 10(d) provides that test data on pesticides are available for disclosure to certain portions of the public, even if they constitute trade secret or confidential information. The constitutionality of this disclosure provision was upheld in *Ruckelshaus v. Monsanto Co.*, 467 U.S. 986 (1984).

Under § 10(d), test data include:

> All information concerning the objectives, methodology, results, or significance of any test or experiment performed on or with a registered or previously registered pesticide or its separate ingredients, impurities, or degradation products, and any information concerning the effects of such pesticide on any organism or the behavior of such pesticide in the environment, including, but not limited to, data on safety to fish and wildlife, humans and other mammals, plants, animals, and soil, and studies on persistence, translocation and fate in the environment, and metabolism . . . .

Even though test data are releasable, the following kinds of information are not ordinarily disclosed if they meet the criteria for trade secret or confidential information:

1. Manufacturing or quality control processes.

2. Details of methods for testing, detecting or measuring the quantity of any deliberately added inert ingredient.
3. The identity or percentage quantity of any deliberately added inert ingredient. This information can be disclosed, however, if EPA finds that its release is necessary to protect against an unreasonable risk of injury to health or the environment.

Test data submitted to support a petition for a tolerance on a pesticide are not disclosable under § 408(f) of the Federal Food, Drug, and Cosmetic Act (FFDCA). These data are given absolute confidentiality until EPA has published a regulation establishing a tolerance for the pesticide, or exempting the pesticide from the tolerance requirement. 40 C.F.R. § 2.308.

Test data submitted in support of an application for registration may not be disclosed under FIFRA § 3(c)(2)(A) until after the pesticide is registered. This section provides that EPA shall make available to the public the data called for in the registration statement within thirty days after registration. The legislative history of this provision indicates that Congress did not intend for such information to be publicly released until after the registration is granted. S. Rep. No. 838 (pt. II), 92d Cong., 2d Sess. 36, *reprinted in* 1972 U.S. Code Cong. & Admin. News 3993, 4057-58. EPA generally follows this practice but has not promulgated a regulation to this effect.

Information concerning production, distribution, sale or inventories of a pesticide is entitled to limited protection under § 10(d)(2). Even if trade secret or confidential, this information may be publicly disclosed in a public proceeding to determine whether a pesticide causes unreasonable adverse effects on health or the environment, if the Administrator determines that disclosure is necessary in the public interest. In many cases, it will be possible to maintain the continued confidentiality of this information through a protective order granted by the officer presiding over the proceeding.

Prohibition on Disclosure to Multinational Producers

Under § 10(g) of FIFRA, data may not be disclosed by EPA to foreign or multinational pesticide producers. This prohibition includes not only test data, but any information submitted by an applicant or registrant. The information cannot be disclosed to any employee or agent of a firm engaged in the production, sale or distribution of pesticides outside the United States, or to any person who intends to deliver the data to such a firm. The purpose of the prohibition is to prevent such producers from using data to obtain foreign registrations without paying compensation.

Anyone who wishes to inspect pesticide information must sign an Affirmation of Non-Multinational Status attesting that he does not seek access to the data for the purpose of providing it to such a firm, and will not intentionally or negligently cause the data to be delivered to such a firm. The prohibition in § 10(g) does not, however, prevent disclosure of pesticide information in a public proceeding held by EPA where the information is relevant to whether a pesticide causes unreasonable adverse effects to health or the environment. EPA is required to keep records of the names of persons who inspect pesticide information. EPA has revised its § 10(g) affirmation form to provide that the publication of test data could be deemed a violation of the Act because it would result in the prohibited delivery of the information to foreign and multinational producers.

Dispute Procedures

Whenever EPA proposes to disclose trade secret or confidential information protected by § 10(b), it must notify the submitter by certified mail and provide thirty days in which to file an action in federal court for a declaratory judgment and an injunction as to whether the information is protectable. FIFRA § 10(c).

The notification procedures of § 10(c) apply only to information which is protected under § 10(b). Nevertheless, EPA has made the thirty-day notice procedure applicable to test data. 43 Fed. Reg. 59,060 (1978). EPA may shorten the notice period to ten days for manufacturing processes and other information entitled to limited protection under § 10(d), if EPA finds that disclosure is necessary to avoid an imminent and substantial risk of injury to the public health.

Disclosure to Contractors, Physicians and Other Qualified Persons

EPA may disclose trade secret or confidential information to government contractors if disclosure is necessary for satisfactory performance of the contract. The contractor is required, however, to take security precautions as prescribed by EPA. FIFRA § 10(e). EPA regulations promulgated under § 12(a)(2)(D) of FIFRA provide that trade secret and confidential information may be disclosed to physicians, pharmacists, hospitals, veterinarians, law enforcement personnel, or government agencies with responsibilities for protection of public health or to other qualified persons, to the extent necessary in order to treat illness or injury or to prevent imminent harm to people, property or the environment.

FIGURE 7-1

FIFRA ARBITRATIONS

	§ 3(c)(1)(D)(ii)	**§ 3(c)(2)(B)(iii)**
Purpose	Award compensation for use of data	Determine allocation of testing costs/terms of joint data development arrangement
Trigger	Offer to compensate	Agreement to develop data jointly, or agreement to share in the cost
Waiting period	90 days after delivery of other compensation	60 days after advising administrator of agreement to develop data jointly, or to share in the cost
Procedures	FIFRA Arbitration Rules (29 C.F.R. pt. 1440)	FIFRA Arbitration (29 C.F.R. pt. 1440)
Judicial review of arbitration decision?	No (except for fraud, misrepresentation, or other misconduct)	No
Sanction for failure to participate/ comply with decision	Cancellation (me-too registrant) (15-day notice) Forfeit right to compensation (data submitter)	Suspension (30-day notice)
Existing stocks provision in cancellation/ suspension order?	Discretionary	Discretionary

FIFRA ARBITRATIONS (continued)

	§ 3(c)(1)(D)(ii)	**§ 3(c)(2)(B)(iii)**
Right to hearing to challenge sanction?	No	Yes (decision within 75 days)
Ability to reinstate registration following sanction	No	Yes

Chapter 8
SPECIAL REVIEW, CANCELLATION, AND SUSPENSION

Congress gave EPA authority not only to register pesticides but also to cancel the registrations of any pesticide that "generally causes unreasonable adverse effects on the environment." FIFRA § 6(b). The definition of "unreasonable adverse effects" encompasses both the "economic, social, and environmental . . . benefits" of using a pesticide as well as the risks. FIFRA § 2(bb). As a result, before a registered pesticide or any of its uses can be cancelled or otherwise restricted, EPA is required to weigh the risks of adverse effects on the environment against the benefits derived from the use of an efficacious pesticide. This regulatory assessment is usually accomplished by means of a Special Review. Because § 6 cancellation or suspension hearings are resource-intensive and time-consuming, EPA attempts to resolve most issues through less formal procedures, and resort to § 6 hearings only when absolutely necessary. The right to request a hearing on proposed cancellations or suspensions, however, is one of the most important protections afforded by FIFRA to pesticide producers and users.

SPECIAL REVIEW

Special Review (formerly called Rebuttable Presumption Against Registration or RPAR review) is EPA's principal procedure for addressing high priority concerns about the safety of pesticides. The purpose of a Special Review is to assess the risks and benefits of a pesticide's uses to determine whether to issue a notice of cancellation, impose additional restrictions as a condition for continued registration, or deny registration. Regulations governing the Special Review process are set forth at 40 C.F.R. part 154. Lower priority concerns are addressed through the Registration Standards program discussed in Chapter 3.

Evolution of the Special Review Process

The Federal Environmental Pesticide Control Act of 1972 (FEPCA), Pub. L. No. 92-516, 86 Stat. 973, added the "unreasonable adverse effects" standard to FIFRA. As part of the FEPCA amendments, Congress directed EPA to review the thousands of pesticides that were registered prior to 1972 and to re-register only those that met the unreasonable adverse effects standard. FIFRA § 4(g)(2)(C).

In the early and mid 1970s, EPA began carrying out this task by initiating formal cancellation or suspension proceedings against widely used but environmentally persistent agricultural pesticides in the organochlorine family. As the lengthy DDT, aldrin/dieldrin, and heptachlor/chlordane hearings made clear, however, it would have been impossible for EPA to elicit risk and benefit information by conducting simultaneous cancellation proceedings on the forty or fifty pesticides that were on the Agency's unofficial "problem" list. EPA needed a way to identify those pesticides which posed a "substantial question of safety" warranting issuance of a § 6(b) cancellation notice. *Environmental Defense Fund, Inc. v. Ruckelshaus*, 439 F.2d 584 (D.C. Cir. 1971). In July 1975, EPA devised the RPAR as an internal administrative mechanism for identifying such pesticides. 40 Fed. Reg. 28,242 (1975).

The Agency learned that adjudicatory hearings are an undesirable way to elicit scientific data about the risks and benefits of a pesticide's uses. In contrast, RPAR reviews offered a relatively informal, non-adjudicatory (but sometimes adversarial) administrative procedure for collecting data on the risks and benefits of a pesticide. The RPAR process quickly grew into much more than a screening device for the issuance of cancellation notices. By the late 1970s and early 1980s, EPA's pesticide program had matured and the Agency moved away from a black-and-white approach to pesticide regulation. In most cases the issue was no longer whether to cancel a registration, but whether new label warnings or restrictions, application precautions, protective clothing requirements or other restrictions would adequately reduce the risk of unreasonable adverse effects. The RPAR became a vehicle for deciding what additional safeguards could be adopted to shift the risk-benefit balance toward continued registration. Thus, the RPAR process facilitated the "more finely tuned control of pesticides" envisioned by Congress when it overhauled FIFRA in 1972. S. Rep. No. 838, 92d Cong., 2d Sess. 5, *reprinted in* 1972 U.S. Code Cong. & Admin. News 3993, 3997.

Special Review has undergone three distinct phases since its advent as the RPAR program in 1975. During the late 1970s, EPA issued approximately thirty-five RPARs. They dragged on interminably, and few were resolved. After the departure of Anne Burford as EPA Administrator in 1983, new management revitalized the program. Since that time, revised Special Review criteria and comprehensive procedures have been adopted (40 C.F.R. part 154), the original docket of RPARs was substantially reduced, and new Special Reviews were initiated. Recent Special Review decisions are discussed in Chapter 9; a list of chemicals in the Special Review program is provided at the end of this chapter.

Statutory Basis for Special Review

The terms "Special Review" and "Rebuttable Presumption Against Registration" do not appear in FIFRA. The original RPAR regulations (40 C.F.R. § 162.11) were promulgated in 1975 as internal administrative criteria for determining adverse effects pursuant to the "unreasonable adverse effects" standard of FIFRA § 6(b). 40 Fed. Reg. 28,242, 28,253 (1975).

Congressional dissatisfaction with the "indictment-like characteristics" of the RPAR process led to the Grassley-Allen Amendment in 1978.[14] That amendment, enacted as § 3(c)(8) of FIFRA, refers to a "public interim administrative review process to develop a risk-benefit evaluation of the ingredients of a pesticide or any of its uses prior to initiating a formal action to cancel, suspend, or deny registration of such pesticide." The conference

14 The name change to "Special Review" was an effort to ameliorate the stigma and misunderstandings which arose from the "Rebuttal Presumption Against Registration" terminology. 50 Fed. Reg. 12,188, 12,191-92 (1985).

report on the 1978 amendments confirms that the "interim administrative review" referred to in § 3(c)(8) is the RPAR (now Special Review) process. S. Rep. No. 1188, 95th Cong., 2d Sess. 35 (1978) (Conference Report on the Federal Pesticide Act of 1978).

Section 3(c)(8) does not expressly authorize or establish a Special Review procedure. It does prohibit the Administrator, however, from initiating a "public interim administrative review" unless it is based on a "validated test or other significant evidence raising prudent concerns of unreasonable adverse risk to man or to the environment." Thus, the statute implicitly recognizes EPA's authority to conduct a Special Review. In 1986 both houses of Congress approved comprehensive amendments that would have elevated Special Review to a full statutory procedure integrated with § 6(b) cancellation proceedings, but the amendments were not enacted into law.[15] There is little doubt, therefore, that in Congress' view, Special Review has a firm statutory basis.

Criteria for Initiating a Special Review

In 1985, EPA adopted revised risk criteria to replace the original RPAR "triggers." 50 Fed. Reg. 49,003 (1985). In EPA's words, the old criteria of 40 C.F.R. § 162.11(a) were "inflexible," "unclear," "unrealistic," and "impractical." 50 Fed. Reg. 12,188, 12,192 (1985). The revised criteria, set forth at 40 C.F.R. § 154.7, "afford the Agency greater flexibility in evaluating potential risks." 50 Fed. Reg. at 49,005.

The principal defect in the old criteria was that they generally did not take into account whether exposure to humans or nontarget organisms was high enough to pose a significant risk. In contrast, the new criteria either explicitly or implicitly mandate consideration of available exposure information. The Agency is specifically directed to

15 The amendments were part of the larger FIFRA bill which ultimately failed to be enacted by the 99th Congress. See Chapter 1. The Special Review provision would have described each step of the process and established time limits for conducting the review.

"consider available evidence concerning both the adverse effect in question and the magnitude and scope of exposure." 40 C.F.R. § 154.7(b).

Under the current criteria, a Special Review may be conducted if the Agency determines that a pesticide's uses (taking into account the pesticide's ingredients, impurities, metabolites, and degradation products):

> (1) May pose a risk of serious acute injury to humans or domestic animals.
>
> (2) May pose a risk of inducing in humans an oncogenic, heritable genetic, teratogenic, fetotoxic, reproductive effect, or a chronic or delayed toxic effect, which risk is of concern in terms of either the degree of risk to individual humans or the number of humans at some risk . . . .
>
> . . . .
>
> (3) May result in residues in the environment of nontarget organisms at levels which equal or exceed concentrations acutely or chronically toxic to such organisms, or at levels which produce adverse reproductive effects in such organisms, as determined from tests conducted on representative species or from other appropriate data.
>
> (4) May pose a risk to the continued existence of any endangered or threatened species . . . .
>
> (5) May result in the destruction or other adverse modification of any . . . . critical habitat for any endangered or threatened species.
>
> (6) May otherwise pose a risk to humans or to the environment which is of sufficient magnitude to merit a determination whether the use of the pesticide product offers offsetting social, economic, and environmental benefits that justify initial or continued registration.

40 C.F.R. § 154.7(a).

These risk criteria give EPA considerable flexibility in determining whether to conduct a Special Review. Under these criteria, the mere potential of a pesticide to produce toxicological effects in humans or animals is not enough to warrant a Special Review. Instead, the Agency must consider "the magnitude and scope of exposure" to assess the significance of the potential risks in deciding whether to initiate a Special Review. 40 C.F.R.

§ 154.7(b). On the other hand, EPA has indicated that it will not necessarily defer a Special Review in instances where exposure data are unavailable. 50 Fed. Reg. at 49,005-06.

Finally, it should be noted that the criteria for initiating a Special Review are *risk* criteria. The Agency will not consider the benefits of a pesticide in deciding whether to initiate a Special Review. 40 C.F.R. § 154.23. EPA's stated rationale for excluding benefits information is that the "pre-Special Review" could become "too elaborate and come to resemble the Special Review process itself." 50 Fed. Reg. at 49,009. Once a Special Review is initiated, however, the benefits of a pesticide play an important role since the process itself represents a risk-benefit review.

Special Review Procedures

Special Reviews are administered by the Special Review Branch of the Registration Division. The Special Review regulations prescribe the procedures used by the Agency in conducting a Special Review. 40 C.F.R. pt. 154, subpt. B. The following is a brief description of how a Special Review is conducted under the regulations:

Petition to Initiate Special Review. A Special Review may be conducted only if the Administrator first determines that a pesticide use meets one or more of the risk criteria set forth at 40 C.F.R. § 154.7 (discussed above). The Administrator may evaluate a pesticide under those criteria either on his own initiative "or at the suggestion of any interested person." 40 C.F.R. § 154.10; 50 Fed. Reg. at 49,011.

Preliminary Notification to Registrants. Congressional guidance in the legislative materials accompanying enactment of FIFRA § 3(c)(8) directed the Administrator to notify affected registrants by private written communication prior to initiating a Special Review. S. Rep. No. 1188, 95th Cong., 2d Sess. 71 (1978) (Conference Report on the Federal Pesticide Act of 1978). The purpose of this private notification is to afford registrants an opportunity to refute the Agency's findings, and thereby "furnish a greater degree of protection for the property rights of pesticide registrants and ameliorate the indictment-like characteristics of the interim public review process." Conference Report at 145; 50 Fed. Reg. 12,188, 12,190 (1985), 50 Fed. Reg. 49,998, 49,008 (1985).

To implement this congressional directive, § 154.21 of the regulations requires the Administrator to send written notification to affected registrants prior to initiating a Special Review. (Because this is supposed to be a *private* notification, it is not published in the *Federal Register.*) The notification advises registrants that a Special Review may be conducted and provides a general description of the information which supports the decision. Registrants have thirty days from receipt of the notification to respond in writing "to dispute the validity of the Agency's conclusions." § 154.21. EPA's policy is to limit the scope and duration of pre-Special Review discussions. Registrants, however, can propose risk reduction measures in response to the pre-Special Review notification, and thereby attempt to avoid public initiation of a Special Review. 50 Fed. Reg. at 49,009-10. A decision not to conduct a Special Review, however, must be published in the *Federal Register* (see next paragraph).

Notice of Special Review. Following the private notification required by § 154.21, the Agency will determine whether to initiate a Special Review based on the available information and the published risk criteria. If the decision is *not* to conduct a Special Review, the Administrator must publish his proposed decision in the *Federal Register* and afford all interested persons a thirty-day comment period prior to announcing a final decision. 40 C.F.R. § 154.23.[16] A decision that a Special Review *should* be conducted must be published in the *Federal Register* along with a statement of reasons. 40 C.F.R. § 154.25(c). The Notice of Special Review identifies each use under review and the risk criteria which have been triggered by each such use.

The regulations do not expressly refer to staff "decision" or "position" documents. The availability of these detailed technical support documents, however, is normally announced in the *Federal Register.* They are a critical part of the Special Review and should

[16] As discussed previously, the benefits of a pesticide are not considered in deciding whether to conduct a Special Review. 40 C.F.R. § 154.7.

be obtained and evaluated by interested parties. When a Notice of Special Review is issued, it is usually supported by a "Position Document 1."

Comment Opportunity. The Notice of Special Review establishes a deadline for submission of written comments from registrants, users, or other interested persons (normally thirty to sixty days). Typically, comments discuss the validity of the Agency's determination that particular risk criteria have been met, the benefits of the pesticide uses under review, and the regulatory actions the Agency should consider taking. 40 C.F.R. § 154.26. In addition to written comments, meetings can be held with Agency staff to exchange information and discuss regulatory options. 40 C.F.R. § 154.27. (See discussion below on Public Participation.) Informal public hearings to gather relevant information also may be conducted. FIFRA § 21(b); 40 C.F.R. § 154.29.

Notice of Preliminary Determination. After consideration of the information presented and the regulatory options, the Administrator will publish a Notice of Preliminary Determination. This notice announces the use-by-use regulatory actions the Agency intends to take as a result of the Special Review and presents a summary of the reasons therefor. 40 C.F.R. § 154.31(a). Measures can range from no action to outright cancellation. Typically, however, the Agency concludes Special R(views by imposing various label changes (such as additional label warnings, use precautions, and protective clothing requirements). EPA also may initiate reclassification for restricted use and a § 3(c)(2)(B) call-in for additional data. A detailed technical support document (commonly referred to as "Position Document 2/3") discusses the rationale for the actions being taken.[17] The Agency solicits public comments on the preliminary determination, which also is referred to the Secretary of Agriculture and the FIFRA Scientific Advisory Panel (SAP) for comment. 40 C.F.R. § 154.31(b). (See discussion below on External Review.)

---

17 This document combines "Position Document 2," which discusses risks, and "Position Document 3," which presents a use-by-use risk-benefit assessment.

Notice of Final Determination. Following receipt and consideration of comments on the preliminary determination, including those of the Secretary of Agriculture and SAP, the Agency announces its final determination. 40 C.F.R. § 154.33. A detailed "Position Document 4" supports the final determination. The use-by-use final determination normally is in the form of a FIFRA § 6(b) notice of intent to cancel or reclassify for restricted use. Even if the final determination is to *continue* registration of a pesticide's uses, albeit with new label requirements and restrictions, the only way the Administrator can impose such changes is through issuance of a cancellation notice. Registrants either must submit timely applications to amend their registrations as specified by the notice, or suffer cancellation. Cancellation does not become effective, however, if the registrant files a timely request for hearing pursuant to FIFRA § 6(b). (See discussion below on Cancellation Proceedings.) Special Review remains a relatively informal process. The regulations specifically authorize the Administrator to collapse the process in order to expedite a decision. 40 C.F.R. § 154.34. (In such cases the Notice of Special Review and Notice of Preliminary Determination are combined and supported by a "Position Document 1/2/3.") Moreover, the regulations state that a Special Review need not be conducted at all prior to issuing a notice to cancel or suspend under FIFRA § 6. Further, even if a Special Review has been commenced, it need not be completed prior to issuing a § 6 notice. 40 C.F.R. § 154.34(a).

External Review

As discussed above, the publication of a notice pursuant to § 6(b) of FIFRA generally concludes a Special Review. Such a notice either may initiate outright cancellation of a pesticide's uses, or may impose new label requirements or other restrictions in lieu of cancellation.

Prior to issuing a § 6(b) notice, the Administrator must solicit the views of the Secretary of Agriculture and the FIFRA Scientific Advisory Panel (SAP).[18] Congress enacted this "external review" requirement in 1975 as a critical check on EPA's authority to ban or restrict the use of registered pesticides. Section 6(b) allows the Administrator to waive external review prior to issuing a suspension notice under FIFRA § 6(c). EPA is required, however, to obtain SAP review of suspension notices after they have been issued. FIFRA § 25(d). In addition, pursuant to FIFRA § 25(a), proposed and final forms of regulations are required to be referred to the Secretary of Agriculture and SAP for comment prior to publication.

According to EPA, the Agency fulfills its duty to obtain external review by sending notices of preliminary determination to the Secretary of Agriculture and SAP at the time they are made available for public comment. 50 Fed. Reg. at 49,011. The Special Review regulations incorporate this procedure. 40 C.F.R. § 154.31(b). A 1984 district court suit challenging the Agency's practice of referring preliminary, but not final, notices of determination to the Secretary of Agriculture and SAP was dismissed on jurisdictional grounds. *American Wood Preservers Institute v. Ruckelshaus*, No. 84-2695 (W.D. La. Sept. 9, 1985).

Review by U.S. Department of Agriculture. The U.S. Department of Agriculture (USDA) has considerable expertise on the use and benefits of pesticides and the impact of pesticide restrictions on the agricultural economy. In determining whether to issue a notice under § 6(b), the Administrator must take into account "the impact of the action proposed in such notice . . . on the agricultural economy." FIFRA § 6(b). At least sixty days prior to making the notice public, the Administrator must provide the Secretary of Agriculture with

18 The FIFRA Scientific Advisory Panel should not be confused with EPA's "Science Advisory Board." The Science Advisory Board was established under the National Environmental Policy Act of 1969 (NEPA) and provides advice on broad scientific issues at the discretion of EPA. 42 U.S.C. § 4365 (1982). Referral of § 6(b) notices to the FIFRA Scientific Advisory Panel is mandatory, not discretionary.

a copy of the notice and analysis. If the Secretary comments within thirty days, the Administrator may not issue the notice without including the Secretary's comments and a response. FIFRA § 6(b).[19] Registrants and user groups are free to submit benefits and other economic information to USDA for consideration in connection with preparation of the Secretary's comments. There are no formal procedures governing the method by which USDA conducts its review. No hearings are held before USDA. The Departmental Coordinator for Pesticides and Pesticide Assessment prepares comments on behalf of the Secretary. He is assisted by various pesticide impact assessment teams in which EPA and state extension personnel participate along with USDA personnel.

Review by SAP. The Administrator also must refer proposed § 6(b) notices to SAP at least sixty days prior to issuance. SAP may comment as to the impact of the proposed action "on health and the environment." FIFRA § 25(d). As a practical matter, SAP reviews the scientific bases for the regulatory actions being proposed by the Agency. The seven scientists who constitute SAP are appointed by the Administrator from a list of nominees provided by the National Institutes of Health and the National Science Foundation. FIFRA § 25(d). An Executive Secretary who is also a member of the Office of Pesticide Programs (OPP) staff administers SAP.

Unlike USDA, SAP holds a public session prior to issuing its recommendations. An announcement of the meeting is published in the *Federal Register.* Data and other information may be submitted in writing by registrants, user groups, and other interested parties for SAP's consideration in advance of the meeting. While these submissions typically focus on risk issues, benefits information also can be submitted. Time available for oral presentations at the public session is extremely limited. After EPA staff summarizes the Agency's findings and proposed actions, registrants, users, and other interested persons are afforded only five minutes each to speak. Obviously, written submissions are essential if

---

[19] Even § 6(b) notices relating to nonagricultural pesticides must be referred to the Secretary of Agriculture. USDA frequently waives comment on such pesticides.

registrants and users hope to make their case before SAP. SAP members may ask questions at the hearing, and a transcript is made.

Following the public meeting, SAP meets in executive session and issues written recommendations. While not binding, SAP's comments carry considerble weight. SAP has become an important forum for pesticide registrants and members of the public to obtain independent peer review of the Agency's scientific conclusions. SAP has attempted to fulfill its role of providing independent review of EPA positions and has taken issue with the Agency staff's scientific methods and findings on several occasions. SAP's meetings and agenda, however, are controlled by the Office of Pesticide Programs (OPP). In accordance with informal procedures suggested by OPP, SAP has limited time to consider each pesticide under review, and focuses on questions drafted by OPP itself. The opportunity for registrant and user presentations at SAP meetings has been substantially curtailed. Such practices seem inconsistent with the very purpose for which SAP was established.

Deadlines

Despite complaints from industry, environmental groups, and Congress that Special Reviews take too long, the Special Review regulations impose no time limits on the Agency. EPA has stated that mandatory deadlines would impair its flexibility in administering the Special Review process. 50 Fed. Reg. at 49,011-12. As a result, the Agency opposed congressional efforts in 1986 to enact amendments imposing an eighteen-month deadline for conducting Special Reviews and various interim deadlines. On the other hand, public comment periods afforded by EPA are typically quite short (thirty to sixty days). This seeming double standard is a source of continuing dissatisfaction within the regulated community.

Public Participation in Special Reviews

When the Reagan Administration assumed control of EPA in 1981, one of the priorities was to conclude the numerous RPARs that were begun during the mid and late

1970s. The Reagan EPA sought to settle several RPARs through "decision conferences" with registrants or user groups. Such a settlement involved an agreement by the registrant to accept new restrictions on a pesticide's uses and to conduct further testing on the pesticide, in return for EPA's concluding the RPAR review without cancellation of registrations.

Environmental groups objected to the decision conference concept and filed suit against EPA. They claimed that RPAR settlements and other key pesticide decisions made during the first two years of the Reagan Administration were unlawfully negotiated with industry "behind closed doors." *Natural Resources Defense Council, Inc. v. EPA*, 99 F.R.D. 607 (D.D.C. 1983).

The post-Burford EPA treated NRDC's suit as "friendly." In a negotiated settlement between EPA and NRDC, the Agency agreed to give pro forma reconsideration to numerous RPAR decisions and to adopt regulations ensuring openness in future reviews. 49 Fed. Reg. 45,486 (1984).

The Special Review regulations were revised to incorporate the terms of the *NRDC* settlement. 50 Fed. Reg. 12,188, 12,195 (1985); 51 Fed. Reg. 10,391 (1986). Specifically, a public docket must be established for each Special Review. The docket must include all documents and communications with the Agency relating to a Special Review, and any pre-Special Review communications from affected registrants. 40 C.F.R. § 154.15. Agency staff *can* hold "private meetings" with registrants or user groups (or other interested persons). 40 C.F.R. § 154.27. However, within ten days of the meeting EPA must draft and place into the public docket detailed minutes of each meeting. § 154.15(b)(7), (d). Moreover, the regulations prohibit the Agency from giving preferential access to any group, or commiting to any particular action concerning a pending decision. § 154.27(b). A procedure for asserting confidentiality of information contained in the public docket is set forth at § 154.15(c).[20]

CANCELLATION PROCEEDINGS

Because the Special Review process is intended to resolve risk concerns without resorting to adjudicatory proceedings, cancellation hearings are the exception rather than the rule. In many instances, a cancellation notice issued at the conclusion of a Special Review will recommend label changes or other risk reduction measures that a registrant may adopt to avoid the cancellation of products. The right to a hearing, however, can be critically important and registrants have sought judicial relief where cancellation hearings were denied to them by EPA. *Environmental Defense Fund, Inc. v. Costle*, 631 F.2d 922 (D.C. Cir. 1980); *Amvac Chem. Corp. v. EPA*, 653 F.2d 1260 (9th Cir. 1980).

Commencement of Cancellation Proceeding

FIFRA § 6(b) authorizes the Administrator to cancel a pesticide if "it appears . . . that a pesticide or its labeling . . . when used in accordance with widespread and commonly recognized practice, generally causes unreasonable adverse effects on the environment." Because FIFRA § 2(bb) defines "unreasonable adverse effects" in a manner that takes benefits into account, cancellation is warranted only if the human or environmental risks of using a pesticide outweigh its benefits. In assessing a pesticide's benefits, the Administrator must consider the impact that the cancellation action would have "on production and prices of agricultural commodities, retail food prices, and otherwise on the agricultural economy." FIFRA § 6(b). At least one court has suggested that the Administrator should accord greater weight to a pesticide's capacity to control disease, and lesser weight to its capacity

20 For a discussion of issues relating to the Special Review process, see U.S. General Accounting Office, Pesticides: EPA's Formidable Task to Assess and Regulate Their Risks, ch. 7 (Apr. 1986).

to protect commercial crops. *Environmental Defense Fund, Inc. v. Ruckelshaus*, 439 F.2d 584 (D.C. Cir. 1971).

Although FIFRA directs EPA to weigh risks against benefits in making cancellation decisions, it allows cancellation only if use of a pesticide "generally" poses unreasonable risks. In *Ciba-Geigy Corp. v. EPA*, 874 F.2d 277 (5th Cir. 1989), EPA decided that the pesticide diazinon should be cancelled because of its potential to kill birds, but made no finding that diazinon use 'generally' caused unreasonable adverse effects to birds. In the first court decision to consider this question, the court of appeals agreed with Ciba-Geigy that the Administrator had "improperly read the word 'generally' out of FIFRA § 6(b)." The court concluded that the word "generally" means " with 'considerable frequency,' though not necessarily 'more likely than not.' " 874 F.2d at 279.

In *Ciba-Geigy Corp. v. EPA*, 801 F.2d 430 (D.C. Cir. 1986), the question was raised whether EPA had the authority to impose label changes by means of the FIFRA misbranding provision (§ 12(a)(1)(F)), rather than by cancellation notice under § 6(b). The validity of the procedure is still in question, however, as the case was settled before the court could reach a decision.

To initiate cancellation, the Administrator issues a notice of intent to cancel, or a notice of intent to hold a hearing to determine whether a registration should be cancelled. FIFRA § 6(b)(1), (2). Prior to issuance of the notice of intent to cancel, the Administrator must submit the proposed notice to the SAP and to the USDA for comment. FIFRA §§ 6(b), 25(d). (See discussion of External Review, above.)

The proposed action announced in the Administrator's notice automatically becomes final and effective after thirty days unless within that time either (1) the registrant makes the necessary corrections to the product or its labeling, if possible, or (2) a request for hearing is made by a person adversely affected by the notice. FIFRA § 6(b). If a timely hearing request is filed, the regulatory action proposed in the notice is held in abeyance pending the outcome of the hearing.

An EPA decision to issue a cancellation notice is not reviewable, because "it merely sets in motion the administrative process that terminates in a reviewable final order." *Environmental Defense Fund, Inc. v. Ruckelhaus*, 439 F.2d at 592. *Dow Chemical Co. v. Ruckelhaus*, 477 F.2d 1317, 1323 (8th Cir. 1973); *Environmental Defense Fund, Inc. v. EPA*, 465 F.2d 528, 534 (D.C. Cir. 1972); *Pax Co. of Utah v. United States*, 454 F.2d 93, 95-97 (10th Cir. 1972).

Objections and a request for hearing usually must be filed with the EPA hearing clerk within thirty days of receipt of the § 6(b) notice of intent to cancel, or within thirty days from the date the § 6(b)(1) notice is published in the *Federal Register*, whichever is later. This will be specified in the cancellation notice.

The Rights of Non-Registrants to a Hearing

Other adversely affected parties, such as user groups, have thirty days from the date the § 6(b)(1) notice is published in the *Federal Register* to file such a request. *Environmental Defense Fund, Inc. v. Costle*, 631 F.2d 922, 937-38. Yet, the rights of non-registrants may not be coextensive with the rights of registrants. The Fifth Circuit reached such a conclusion in *McGill v. EPA*, 593 F.2d 631, 636-37 (5th Cir. 1979), holding that non-registrant pesticide users did not have the right to disturb a settlement agreement reached by EPA and the registrant, which indefinitely suspended a cancellation hearing.

Denials of Registrations, Changes in Classification, and Conditional Registrations

Section 6(b) cancellation procedures also apply to a denial of an application for registration and to a change in classification from general to restricted use. FIFRA § 3(c)(6), (d)(2). When an application for a registration is denied, a hearing may be requested within thirty days by the applicant "or other interested person with the concurrence of the applicant." FIFRA 3(c)(6); 40 C.F.R. § 162.7(f)(4). In the case of a notice of intent to reclassify for restricted use, a hearing may be requested within thirty days by the registrant "or other interested person with the concurrence of the registrant." FIFRA § 3(d)(2).

A different provision applies to cancellation of *conditional* registrations. FIFRA § 6(e). The Administrator is authorized to issue a notice of intent to cancel a conditional registration issued under § 3(c)(7) of the Act "at any time" he determines that the registrant "has failed to initiate and pursue appropriate action toward fulfilling any condition imposed," or "at the end of the period provided for satisfaction of any condition imposed, [if] that condition has not been met." FIFRA § 6(e)(1). Continued sale and use of existing stocks is permitted when provided for in the § 6(e) cancellation notice.

Cancellation of a conditional registration becomes final and effective at the end of thirty days unless a person adversely affected by the notice requests a hearing. FIFRA § 6(e)(2). The statute provides that the "only matters for resolution" at the § 6(e) hearing are (1) whether the registrant has initiated and pursued appropriate action to comply with the conditions of registration within the time provided, (2) whether the conditions have been satisfied within the time provided or an extension has been granted, and (3) whether the Administrator's determination regarding disposition of existing stocks is consistent with the Act. FIFRA § 6(e)(2); 40 C.F.R. § 162.177.

An administrative hearing also can be requested when an application for conditional registration is denied. 40 C.F.R. §§ 162.167(e), 162.7(F)(4). As with denial of an application for full registration, a hearing may be requested by the applicant, or by another interested party with the applicant's concurrence, within thirty days of the notice of denial.[21]

21 The Agency's refusal to cancel or suspend registrations or impose use restrictions in the absence of a hearing is reviewable in federal district court. FIFRA § 16(a).

Other Grounds for Cancellation

The Eighth Circuit has ruled that the Endangered Species Act (ESA) may be a basis for cancellation of pesticide registration. *Defenders of Wildlife v. Administrator*, 882 F.2d 1294 (8th Cir. 1989). After EPA refused to cancel certain uses of strychnine, various environmental groups sued, arguing that the incidental killing of various endangered species through continued use of strychnine violated the ESA, the Bald and Golden Eagle Protection Act, and the Migratory Bird Treaty Act. They argued that the incidental killing of various endangered species through continued use of strychnine violated these statutes. The Eighth Circuit noted that, unlike the other two wildlife protection acts, the ESA allows private citizens to sue to enforce its provisions. 882 F.2d at 1299-1300, 1301-02. The court found that EPA's continuing registration of strychnine constituted an illegal "taking" of endangered species under the ESA and upheld a district court injunction prohibiting the registration of strychnine, but said that the injunction should be lifted

> [i]f the EPA can show that it has now obtained authorization for the incidental takings [from the Fish and Wildlife Service as required by the ESA] and has acted in compliance with the requirements of the taking statement.

Defenders of Wildlife, 882 F.2d at 1301.

In *Merrell v. Thomas*, 807 F.2d 776 (9th Cir. 1986), *cert. denied*, 484 U.S. 848 (1987), the court decided that the National Environmental Policy Act (NEPA) could not be used to invalidate federal pesticide registrations. Merrell argued that the EPA violated NEPA by failing to prepare a site-specific environmental impact statement (EIS) for each pesticide registration. The court held that "[t]o apply NEPA to FIFRA's registration process would sabotage the delicate machinery that Congress designed to register new pesticides," *Merrell*, 807 F.2d at 779, and refused to extend NEPA to block FIFRA registration.

Rules of Practice

Cancellation hearings authorized by FIFRA §§ 6(b) and 6(e) are conducted in accordance with § 6(d). Rules of Practice governing such proceedings are set forth at

40 C.F.R. part 164. These rules were adopted in 1973 and do not take into account Special Review and other subsequent regulatory changes.

A request for hearing must set forth "clearly and concisely . . . objections and the basis for each objection, including relevant allegations of fact concerning the pesticide under consideration." The request for hearing must identify the specific uses for which a hearing is sought. Objections concerning any uses for which a hearing is not requested are waived. A registrant requesting a hearing should include copies of labels for all products for which a hearing is sought. 40 C.F.R. § 164.22.

FIFRA does not afford a hearing on the sufficiency of the actions announced in a notice of intent to cancel. In *Environmental Defense Fund, Inc. v. Costle*, 631 F.2d 922 (D.C. Cir. 1980), the court of appeals held that a registrant or other adversely affected party may initiate a hearing only for the purpose of contending that the cancellation (or other restriction) of a particular use is not justified. Contentions that the Agency has not imposed sufficient restrictions, or that it was wrong to restrict rather than cancel a use, are not the proper subject matter of § 6(b) hearings.

Although the D.C. Circuit held that the judicial review of a denial of a hearing request lies in the courts of appeal, not the district courts, 631 F.2d at 925-32, a district court has held otherwise. *Amvac Chemical Corp. v. EPA*, 653 F.2d 1260, 1265 (9th Cir. 1980) (EPA decision not to hold a public hearing must be presented first to a district court and not originally to a circuit court).

After one or more requests for a hearing are filed, an EPA administrative law judge (ALJ) (also known as a hearing examiner) is assigned. 40 C.F.R. § 164.20(c). The ALJ normally consolidates into a single proceeding all requests for a hearing filed in response to a § 6(b)(1) notice. 40 C.F.R. § 164.32. He also may entertain motions for leave to intervene, which "will be freely granted but only insofar as such leave raises matters which are pertinent to and do not unreasonably broaden the issues already presented." 40 C.F.R. § 164.31(c). An intervenor need not qualify as an "adversely affected" party.

When the Administrator publishes a § 6(b)(2) notice of intent to hold a hearing, he includes a statement of issues to be heard and sets a time in which any party wishing to participate in the hearing may file a written response to the statement of issues. 40 C.F.R. § 164.23(a). Such a response must include a statement describing the interest of the party in the proceeding. 40 C.F.R. § 164.24. After responses are filed, an ALJ is assigned, and, thereafter, most of the same procedures that govern § 6(b)(1) proceedings apply.

Prehearing Procedures and Discovery

The presiding ALJ is vested with the power "to take actions and decisions in conformity with statute or in the interests of justice." 40 C.F.R. § 164.40(d). Prior to commencement of the evidentiary hearing, he may convene a pre-hearing conference among the parties to consider preliminary matters such as simplification of issues, prehearing discovery, stipulations of fact, issuance of subpoenas, and the schedule for and structure of the hearing. 40 C.F.R. § 164.50(a). The ALJ also can make provision for *in camera* (in chambers) submission and protection of confidential evidence. 40 C.F.R. § 164.70. He also may rule on preliminary legal issues.

Prior to the hearing, the parties must exchange witness lists, summaries of expected testimony, and lists of documents and exhibits to be introduced into evidence. 40 C.F.R. § 164.50(b). Unlike district court cases, other discovery such as interrogatories and depositions may be conducted only with leave of the ALJ. Such discovery may be permitted if it will not "in any way unreasonably delay the proceeding," "is not otherwise obtainable," and "has significant probative value." 40 C.F.R. § 164.51(a). Registrants may also attempt to obtain Agency materials before and during the discovery period by using the Freedom of Information Act.

The ALJ also is empowered under FIFRA § 6(d) to compel by subpoena attendance of witnesses or the production of documentary evidence. 40 C.F.R. § 164.70(a). United States district courts are vested with jurisdiction under FIFRA § 6(d) to enforce such subpoenas.

In accordance with § 6(d), any party may request the ALJ to refer "questions of scientific fact" to the National Academy of Sciences for resolution. After considering the views of the parties, the judge determines whether submission of such questions is desirable, and if questions are to be submitted, he prepares and transmits them. 40 C.F.R. § 164.50(e). The answers provided by the Academy (theoretically due within sixty days of referral) are made part of the public record. § 164.50(e)(3).

The Hearing

Section 6(d) of FIFRA states that a hearing initiated under § 6(b) "shall be held after due notice for the purpose of receiving evidence relevant and material to the issues." The hearing is considered an "adjudicatory proceeding" governed by §§ 5, 7, and 8 of the federal Administrative Procedure Act (5 U.S.C. §§ 554, 556, 557).

The respondent (i.e., the EPA Assistant Administrator for Pesticides and Toxic Substances) has the "burden of going forward to present an affirmative case." 40 C.F.R. § 164.80(a). However, the "ultimate burden of persuasion" rests with the proponents of the registrations at issue. 40 C.F.R. § 164.80(b); *Environmental Defense Fund, Inc. v. EPA*, 548 F.2d 998, 1004 (D.C. Cir. 1976), *cert. denied*, 431 U.S. 925 (1977); *Environmental Defense Fund, Inc. v. EPA*, 510 F.2d 1292, 1297-98 (D.C. Cir. 1975); *Environmental Defense Fund, Inc. v. Ruckelshaus*, 439 F.2d 584, 593 (D.C. Cir. 1971).

The ALJ must admit "all relevant, competent and material evidence, except evidence that is unduly repetitious." 40 C.F.R. § 164.81(a). This may include pertinent information developed during the Special Review. The usual practice is to admit direct testimony in written form and to make witnesses available for oral cross-examination. 40 C.F.R. § 164.50(a)(7).

Interlocutory appeals to the Administrator may be filed only by leave of the ALJ. In deciding whether to certify a matter for interlocutory appeal, the ALJ must determine whether the order or ruling at issue "involves an important question of law or policy about which there is substantial ground for difference of opinion"; whether an immediate appeal "will materially advance the ultimate termination of the proceeding"; and whether "review after the final judgment is issued will be inadequate or ineffective." 40 C.F.R. § 164.100.

The Initial Decision

After the hearing is completed, the parties submit proposed orders, findings of fact, and conclusions of law. The ALJ then prepares an "initial decision." 40 C.F.R. § 164.90. In the case of a § 6(b)(1) hearing, the ALJ may adopt, modify, or reject the proposed regulatory actions at issue. If a § 6(b)(2) hearing has been held the ALJ must recommend specific regulatory actions for the uses at issue. Unless the Administrator takes further action on the initial decision, it becomes final. §§ 164.90(b), 164.101(b). FIFRA requires the Administrator to "consider restricting a pesticide's use or uses as an alternative to cancellation." FIFRA § 6(b).

Appeals

Following issuance of the ALJ's initial decision, any party may file an appeal with the Administrator. 40 C.F.R. § 164.101. Oral arguments before the Administrator are not normally held, but the issues are fully briefed. The ALJ's report is "part of the record and is entitled to consideration but does not preclude the Administrator from making contrary findings when supported by the record as a whole." *Southern National Manufacturing Co. v. EPA*, 470 F.2d 194, 197 (8th Cir. 1972); *Environmental Defense Fund, Inc. v. EPA*, 489 F.2d 1247, 1253 (D.C. Cir. 1973). The Administrator then renders his final decision and order. 40 C.F.R. § 164.103.

Under § 16(b) of the Act, the Administrator's final decision may be appealed within sixty days to the United States court of appeals for the circuit in which the appellant resides or has a place of business. FIFRA § 6(f). The Act states that the Administrator's order shall be sustained "if it is supported by substantial evidence when considered on the record as a whole." FIFRA § 16(b). In this context, "substantial evidence" means "such relevant evidence as a reasonable mind might accept as adequate to support a conclusion."

Environmental Defense Fund, Inc. v. EPA, 489 F.2d 1247, 1251 (D.C. Cir. 1973). In addition, since § 16(b) mandates consideration of the "record as a whole," the "substantiality of evidence must take into account whatever in the record fairly detracts from its weight . . . [but does not mean that] a court may displace [EPA's] choice between two fairly conflicting views." *Environmental Defense Fund, Inc.*, 489 F.2d at 1251. Because of this highly deferential standard of review and the complexity of the scientific issues in contention, appellate review would normally offer little hope of reversal, except in those cases where the administrative proceedings were "dominated by an erroneous interpretation of agency authority." *S.L. Cowley & Sons Manufacturing Co. v. EPA*, 615 F.2d 1312, 1314 (10th Cir. 1980).

Voluntary Cancellation

FIFRA § 6(f) was amended in 1990 to provide procedural safeguards for the voluntary cancellation of a pesticide registration. Food, Agriculture, Conservation, and Trade Act of 1990, Pub. L. No. 101-624, sec. 1494, 104 Stat. 3359, (Nov. 28, 1990). Before granting the registrant's request, the Administrator must publish in the *Federal Register* the notice of the request for cancellation and provide a thirty-day period in which the public may comment on the proposed cancellation. Where the voluntary cancellation applies to a pesticide that is registered for minor agricultural use and where the Administrator determines that the cancellation of the registration would adversely affect the availability of the pesticide for use, the Administrator must undertake reasonable efforts to inform users of the pesticide of the request for cancellation and is precluded from acting on the request until the expiration of a ninety-day waiting period which begins on the date of publication of the notice in the *Federal Register*. FIFRA § 6(f)(1)(c)(ii). In the case of a minor use pesticide that is awaiting the completion of the ninety-day waiting period, the registrant may notify the Administrator of an agreement between the registrant and another person regarding the transfer of the registration of the pesticide in lieu of cancelling or amending the registration to terminate its use. FIFRA § 6(f)(3)(B). Upon transfer, the new registrant of the pesticide is required to assume the outstanding data and other requirements which are pending at the time of transfer.

SUSPENSION PROCEEDINGS

Suspension Based on Imminent Hazard

At any time (including during a § 6(b)(1) cancellation proceeding), the Administrator may initiate action in accordance with § 6(c) of the Act to suspend a pesticide registration. FIFRA authorizes suspension when the Administrator determines that suspension "is necessary to prevent an imminent hazard during the time required for cancellation . . . proceedings." FIFRA § 6(c)(1).

> [T]he term "imminent hazard" is not limited to a concept of crisis: "It is enough if there is a *substantial likelihood* that serious harm will be experienced during the year or two required in any realistic projection of the administrative process."

Environmental Defense Fund, Inc. v. EPA, 510 F.2d 1292, 1297 (D.C. Cir. 1975) (quoting *Environmental Defense Fund, Inc. v. EPA*, 465 F.2d 528, 540 (D.C. Cir. 1972)). A finding of imminent hazard results from an administrative determination that the risks appear to outweigh the benefits of a pesticide during the cancellation proceedings. *Dow Chemical Co. v. Blum*, 469 F. Supp. 892, 898-99 (E.D. Mich. 1979).

Before a suspension order can take effect, however, affected registrants have the right to an expedited hearing "on the question of whether an imminent hazard exists." FIFRA § 6(c). The suspension does not go into effect while this hearing is pending, unless the Administrator determines that "an emergency exists that does not permit him to hold a hearing before suspending" a registration. FIFRA § 6(c)(3). In this case he may issue an emergency suspension order in advance of notification to affected registrants. FIFRA § 6(c)(3). Even though a suspension order will have issued, the registrant still has the right to a hearing to contest the suspension. No order of suspension can be issued unless the

Administrator has issued or at the same time issues a notice of intent to suspend pursuant to § 6(b). FIFRA § 25(d) directs the Administrator to "promptly submit" suspension actions to SAP "for comment, as to the impact on health and the environment." Compare FIFRA § 6(b).

Final orders of suspension issued pursuant to FIFRA § 6(c)(1) are reviewable by the United States courts of appeals. A refusal to suspend or a failure to act on a request for a suspension is likewise reviewable. *Wellford v. Ruckelshaus*, 439 F.2d 598, 600 (D.C. Cir. 1971) (citing *Environmental Defense Fund, Inc. v. Hardin*, 428 F.2d 1093 (D.C. Cir. 1970)). Emergency suspension orders issued pursuant to FIFRA § 6(c)(3) are reviewable by the district courts on the issue of whether the suspension order was arbitrary, capricious or an abuse of discretion. FIFRA § 6(c)(4).

Suspension for Failure to Submit Data

The Act also authorizes the Administrator to issue a notice of intent to suspend for noncompliance with a request for submission of additional data in support of existing registrations. FIFRA § 3(c)(2)(B)(iv). See discussion in Chapter 6. The suspension becomes effective at the end of thirty days unless during that time a request for hearing is made by a person adversely affected by the notice, or the registrant complies fully with the requirements that served as the basis for the notice. If a hearing is requested, the only issues for resolution are (1) whether the registrant has failed to take the action that served as the basis for the notice, and (2) whether the Administrator's determination as to disposition of existing stocks is consistent with the Act.

Suspension Procedures

Except for emergency suspension orders issued pursuant to § 6(c)(3), the Administrator may not suspend a pesticide registration unless he "has issued or at the same time issues notice of his intention to cancel the registration or change the classification of the pesticide." FIFRA § 6(c)(1). In 1980, Congress added a provision to § 25(d) of the Act that requires the Administrator to submit to the SAP for comment any non-emergency suspension actions taken under FIFRA § 6(c)(1).

The rules governing suspension hearings are set forth at 40 C.F.R. part 164. The Administrator must first notify affected registrants of his intent to suspend the registrations at issue and of his findings on the need to prevent an "imminent hazard." 40 C.F.R. § 164.120. The suspension cannot go into effect, however, if a request for an expedited hearing is made within five days of a registrant's receipt of the notice to suspend. FIFRA § 6(c)(2). The request for hearing must set forth objections to the Administrator's findings on the question of "imminent hazard." 40 C.F.R. § 164.121(a)(3). Unlike requests for hearing on § 6(b) cancellation notices, only registrants, not users or other adversely affected parties, can request a suspension hearing.

The presiding officer at an expedited hearing normally is, but need not be, an ALJ. FIFRA § 6(c)(2). The hearing is supposed to begin within five days after the filing of the request for hearing, but can be extended by agreement with EPA. The sole issue to be resolved is "whether an imminent hazard exists." 40 C.F.R. § 164.121(c). As in the case with § 6(b) hearings, the Agency has the burden of going forward with an affirmative case but the ultimate burden of persuasion rests with the proponents of registration.

After the hearing is completed, the presiding officer prepares "recommended findings and conclusions and a statement of the reasons on which they are based." 40 C.F.R. § 164.121(j)(2). The Administrator then issues a final decision and order which may accept or reject in whole or in part the presiding officer's recommendations. 40 C.F.R. § 164.122. The Administrator is obligated "to articulate the criteria that he develops in making each individual decision" and the failure to do so is grounds for reversal. *Environmental Defense Fund, Inc. v. Ruckelshaus*, 439 F.2d 584, 596 (D.C. Cir. 1971). Final orders of suspension and denial of such orders are reviewable in the courts of appeals under § 16(b) of FIFRA, even while cancellation proceedings are pending. As with cancellation proceedings, the Administrator's order need only be supported by "substantial evidence" and thus, it is "enough at this stage that the administrative record contain respectable scientific authority supporting the Administrator." *Environmental Defense Fund, Inc. v. EPA*, 510 F.2d 1292, 1298 (D.C. Cir. 1975) (quoting *Environmental Defense Fund, Inc. v. EPA*, 465 F.2d 528, 537 (D.C. Cir. 1972)).

Emergency Suspension Procedures

If the Administrator determines that an emergency exists in accordance with FIFRA § 6(c)(3), he can issue a suspension order in advance of notifying the registrants or holding a hearing. "Emergency" has been defined as "a substantial likelihood that serious harm will be experienced during the three or four months required in any realistic projection of the administrative suspension process." *Dow Chemical Co. v. Blum*, 469 F. Supp. 892, 902 (E.D. Mich. 1979). Emergency suspensions have been sought in proceedings involving 2,4,5-T, EDB, and dinoseb. If a registrant subsequently requests a suspension hearing, the emergency suspension order will remain in effect (unless vacated by a district court) pending issuance of the final order of suspension. 40 C.F.R. § 164.123.

An emergency suspension order is immediately reviewable in the district courts "solely to determine whether the order of suspension was arbitrary, capricious or an abuse of discretion, or whether the order was issued in accordance with the procedures established by law." FIFRA § 6(c)(4). *Dow Chemical Co.*, 469 F. Supp. at 902-903 (emergency suspension of 2,4,5-T).

In a case involving the emergency suspension of dinoseb, the Ninth Circuit interpreted the § 6(c)(4) limit on the scope of district court review of EPA emergency suspension orders. The district court had found that EPA insufficiently weighed the risks and benefits in its determination of whether dinoseb posed an imminent hazard because the Agency failed to review the economic impact of cancelling dinoseb on regional farmers and processors of crops for which there were no acceptable substitutes. *Love v. Thomas*, 668 F. Supp. 1443, 1450 (D. Or. 1987), *aff'd in part, rev'd in part*, 858 F.2d 1347 (9th Cir. 1988), *cert. denied*, 109 S. Ct. 1932 (1989). The district court issued a modified suspension order allowing the use of dinoseb in Oregon, Washington and Idaho on certain crops as long as a list of precautions was met.

On appeal, the Ninth Circuit agreed that EPA's risk/benefit review was inadequate, but vacated the modified suspension order on the ground that the district court's powers are limited to "stay[ing] the effectiveness of the suspension order, pending the Administrator's final decision with respect to cancellation or change in classification." *Love v. Thomas*, 858 F.2d 1347, 1364 (9th Cir. 1988), *cert. denied*, 109 S. Ct. 1932 (1989), *quoting* FIFRA § 6(c)(4).

In a related proceeding, after all remaining dinoseb registrants reached a settlement with EPA providing for the cancellation of their registrations, food processors and growers challenged the cancellation in federal district court, claiming that EPA had not developed an administrative record sufficient to justify its decision to cancel. *Northwest Food Processors Association v. Reilly*, 886 F.2d 1075, 1076-77, 1079, (9th Cir. 1989), *cert. denied*, 110 S. Ct. 3239 (1990). The Ninth Circuit ruled that "once . . . the remaining . . . registrants agreed to the cancellation of their registrations, the Administrator had no obligation to conduct further analyses or proceedings pertaining to cancellation," and that "FIFRA does not give nonregistrant users . . . the right to prevent a settlement and force futher proceedings, once the registrants have agreed to abandon their registrations." 886 F.2d at 1079.

INDEMNIFICATION OF REGISTRANTS FOR SUSPENSION ACTIONS

Characterized by one congressman in 1971 as a potentially "broad and far reaching" policy (H.R. Rep. No. 511, 92d Cong., 1st Sess. 69 (1971)), § 15 of FIFRA as orginally added in 1972 provided, with certain exceptions, that the Administrator of EPA "shall" indemnify owners of pesticides whose registrations are suspended and subsequently cancelled by EPA on the ground that usage would create an "imminent hazard." This statutory right extended to "owners" of pesticides, including manufacturers, formulators, distributors, retailers, and consumers. In *Gro-Green Products, Inc. v. United States*, 3 Cl. Ct. 639 (1983), the United

States Claims Court held that the indemnity provisions of FIFRA applied to a company which had held a state registration and had applied for a federal registration.

Indemnity under FIFRA was the outcome of stormy congressional debate. In 1972, many of its congressional critics thought that government indemnification for losses suffered by owners of products found to be hazardous was unwarranted. Opponents of § 15 argued that enactment would open a floodgate of similar demands for indemnification under other regulatory statutes such as the Federal Food, Drug, and Cosmetic Act. Initially approved by the House and rejected by the Senate, indemnity was the product of compromise by the conference committee.

In October 1988, § 15 of FIFRA was amended, effective April 1989. Under FIFRA § 15(a), payments may still be made generally to pesticide "owners" including manufacturers, formulators, distributors, retailers and consumers, as prior to the 1988 amendment, however two added subsections of § 15(a) require EPA to directly request appropriations for such payments from the House and Senate appropriations and agriculture committees, which must approve such appropriations before these payments may be made. FIFRA § 15(a)(3), (4). Congressional concern over EPA's limited resources and the policital unpopularity of indemnifying manufacturers makes it difficult for registrants and other distributors to receive indemnification payments.

Indemnification payments can also be made from a more readily available Justice Department "judgment fund," but EPA must generally indemnify only end users of pesticides from this fund — those who owned the pesticide for purposes of applying it, (FIFRA § 15(b)(1)) — and others that cannot receive reimbursement from a "seller" because of "bankruptcy of the seller and the seller's resulting inability to provide such reimbursement." FIFRA § 15(b)(2)(B)(iv)(II). A "seller" is any "registrant, wholesaler, dealer, or other distributor." FIFRA § 15(b)(2)(A). A seller is responsible for reimbursing those who purchase pesticides from it, other than end users, unless the seller provides written notice that it will not do so at the time of sale. FIFRA § 15(b)(2)(A). If a seller does provide such written notice, the purchaser is ineligible for reimbursement from EPA.

Any indemnities that are paid generally cover the same losses as before the 1988 amendment. Indemnities are paid for losses resulting from the inability to use or sell stocks existing as of the date of the notice of intent to suspend. If the registrant of the pesticide requests a hearing, the suspension order is not effective unless the Administrator issues an emergency suspension order, and existing stocks may be used and distributed until EPA issues a final order. Additionally, the Agency may invoke the "special rule" provision of § 15(c)(2) (formerly § 15(b)(2)), which authorizes a reasonable time "for use or other disposal" of the suspended pesticide. Because stocks existing as of the date of the Administrator's notice of intent to suspend under FIFRA § 6(c)(1) may be exempted from or depleted before the final suspension order, the need for indemnity does not usually arise in this situation. Rather, the need emerges more often as a consequence of the Administrator's issuance of an emergency suspension order under FIFRA § 6(c)(3). Such an order immediately suspends the use or distribution of the product.

The statute provides that the "amount of the indemnity payment . . . shall be determined on the basis of the cost of the pesticide." FIFRA § 15(c)(1). The legislative history sheds little light on the meaning of this term, except for the House amendment that the pesticide's cost cannot exceed its fair market value. Nevertheless, it is safe to surmise that the meaning of "costs" and "fair market value" is in a state of much confusion, whether viewed from the perspective of accountant or lawyer. The 1988 amendment added the phrase "other than the cost of transportation, if any" to modify the definition of cost of a pesticide.

STATUS OF CHEMICALS IN THE SPECIAL REVIEW PROGRAM[22]

Special Reviews in Process

Alachlor

Aldicarb

Amitrole

Carbofuran

2,4-D [also 2,4-DB and 2,4-DP]

Daminozide

1,3-Dichloropropene

Dichlorvos (DDVP)

EBDCs

Ethylene Oxide

Inorganic Arsenicals (non-wood use)

Pentachlorophenol (non-wood use)

TPTH (Triphenyltin hydroxide)

Tributyltins (antifoulant use)

Special Review Completed

Amitraz

Benomyl

Cadmium

Captafol

Captan

---

22 EPA Office of Pesticide Programs, Report on the Status of Chemicals in the Special Review Program (Dec. 1989) (supplemented through Jan. 9, 1990). An update to this list will be available from the Office of Pesticide Programs Field Operations Division, Public Information Branch.

Carbon Tetrachloride

Chlorobenzilate

Coal Tar and Creosote (non-wood use)

Creosote (wood use)

Cyanazine

DBCP (Dibromochloropropane)

Diallate

Diazinon

Dicofol

Dimethoate

Dinocap

Dinoseb

EBDCs

Endrin

EPN (Ethyl (p-nitrophenyl)thiobenzene phosphonate)

Ethalfluralin

Ethylene Dibromide (EDB)

Goal (Oxyfluorfen)

Inorganic Arsenicals (wood use)

Lindane

Linuron

PCNB (Pentachloronitrobenzene)

Pentachlorophenol

Pronamide

Sodium Fluoroactetate (Compound 1080)

Strychnine/Strychnine Sulfate

2,4,5-T/Silvex

Thiophanate-Methyl

Toxaphene

2,4,5-Trichlorophenol (TCP)

Trifluralin

Final Determinations on Pre-Special Reviews

Cacodylic Acid and Salts

Carbaryl

Chlorinated Isocyanurates

Chloroform

Diflubenzuron (Dimilin)

Fluoroacetamide (Compound 1081)

Maleic Hydrazide

Methanearsonates (includes Amine Methanearsonate, Calcium Acid Methanearsonate (CAMA), Monodosodium Methanearsonate (MSMA), Disodium Methanearsonate (DMSA))

Naled

OBPA (10,10'-Oxybisphenoxarsine)

Paraquat Dichloride

Piperonyl Butoxide

Rotenone

Terbutryn

Triallate

S,S,S-Tributyl Phosphorotrithioate

Tributyl Phosphorotrithioite

Trichlorfon

Final Actions - Voluntary Cancellations

Acrylonitrile

Arsenic Trioxide

Benzene

BHC

Captafol

Carbon Tetrachloride

Chloranil

Chlordane

Chlordimeform

Copper Acetoarsenite

Copper Arsenate

Cyhexatin

Endrin

EPN (Ethyl(p-nitrophenyl)thiobenzene phosphonate)

Erbon

Heptachlor

Monocrotophos

Monuron

Nitrofen (TOK)

OMPA (Octamethylpyrophosphoramide)

Perthane

Phenarsazine Chloride

Ronnel

Safrole

Sodium Arsenite

Strobane

2,4,5-Trichlorophenol

Trysben

Cancellations Issued Prior to RPAR/Special Review Process

Aldrin

Chlordane/Heptachlor

Chlordecone (Kepone)

DDD (TDE) (Dichlorodiphenyl dichloroethane)

DDT (Dichlorodiphenyl trichloroethane)

Dieldrin

Mirex

Chapter 9
ASSESSING RISKS AND BENEFITS

THE STATUTORY STANDARD

In order to register a pesticide, EPA must determine that, among other things, the pesticide "will perform its intended function without unreasonable adverse effects on the environment; and . . . when used in accordance with widespread and commonly recognized practice it will not generally cause unreasonable adverse effects on the environment." FIFRA § 3(c)(5)(C), (D). After registration, the Administrator must make this same determination whenever a "substantial question of safety" has been raised about the pesticide. *Environmental Defense Fund, Inc. v. Ruckelshaus*, 439 F.2d 584, 593 (D.C. Cir. 1971).

FIFRA § 2(bb) defines "unreasonable adverse effects on the environment" as "any unreasonable risk to man or the environment, taking into account the economic, social, and environmental costs and benefits of the use of any pesticide." In the registration regulations, EPA has specified criteria for determining unreasonable adverse effects in order to establish the pesticide's appropriate use classification. 40 C.F.R. § 152.170. In the Special Review regulations, EPA has specified unreasonable adverse effects criteria for determining when a substantial question of safety is raised about a pesticide registration. 40 C.F.R. § 154.7.

To determine whether a pesticide may cause unreasonable adverse effects on the environment, the Administrator, in current practice, uses risk assessment and risk management techniques. Risk assessment is a multistage process: First, the Agency considers the hazard posed by the pesticide; that is, its inherent toxicity or capacity to cause adverse toxicological effects. Second, EPA considers the exposure to the pesticide because, without exposure, even the most hazardous pesticide would pose no risk. Third, the Agency assesses the risk — the degree of probability that the identified adverse effect will result given the duration, level, and type of exposure to the pesticide.

After quantifying the risk, EPA then applies risk management: The Agency considers whether, by making changes in labeling or by restricting or limiting particular uses, it can reduce a risk to reasonable proportions. During this risk management stage, EPA considers the use-by-use benefits of the pesticide and the cost effectiveness of the available control measures, including alternative pesticides. The Agency's guidelines for conducting each stage of the risk assessment/risk management process are elaborated below. In order to fully appreciate these guidelines, however, it is important to understand the meaning and application of the unreasonable risk standard as it has evolved through several statutory amendments and numerous regulatory and enforcement decisions.

In 1972, Congress adopted the unreasonable adverse effects standard, and left the recently created EPA to interpret and enforce it. By 1975, Congress had to address "the often-stated concern that EPA has not adequately considered the impact of its actions on agriculture" and that "more effective consultation is necessary." S. Rep. No. 452, 94th Cong., 1st Sess. 8 (1975). Congress accomplished this by amending FIFRA § 6(b) to require that EPA give the Secretary of Agriculture advance notice of proposals to cancel or change the classification of a pesticide. At the same time, Congress amended FIFRA § 25(d) to create the FIFRA Scientific Advisory Panel (SAP) and to require that EPA ask the SAP to review the scientific basis for proposed cancellation actions. Pub. L. No. 94-140, 89 Stat. 751 (1975).

As a result of the 1975 amendments, the Administrator had to consider specifically the impact of a proposed cancellation "on production and prices of agricultural commodities, retail food prices, and otherwise on the agricultural economy," and to weigh the alternative of restricting a pesticide's use or uses before initiating a cancellation proceeding. In 1978, Congress again amended FIFRA by adding § 3(c)(8), which requires that the Agency initiate interim administrative (special) reviews of pesticide registrations only on the basis of "a validated test or other significant evidence raising prudent concerns of unreasonable adverse risk to man or to the environment."

With each of these amendments, Congress attempted to refine the term "unreasonable adverse effect" and to give EPA guidance in interpreting and applying this standard. In so doing, Congress made clear that FIFRA embodies a risk/benefit standard, that EPA must base risk/benefit decisions on reliable and sound scientific analyses and data, that EPA's risk/benefit decisions must take into account the social and economic consequences, and that EPA first must consider restrictions and other limitations on a pesticide that will reduce the perceived risk before initiating cancellation proceedings.

EPA'S APPLICATION OF THE STANDARD

Human Hazard

For potential human health hazards, the evolution of EPA's approach to risk/benefit analyses from 1970 to the present can be charted through its proceedings against the organochlorine insecticides: DDT, aldrin/dieldrin, heptachlor/chlordane, lindane and dicofol. In each case, the central safety question, namely possible human cancer risk, was raised by chronic bioassay data showing liver tumors in mice. The Agency's regulatory approach and responses changed markedly from early 1972 cancellation decisions on DDT to the 1986 Special Review decision on dicofol. These several proceedings show a gradual shift in approach from early emphasis upon hazard identification, through subsequent emphasis upon risk assessments, to the current vogue of risk management.

The Chlorinated Hydrocarbon Insecticides. In 1971, EPA initiated proceedings to cancel DDT on the basis of tumors in four mouse studies. Although the Delaney Clause (see Chapter 5) did not apply directly to a cancellation proceeding under FIFRA, the United States Court of Appeals for the D.C. Circuit on remand to EPA stressed that Delaney placed "a heavy burden on any administrative officer to explain the basis for his decision to permit the continued use of a chemical known to produce cancer in experimental animals." *Environmental Defense Fund, Inc.*, 439 F.2d at 596 n.41 (citation omitted).

After considering the animal test data, the presiding administrative law judge (ALJ) found that DDT was not a human carcinogen and should be retained for its essential uses. Administrator Ruckelshaus, however, overturned the ALJ's decision and banned DDT. In the Administrator's judgment, the mouse studies showed an unreasonable — albeit remote and unqualifiable — human carcinogenic risk from continued DDT use. 37 Fed. Reg. 13,369,

13,375 (1972). Thus, hazard identification characterized the DDT decision: the Administrator drew no distinctions between benign and malignant tumors, gave little consideration to the levels and types of exposures, but rather focused on the persistence of DDT and his inability to disprove the existence of a cancer risk. Significantly he made essentially no risk assessment.

The 1974 proceeding against aldrin/dieldrin again turned on hazard identification rather than risk assessment. Because aldrin/dieldrin caused liver tumors in five strains of mice, the Administrator concluded that these pesticides presented a "high risk of causing cancer in man." 39 Fed. Reg. 37,246, 37,259 (1974) (footnote omitted). Indeed, the Administrator suspended all aldrin/dieldrin pesticide registrations and questioned whether he could accept any continued use of these pesticides "however significant or beneficial in social or economic terms." 39 Fed. Reg. at 37,270. With respect to the human relevance of mouse liver tumors, the Administrator stated: "[A] carcinogenic reaction in any species of test animal must be considered sufficient to describe the test compound as a carcinogen and so a threat to human health." 39 Fed. Reg. at 37,270.

In 1976, however, EPA adopted interim cancer guidelines, which remained in effect for a decade. There, for the first time, EPA recognized that the Delaney "risk eradication" approach to regulating suspect carcinogens was not always appropriate because "in many areas risks cannot be eliminated completely without unacceptable social and economic consequences." 41 Fed. Reg. 21,402-03 (1976). Under these interim cancer guidelines, a positive carcinogenicity bioassay did not automatically trigger a regulatory response. Rather, the guidelines required the Agency to make a "detailed and independent" assessment "concerning the weight of evidence that an agent is a potential human carcinogen and, if so, how great an impact it is likely to have on public health." 41 Fed. Reg. at 21,404. As to hazard, the interim cancer guidelines required "considerations of the quality and adequacy of the data and the kinds of responses induced by the suspect carcinogen." As to risk, they required extrapolation from animal data to man based upon estimates of "current and anticipated levels of exposure." 41 Fed. Reg. at 21,404.

The impact of the interim cancer policy was soon apparent. In 1976, the Administrator suspended most heptachlor/chlordane registrations (as he had done with aldrin/dieldrin registrations) on the basis of several mouse bioassays showing liver lesions. The ALJ had declined to suspend, however, because the controversy over these lesions was troublesome in "the absence of meaningful metastisis or invasion and . . . should be reviewed by others more qualified with respect to pathology than we for a more definitive appraisal." 41 Fed. Reg. 7552, 7567 (1976). In the subsequent cancellation proceeding, the ALJ referred both the pathology and the human cancer risk questions to a committee of the National Academy of Sciences (NAS) under FIFRA § 6(d). The NAS committee described its diagnostic criteria for mouse liver cancer, independently evaluated the liver slides from the bioassays, and rendered diagnoses which differed significantly from pathology experts who testified for the Agency. The NAS committee did find a dose-dependent increase in mouse liver cancer and a possible human cancer risk, but declined to assess the degree of risk "because of the uncertainties in the available data and in the extrapolation of carcinogenicity data from laboratory animals to man." Following the precedent-setting NAS committee report, the parties agreed to dismiss the proceeding on terms which provided for gradual phase-out over several years of the non-termiticide uses of these pesticides.

In 1977, EPA proposed to cancel virtually all uses of lindane because of two lifetime and two subchronic mouse studies showing liver tumors in high doses. In its 1983 review of EPA's Position Document 2/3,[23] however, the FIFRA SAP concluded: "Available data suggest that lindane is at worst a weak animal carcinogen." 48 Fed. Reg. 48,512, 48,518 (1983). Furthermore, the SAP noted that the amounts produced and resulting exposures were too low to present an unreasonable risk, and recommended retention of the principal uses of lindane. 48 Fed. Reg. at 48,518-19. Acknowledging the controversy over "whether or not increased mouse liver tumors are predictive of human carcinogenicity, especially in the absence of positive mutagencity data, or induction of primary tumors at other sites," EPA

---

[23] See pages 176-77.

reconsidered and sided with the SAP. After performing an exposure and a risk analysis, EPA found that the cancer risk was not unreasonable in light of the benefits of lindane, and retained most uses with additional restrictions and safety precautions. Lindane Position Document 4, at 39 (Oct. 1983).

In November 1984, EPA proposed guidelines for carcinogen risk assessment using a system modeled on that of the International Agency for Research on Cancer (IARC) for assessing the significance of animal studies and for classifying the overall weight of the evidence for carcinogenicity of a particular agent. 49 Fed. Reg. 46,294 (1984). The guidelines classified evidence from studies in experimental animals into one of five groups: sufficient, limited, inadequate, no evidence, or no data. To be sufficient, the evidence needs to show an increase of malignant tumors, or combined malignant and benign tumors, in multiple species or strains. Significantly, the Agency would not lump benign tumors with malignant tumors when classifying evidence if the benign tumors lacked the potential to progress to malignancies. Evidence showing only mouse liver tumors, moreover, could be downgraded from sufficient to limited. In cases where there is only limited evidence of carcinogenicity (such as positive effects in a single species, strain, or experiment, or an increase only in the benign tumors), the Agency even would consider omitting a human risk assessment.

In 1985, on the basis that dicofol had produced liver tumors in a mouse bioassay and also that it contained DDT related impurities (DDTr), the Agency initially proposed to cancel dicofol because of its carcinogenic risk. On its own initiative, however, the Agency reevaluated the initial diagnoses of mouse liver cancer and on the basis of this reevaluation concluded that the evidence was at most "suggestive": the positive tumorigenic response was only in one sex (male), one species (mouse), and one tissue (liver); the tumors were predominantly benign with no evidence of invasion or metastisis; the malignant tumors were not dose related; and the short-term mutagenicity tests were inconclusive. 51 Fed. Reg. 19,508, 19,514-15 (1986). Earlier, the SAP had concluded that the evidence for the carcinogenicity of dicofol was weak. The Agency not only agreed with the SAP conclusion but said it could place no confidence in any quantitative estimates of risk because the animal data were so weak. 50 Fed. Reg. 33,008, 33,012, 33,015 (1985).

The 1988 *De Minimis* Policy. As detailed more fully in Chapter 5, a pesticide intended for use on a raw agricultural commodity must have a tolerance or a tolerance exemption under § 408 of the Federal Food, Drug, and Cosmetic Act (FFDCA) prior to registration. Furthermore, if during processing of the raw agricultural commodity, the pesticide residue concentrates above the § 408 tolerance or if the pesticide itself is used in or around processed foods, the pesticide must have a food additive regulation or tolerance under FFDCA § 409. While § 408 employs a risk benefit standard, § 409 has an all-or-nothing standard for carcinogens — § 409 prohibits EPA from establishing a tolerance for any substance which induces cancer in man or animals.

The different standards for suspect carcinogens under §§ 408 and 409 can produce anomalous results. Under § 409, EPA may need to deny a tolerance for a low risk pesticide simply because the residue marginally concentrates in processing.

Conversely, under § 408 EPA may have to grant a tolerance for a risky pesticide simply because the residue does not concentrate. Furthermore, because the Delaney Clause (see Chapter 5) effectively eliminates the lower risk pesticides from the marketplace, EPA must take into account that the safer alternative product is not available when considering whether to cancel the higher risk pesticide. Such counter-productive results, in terms of human safety, have been termed the *Delaney Paradox.*

To mitigate such unwanted effects of Delaney, EPA developed its *de minimis* policy. Under the *de minimis* policy, "where the human dietary risk from residues of the pesticide is at most negligible" (one in a million), EPA will not consider the Delaney Clause a bar to issuing a tolerance under § 409. 53 Fed. Reg. 41,104, 41,104 (1988). While EPA's early decisions on the chlorinated hydrocarbon insecticides seemed to apply the spirit of the Delaney Clause even when it was not directly applicable, EPA now has taken steps such as its *de minimis* policy to limit Delaney's applicability.

Even this cursory review of Agency decisions on suspect chemical carcinogens shows a definite shift in approach from hazard identification in the early 1970s to risk assessment in the late 1970s and early 1980s and now to its emphasis on risk management. The transition in Agency thinking is attributable to several factors: the involvement of independent scientists in the evaluation of data through participation of the SAP; the openness of the Agency to advances in knowledge about chemical carcinogenesis and cancer risk assessment; the broadened focus of the Agency upon classes of pesticides posing similar risks through the Special Review and Registration Standard processes; and the increased recognition of economic factors through USDA involvement.

The Agency evolution in approach is captured in the comments of EPA's first Administrator, William D. Ruckelshaus, upon his return to that position a decade later:

> Ten years ago . . . I believed it would become apparent to all that we could virtually eliminate the risks we call pollution if we wanted to spend enough money. . . .
>
> This time around . . . I am determined to improve our country's ability to cope with the risk of pollutants over where I left it 10 years ago. . . .
>
> . . . .
>
> . . . [The Agency has] a responsibility for assessing the risk and weighing it, not only against the benefits of continued use of the substance under examination, but against the risks associated with substitute substances . . . .

Ruckelshaus, *Science, Risk, and Public Policy*, 221 Science 1027 (1983).

Ecological Hazard

FIFRA also requires EPA to determine whether a pesticide poses a risk to the environment which in turn necessitates "an environmental risk assessment or more specifically, an ecological risk assessment." EPA Hazard Evaluation Division, Standard Evaluation Procedure, Ecological Risk Assessment, EPA-540/9-85-001 (June 1986).

> The term "ecological" better describes the broad scope of adverse effects that are of concern from the use of pesticides. In the past, the adverse effects of greatest concern have been mortality to single species of non-human, non-target organisms, including endangered or threatened species. The Ecological Effects Branch (EEB) in the [Environmental Fate and Effects Branch] in the Office of Pesticide Programs (OPP) has primary responsibility for developing environmental risk assessments for pesticides. It should be noted that the state-of-the-art in ecological risk assessment is rapidly evolving and EEB is striving to broaden our risk assessment concerns to include not only single species and populations of wildlife, aquatic organisms, plants, and beneficial insects, but also community and ecosystem level concerns, as well.

Ecological Risk Assessment at 1.

The DDT and dicofol cases also exemplify the evolution of EPA's approach to ecological risks from the 1970s to the present. Each case involved a possible chronic avian risk from egg shell thinning.

In 1971, EPA initiated cancellation proceedings for all uses of DDT in part because of the risk it posed to fish and birds. While the ALJ determined that DDT did not cause unreasonable adverse affects on wildlife, the EPA Administrator considered that the laboratory studies showing shell thinning and DDT persistence in the environment were sufficient predicates for a ban. The Administrator found that laboratory studies show "acute and chronic effects of DDT on avian animal species and suggest that DDT impairs their reproductive capabilities." 37 Fed. Reg. at 13,371 (footnote omitted). Having found these effects in the laboratory, the Administrator then observed that "once dispersed, DDT is an uncontrollable, durable chemical that persists in the aquatic and terrestrial environments." 37 Fed. Reg. at 13,370. Because DDT can persist, he found that DDT did pose an unquantified yet nonetheless unreasonable long term "adverse impact . . . on fish and birdlife." 37 Fed. Reg. at 13,371. There was no effort to relate environmental levels to the effective levels in laboratory studies or to assess actual effects on avian populations — environmental presence and persistence were sufficient to create an unreasonable risk.

By contrast, when assessing the avian risks posed by dicofol because of its DDTr contaminants, EPA focused on environmental exposure levels. In the Special Review of dicofol, EPA's ecological concerns centered upon effects of DDTr contaminants in dicofol on populations of endangered species. To address these concerns, EPA quantitatively assessed the levels the DDTr contaminants in the product and in the environment, and then examined the use patterns of the pesticide and the likely exposure to certain endangered species. The Agency thereafter engaged in risk management: it lowered the standard for the acceptable concentration of the DDTr contaminants, and effectively required pesticide manufacturers to fund population recovery programs for the threatened endangered species.

Although the ecological effect of concern was essentially the same for both DDT and dicofol, namely the chronic effect of DDT in endangered predator birds at the top of the food chain, the dicofol case shows a significant advance in the agency's approach. With dicofol, EPA turned from simple hazard identification to an extensive exposure assessment, risk quantification, and ultimately risk management.

ELEMENTS OF RISK ASSESSMENT AND RISK MANAGEMENT

Risk Assessment

Risk assessment is the process by which EPA evaluates the full data base of a pesticide in order to determine if the pesticide presents an environmental or human health hazard, which through foreseeable exposures to its use could constitute an unacceptable risk. The elements of risk assessment are: (1) hazard identification (determining the existence and nature of adverse effects associated with exposure to a substance); (2) dose-response evaluation (the relationship between the dose administered and the incidence of the adverse effect); (3) exposure assessment (determining the human or environmental exposure to the substance from the pesticide uses in question); and (4) risk characterization (combining results of exposure assessment and dose-response assessment to quantify risk).

Hazard Identification. This element of risk assessment is a qualitative process to determine whether exposure to a pesticide leads to an increase in occurrence of a particular adverse effect. This information can be obtained from scientific studies including animal studies, epidemiologic studies, and studies of the physical and chemical properties of the pesticide. Under guidelines developed by EPA, the exercise involves weighing the available evidence and deciding whether a substance exhibits a particular adverse health effect.

Dose-Response Evaluation. This risk assessment element is a quantitative process to determine the dose level at which a pesticide first produces an adverse effect and whether the adverse effect is altered by increasing doses.

Because human cancer risk at the very low levels of human exposure cannot be measured directly either by animal experiments or by epidemiologic studies, it presents a special problem. A number of mathematical models have been developed to extrapolate from high-dose animal experiments to the low-exposure levels expected from human contact with the chemical in the environment. There are two general types of models — threshold and non-threshold. A threshold model assumes that a certain dose exists below which no adverse effects occur. A non-threshold model assumes that for any dose larger than zero, adverse effects will occur. If a threshold model is used, then the dose at which no adverse effects occur may be calculated and is termed the "no observable effect level" (NOEL).

Exposure Assessment. This is a quantitative determination of the magnitude, frequency, duration, and route of exposure to a pesticide. EPA has promulgated guidelines which provide a general approach and framework for performing human and nonhuman exposure assessments. 51 Fed. Reg. 34,042, 34,043 (1986).

Risk Characterization. This element combines the exposure assessment and the dose-response assessment to estimate some measure of the risk of the identified adverse effect to a given population. Risk characterization should include: a summary of the strengths and weaknesses in each component of the assessment, major assumptions, scientific judgments and, to the extent possible, estimates of the uncertainties. 51 Fed. Reg. at 34,036.

Risk Management

In risk management, EPA balances the health or environmental risks associated with exposure to a pesticide, determined through risk assessment, against the benefits associated with the uses of the pesticide, and chooses a regulatory approach. The goal of risk management is to obtain the maximum benefits from use of a pesticide coincident with the least risk. EPA uses several different approaches in risk management.

Cost/Benefit Analysis. This approach weighs the costs of control, explicitly and directly, against the monetized benefits of control — the avoidance of disease and the attainment of other social goods. Optimal use of the benefit/cost analysis occurs when all factors affected by a decision can be accurately represented in dollars.

Risk/Benefit Analysis. This approach balances the economic benefits of a polluting activity against the associated risks to health and the environment. For example, the benefits of using a pesticide are explicitly weighed against the risks generated by the pesticide's use. Note that benefits do not refer to the benefits from regulation, but rather to the benefits from the use of the chemical.

Cost-Effectiveness Analysis. Unlike the other two approaches, cost-effectiveness analysis begins by accepting the desirability of a particular control action. It does not weigh risks against benefits, or monetize benefits; it only looks for the lowest cost to achieve a given goal, such as the achievement of a protective standard.

EPA RISK ASSESSMENT GUIDELINES

Cancer Risk Assessment Guidelines

Human carcinogens are identified primarily through long-term animal studies and epidemiologic investigations. EPA's guidelines establish a system for assessing the significance of such studies and for evaluating the overall weight of the evidence for carcinogenicity for a particular agent. 51 Fed. Reg. 33,992 (1986). Under the guidelines, one of five evidentiary weights will be given to a study: sufficient, limited, inadequate, no evidence, or no data. 51 Fed. Reg. at 33,999. Only the first two evidentiary categories trigger the need to perform quantitative risk assessment under the guidelines.

For epidemiologic studies, data which indicate a causal relationship between incidence of human cancer and exposure to the agent constitute *sufficient* evidence of carcinogenicity. Where a causal interpretation is credible but alternative explanations such as "chance, bias, or confounding" cannot be adequately excluded, the evidence of carcinogenicity is *limited.*

To constitute *sufficient* evidence of carcinogenicity, animal studies must demonstrate an increased incidence of malignant tumors or of combined malignant and benign tumors (1) in multiple species, (2) in multiple experiments, or (3) to an unusual degree in one experiment. Increased incidence of benign tumors alone in most cases will be considered *limited* evidence of carcinogenicity. In addition, where a study suggests a carcinogenic effect but (1) involves only a single species or experiment, or (2) is restricted for a variety of reasons relating to study protocol, the evidence will be classified as *limited.* 51 Fed. Reg. at 33,999.

Based upon the weight-of-evidence determination, the agent is placed in one of five categories: human carcinogen (Group A) (sufficient epidemiologic evidence); probable human carcinogen (Group B) (limited epidemiologic evidence, sufficient animal study evidence); possible human carcinogen (Group C) (limited animal study evidence, no epidemiologic data); not classified (Group D) (inadequate evidence or no data available); and no evidence of carcinogenicity for humans (Group E). Group E is reserved "for agents that show no evidence for carcinogenicity in at least two adequate animal tests in different species or in both adequate epidemiologic and animal studies." 51 Fed. Reg. at 34,000.

Agents in Groups A and B are considered appropriate for quantitative risk assessments while those in Groups D and E generally are not. For those in Group C, a quantitative risk assessment will generally be considered appropriate, but these judgments will be made on a case-by-case basis.

As explained earlier, quantitative risk assessment involves coupling dose-response estimates with exposure assessments. With regard to dose-response assessment, the carcinogen guidelines express a general preference for epidemiologic studies over animal studies, and provide a ranking scheme if animal studies are the only available data. In addition, the guidelines suggest the use of certain mathematical extrapolation models. With regard to exposure assessment, the guidelines reference EPA's Exposure Assessment Guidelines.

Mutagenicity Assessment Guidelines

EPA developed these guidelines in order to establish a system to evaluate the significance of mutagenicity studies and determine how likely an agent is to present a mutagenic risk to humans. 51 Fed. Reg. at 34,006.

To determine whether an agent is a human mutagen, EPA prefers the use of mammalian germ cell assays. EPA will classify the test results under one of eight evidentiary categories. Category 1, the highest evidentiary level of mutagenicity, consists of positive data derived from human germ cell mutagenicity studies, while the lowest level is considered to be inadequate evidence relating to mutagenicity or to chemical interaction with mammalian germ cells. Negative mutagenicity studies are classified in category 7 while the remainder of positive results, depending on various factors (such as assay systems used and end points observed for interaction with mammalian germ cells), occupy categories 2 through 6. 51 Fed. Reg. at 34,010. Where an agent demonstrates "mutagenic potential" under this weight-of-evidence scheme, "one may decide to proceed with an evaluation of the quantitative consequences of mutation following expected human exposure." 51 Fed. Reg. at 34,009.

Quantitative assessment of mutagenic risk involves two steps: determination of (1) the heritable effect per unit of exposure (dose-response) and (2) the relationship between mutation rate and disease incidence. With regard to dose-response evaluation, the guidelines reference several "presently accepted" *in vivo* procedures found in the literature. 51 Fed. Reg. at 34,010. In addition, EPA provides general remarks as to the selection of the best data, appropriate extrapolation models, and future approaches. With regard to exposure assessment, the guidelines again reference EPA's Exposure Assessment Guidelines as the preferred protocol.

Guidelines for Health Risk Assessment of Chemical Mixtures

While some potential environmental hazards involve significant exposure to only a single compound, most instances of environmental contamination involve concurrent or sequential exposures to a mixture of compounds that may induce similar or dissimilar effects over exposure periods ranging from short-term to lifetime. EPA promulgated guidelines to establish a general approach for evaluating data on the chronic and subchronic effects of chemical mixtures. 51 Fed. Reg. at 34,014.

The guidelines define a mixture as "any combination of two or more chemical substances regardless of source or of spatial or temporal proximity." 51 Fed. Reg. at 34,015. EPA concedes that no single approach can be recommended to assess the risks of multiple chemical exposures. Nevertheless, a general approach to risk assessment can be recommended depending on the type of mixture, the known toxic effect of its components, the availability of toxicity data on the mixture or similar mixtures, the known or anticipated interactions among components of the mixture, and the quality of the exposure data. In this regard, the guidelines provide a procedural flow chart for assessing risks associated with chemical mixtures. Pursuant to the flow chart, the risk assessment protocol to be followed depends upon whether the data available relate to (1) the mixture of concern, (2) a similar mixture, or (3) the individual components of the mixture.

Guidelines for Health Assessment of Suspect Developmental Toxicants

Developmental toxicology is

> [t]he study of adverse effects on the developing organism that may result from exposure prior to conception (either parent), during prenatal development, or postnatally to the time of sexual maturation. Adverse developmental effects may be detected at any point in the life span of the organism. The major manifestations of developmental toxicity include: (1) death of the developing organism, (2) structural abnormality, (3) altered growth, and (4) functional deficiency.

EPA, Guidelines for the Health Assessment of Suspect Developmental Toxicants, 51 Fed. Reg. 34,028, 34,030 (1986).

A teratogen is defined as a "[s]ubstance that causes malformation or serious deviation from normal development of embryos and fetuses." EPA, Glossary of Environmental Terms and Acronym List (Dec. 1989). The term "fetotoxicity" describes the property of a chemical that causes a wide variety of embryonic and/or fetal divergencies from the normal which cannot be classified as birth defects per se. The differences between teratogens and fetotoxins are at times minimal, and EPA is moving away from the use of such terms, preferring to classify such effects under the broader category of developmental toxicology. EPA has promulgated guidelines to evaluate potential developmental toxicity associated with human exposure to environmental toxicants. 51 Fed. Reg. at 34,029.

The most commonly used protocol for assessing developmental toxicity in laboratory animals involves the administration of a test substance to pregnant animals during the period of major organogenesis, evaluation of maternal responses throughout pregnancy, and examination of the dam and uterine contents just prior to term. 51 Fed. Reg. at 34,029. The guidelines set forth several possible indicators of maternal toxicity in a developmental toxicity study which include: mortality, gestation index, body weight, body weight change, organ weights, food consumption, and gross necropsy.

Developmental effects include not only death, structural abnormalities, and altered growth, but also subtle alterations in the functional competence of an organ or a variety of organ systems that may result from exposure during various critical developmental periods between conception and sexual maturation. Often, these functional defects are observed at levels below those at which gross malformations are evident. Those agents which produce developmental toxicity at a dose that is not toxic to the maternal animal are of greatest concern because the developing organ appears to be selectively affected or more sensitive than the adult.

While there are many methods for extrapolating developmental toxicity risks from test animals to humans, the most widely used and accepted are the Margin-of-Safety (MOS) and the closely related "safety factor" approaches. EPA's guidelines recommend the MOS approach for developmental toxicity risk assessments, although EPA does alternatively use the "safety factor" approach. While the MOS approach is a direct comparison (ratio) between the NOEL and the estimated human exposure, the safety factor approach divides the NOEL by an arbitrary numerical value (the safety factor) to reach a "safe" level of exposure.

EPA typically will use two different exposure calculations. One calculation is the exposure an employee handling the pesticide would receive on any given day, and the other is the exposure a consumer of products treated with the pesticide would receive on any given day. Exposure data are not usually extrapolated to chronic or lifetime exposure levels for developmental toxicity risk assessments. Generally, the Agency will take the NOEL of the most sensitive species divided by the maximum human exposure from all routes and require a MOS in excess of 100 before approving registration of a pesticide.

While the EPA has not fixed a numerical MOS for developmental effects below which they would automatically reject registration of a pesticide and above which registration of the pesticide would be a forgone conclusion, EPA generally requires the MOS to be at least

100. The MOS will vary with the severity of the effect, and EPA will require a high MOS for a chemical that causes severe effects in the animal studies, such as loss of limbs, while accepting a lower MOS for relatively mild adverse effects, such as wavy ribs.

Guidelines for Exposure Assessment

These guidelines provide a general approach for carrying out human or nonhuman exposure assessments for a chemical. Exposure assessment is the determination or estimation (qualitative or quantitative) of the magnitude, frequency, duration, and route of exposure. 51 Fed. Reg. at 34,043. Exposure assessments are generally combined with dose-response calculations to perform risk assessments. The guidelines identify five major topics which should be addressed in an exposure assessment: source of pollutant, exposure pathways, measured or estimated concentrations and duration, exposed population, and integrated exposure analysis. 51 Fed. Reg. at 34,043.

An exposure "source" is the point at which a substance is believed to enter the environment. (Points of entry may be indoors as well as outdoors; environments include indoor settings such as offices as well as outdoor environments.) The environmental releases of a pollutant can be described in terms of geographic and temporal distribution, as well as the receiving environmental media. 51 Fed. Reg. at 34,043.

"Exposure pathways" of a pollutant must be identified to show how an agent moves from the source to the exposed population. The environmental pathways and fate analysis follows the substance from its point of initial environmental release, through the environment, to its ultimate fate. 51 Fed. Reg. at 34,047.

In order to estimate the concentration and duration of likely exposure, the guidelines suggest the following types of measurements: sampling of stacks or discharge pipes for emissions to the environment, testing of products for chemical or radionuclide content, testing of products for chemical or radioactive releases, sampling of appropriate points within a manufacturing plant to determine releases from industrial processors, sampling of potentially exposed populations using personal dosimeters, and sampling of solid waste for chemical or radionuclide content. Such measurements can be used to calibrate or extrapolate models or calculations to assess environmental distributions. 51 Fed. Reg. at 34,048.

Concentrations of agents should be estimated for all environmental media that might contribute to significant exposures. Generally, the environmental concentrations are estimated from measurements, mathematical models, or a combination of the two. If environmental measurements are not limited by sample size or inaccuracies, then exposure assessments based on *measurements* are preferred over estimates based on *models*. 51 Fed. Reg. at 34,048.

The exposed population will generally be identified using the previously performed source and fate studies. Census and other survey data also may be used to identify and describe the population exposed to various contaminated environmental media. It may be appropriate to describe the exposed population by other characteristics such as species, subspecies-age-sex distribution, and health status.

"Integrated exposure analysis" combines the estimation of environmental concentrations (sources and fate information) with a description of the exposed population to yield exposure profiles. Data should be provided on the size of the exposed populations; duration, frequency, and intensity of exposure; and routes of exposure. Exposures should be related to all sources. 51 Fed. Reg. at 34,048.

Standard Evaluation Procedure: Ecological Risk Assessment

EPA guidelines establish standard procedures for evaluating ecological risks. EPA Hazard Evaluation Division, Standard Evaluation Procedure, Ecological Risk Assessment, EPA-540/9-85-001 (June 1986) (Ecological Risk Assessment). The procedures were "designed to ensure comprehensive and consistent treatment of major scientific topics," and "were developed to explain specifically the principles of scientific evaluation within the Office of Pesticide Programs." Ecological Risk Assessment at preamble.

Ecological risk assessment on pesticide use is defined as "estimating the likelihood or probability that adverse effects (e.g., mortality to single species of organisms, or reductions in populations of non-target organisms due to acute, chronic, and reproductive effects, or disruption in community and ecosystem level functions) will occur, are occurring, or have occurred." Ecological risk assessment is a function of the "toxicological hazard" and "environmental exposure" of the pesticide.

Toxicological hazard is defined as the "intrinsic quality of a pesticide to cause an adverse effect under a particular set of circumstances." Ecological Risk Assessment at 2. These data would include laboratory LC_{50} values for aquatic species and effect levels for mammal reproduction tests. An LC_{50} is a statistically derived number that is the median lethal concentration of a substance that can be expected to cause death in 50 percent of test animals. To assess the potential toxicological hazard posed by pesticide use, EPA uses a tiered approach. For example, in the assessment of toxicity in aquatic organisms, EPA uses four levels. Level one involves LC_{50} determinations on two species of fish and one invertebrate organism. Level two involves a variety of studies including LC_{50} determinations on several species of fish and invertebrates, bioaccumulation determinations in fish and a life cycle test in an invertebrate organism. Level three involves full life-cycle fish studies. Level four involves population effects studies and simulated or actual field studies. The amount of testing required is dependent on several factors, including the pesticide's persistence in the environment, toxicity level, whether the pesticide is applied directly to water, and whether the pesticide will likely be transported to water areas. Ecological Risk Assessment at 15-16.

Environmental exposure has two components: (1) the estimated environmental concentration (EEC), which is the amount of pesticide residue that will be present in the environment and available to a non-target organism; and (2) the "numbers, types, distribution, abundance, dynamics, and natural history of non-target organisms which will be in contact with these residues." Ecological Risk Assessment at 2. To assess environmental exposure, EPA again uses a tiered approach. For example, assessment of environmental exposure to aquatic organisms involves four levels of testing. Level one considers the "worst-case" scenario (i.e., direct application to water) and assumes that all of the pesticide will enter and mix with the body of water. If the determined EEC is less than the acute or chronic aquatic regulatory risk criteria (detailed below), EPA "presumes that there will be no risk to aquatic organisms." Ecological Risk Assessment at 20. If the EEC is equal to or greater then the risk criteria, EPA proceeds to level 2.

In level 2, surface runoff and spray drift are the major mechanisms considered that contribute to pesticide loss to the aquatic environment. EPA estimates that approximately 10 percent of the sprayed pesticide will reach the aquatic environment. Chemical and site dependent variables are factored in for surface runoff estimates. The resultant EEC is compared to the risk criteria, and if there is still potential risk, EPA proceeds to level 3. In level 3, EPA utilizes exposure models and factors in the principal "mechanisms of pesticide transport," including leaching and adsorption. EPA may also request actual field residue monitoring studies (level 4), and registrants themselves often submit field residue studies to "verify or refute the model estimates." Ecological Risk Assessment at 21.

The correlation of the environmental exposure and hazard (toxicity) in order to evaluate the risk to the environment is now routinely done with a model utilizing regulatory risk criteria. EPA compares the EEC (amount in the environment) and an effect level (i.e., LC_{50}) which is then compared to the regulatory risk criteria. The regulatory risk criteria, specified in 40 C.F.R. parts 129 and 154, fall into three categories: (1) presumption of no risk; (2) presumption of risk that may be mitigated by restricted use; and (3) presumption of unacceptable risk subdivided into criteria for non-endangered and endangered species. The size of the safety factor varies, as mammals and birds have the lowest safety factors, while aquatic organisms have the highest. Ecological Risk Assessment at 4.

For example, for granular products with a potential hazard to birds, the regulatory risk criteria specify that if the EEC is one-fifth or less of the LC_{50} then there is a presumption of no risk. If the EEC is greater than one-fifth but less than the LC_{50}, then there is a presumption of risk that may be mitigated by restricted use. But, if the EEC is greater than the LC_{50}, then there is a presumption of unacceptable risk. Ecological Risk Assessment at 4.

EPA has recognized that the regulatory risk criteria ratio model has several weaknesses: "(1) it does not adequately account for effects of incremental dosages; (2) it does not compensate for differences between laboratory tests and field populations; (3) it cannot be used for estimating indirect effects of toxicants . . . ; (4) it has an unknown reliability; (5) it does not quantify uncertainties; and (6) it does not adequately account for other ecosystem effects." Ecological Risk Assessment at 8.

Pesticide Assessment Guidelines

These guidelines present EPA's recommended experimental designs to assess a wide variety of pesticidal factors including product chemistry, photolysis studies, and hazard evaluation to human and domestic animals. For example, in Subdivision F, Hazard Evaluation: Human and Domestic Animals, EPA presents its "recommended experimental designs for a set of toxicology and associated studies in experimental animals. The studies were designed to assess the toxicity of pesticidal chemicals in order to evaluate their potential for harm to humans or domestic animals under proposed use conditions." 55 Fed. Reg. 38,578 (1990).

INFLUENTIAL POLICIES OF OTHER AGENCIES

OSTP Policy on Cancer Risk Assessment

The Policy on Cancer Risk Assessment, published by the White House Office of Science and Technology Policy (OSTP), contains guidance for federal agencies performing cancer risk assessments. 50 Fed. Reg. 10,372 (1985). OSTP has developed thirty-one general principles under six different categories relevant to cancer risk (mechanisms of carcinogenesis, short-term testing, long-term testing, epidemiology, exposure assessment, and risk assessment). No particular quantitative details of risk assessment are presented in this policy.

In the risk assessment area, the policy discusses the elements of hazard identification, dose-response assessment (including models for extrapolation of data from animal to human low dose risks), and exposure assessment. 50 Fed. Reg. at 10,437. OSTP defines risk assessment as "a complex mixture of currently available scientific data, assumptions and judgments based on prevailing scientific thought, and policy decisions." 50 Fed. Reg. at 10,441.

NTP Guidelines

In 1978, the Department of Health, Education and Welfare (now the Department of Health and Human Services) established the National Toxicology Program (NTP) to strengthen and improve the area of toxicology testing in order to provide a resource for government agencies that rely on the results of toxicology studies to make regulatory decisions. In June 1983, NTP developed guidelines to evaluate animal carcinogenicity studies in a consistent framework. The guidelines were reviewed and modified by NTP in 1986. 51 Fed. Reg. 11,843 (1986).

The following evidentiary categories were developed by NTP:

> **Clear Evidence of Carcinogenic Activity** is demonstrated by studies that are interpreted as showing a chemically related increased incidence of malignant neoplasms, studies that are interpreted as showing a dose-related . . . marked increase of benign neoplasms if there is an indication from this or other studies of the ability of such tumors to progress to malignancy, or studies that exhibit an increased incidence of a combination of malignant and benign neoplasms where each increases with dose.
>
> **Some Evidence of Carcinogenic Activity** is demonstrated by studies that are interpreted as showing a chemically related increased incidence of neoplasms (malignant, benign, or combined) in which the strength of the response is less than that required for clear evidence.

> **Equivocal Evidence of Carcinogenic Activity** is demonstrated by studies that are interpreted as showing a chemically related marginal increase of neoplasms.
>
> **No Evidence of Carcinogenic Activity** is demonstrated by studies that are interpreted as showing no chemically related increases in malignant or benign neoplasms.
>
> **Inadequate Study of Carcinogenic Activity** demonstrates that because of major qualitative or quantitative limitations, the studies cannot be interpreted as valid for showing either the presence or absence of a carcinogenic effect.

51 Fed. Reg. at 11,844.

Through 1989, the NTP guidelines have been used to evaluate over 162 toxicology and carcinogenicity studies. It should be noted that these guidelines are used to evaluate individual scientific studies and not, as with the EPA and OSTP guidelines, to assess the overall weight of the evidence on carcinogenicity from a number of studies. NTP's evaluations are then published in the Annual Report.

In *Synthetic Organic Chemical Manufacturers Association v. Department of Health and Human Services*, No. 89-0884-LC (W.D. La. Aug. 28, 1989), *aff'd*, No. 89-4677 (5th Cir. Sept. 28, 1989), an industry coalition challenged NTP's procedures used to classify dichlorobenzene, arguing that NTP failed to consider all relevant studies. Although the court refused to enjoin the listing of dichlorobenzene or the publication of the Annual Report, the court did hold that the "adoption of the Classification Procedures and criteria," slip op. at 4, as well as "the issuance of the Annual Report," slip op. at 5, constitute agency action subject to judicial review. This changed the status of the Annual Report from an informational listing to a regulatory tool.

For further information regarding other agencies' cancer policies, see the discussion on OSHA's generic cancer policy in Chapter 17.

COMPUTER MODELS

EPA evaluates the potential risk of a chemical through the use of models. For example, a suspected carcinogen will be given in large doses to test animals, and the agency will use the linearized multistage model (LMS) to achieve real life estimates on the carcinogenic potential of the pesticide. EPA uses many different models and even makes some of them available to registrants. Thus, a registrant can use EPA's models to predict any potential problem. Also, private companies develop and sell models that are not only the same as EPA's model, but also allow for alternative assumptions to counter the conservative worst-case assumptions that are generally found in EPA models.

RECENT DECISIONS

The following recent decisions illustrate EPA's current approach to risk assessment and management:

Human Health Concerns

Dinoseb. Dinoseb, a herbicide, desiccant, fungicide and insecticide, was initially registered in 1948. In October 1986, EPA issued an emergency suspension order and notice of intent to cancel and deny all registrations for pesticide products containing dinoseb because of developmental toxicity concerns. 51 Fed. Reg. 36,634 (1986). The hazard evaluation of dinoseb focused principally on results from animal teratology studies, from which EPA concluded that dinoseb may cause birth defects in females exposed to dinoseb during pregnancy.

Exposure studies indicated that dinoseb was readily absorbed through the skin of animals and humans. 51 Fed. Reg. at 36,651. EPA assessed both worker and dietary exposures from use of dinoseb. Using its surrogate exposure model, EPA estimated that workers were exposed through applying dinoseb by a ground boom sprayer, aerial applicator, or hand-held sprayers. Exposure levels varied depending upon the method of application.

EPA estimated that the number of people exposed to dinoseb annually included 2,200 females. Additionally, EPA assessed indirect exposure to farmworkers or bystanders through spray drift during application and through washing contaminated clothes in the household laundry. EPA also considered dietary exposure through consumption of treated products. In addition, EPA considered public exposure risk from drinking contaminated ground water.

EPA estimated that the margin of safety (MOS) for workers applying dinoseb was less than 100, and in some instances less than one. The MOSs for consuming treated crops and drinking contaminated ground water were established at 2,703 and 2,452, respectively.

Even when certain "worse case" assumptions regarding dietary exposure levels were factored into EPA's analysis, the MOS for the risk of birth defects from eating food crops treated with dinoseb was found to be ample — over 2,700. Similarly, from consumption of contaminated drinking water, the MOS was roughly 2,450. For worker exposure, however, in many instances the estimated worker exposure levels were equal to or greater than the NOEL, leaving essentially no MOS for the offspring of pregnant workers handling the pesticide.

EPA examined the benefits of dinoseb and determined that a suspension could result in first year losses of $80 to $90 million, due primarily to increased control costs and expected yield losses. No significant market and consumer impacts were expected except for possible peanut price increases. EPA considered a range of risk management measures, including additional protective clothing, protective farm equipment, lower application rates, and product reformulation. EPA determined, however, that none of these measures could reduce the occupational risks to reasonable levels. Therefore, although the benefits of dinoseb were substantial in terms of monetary value, EPA concluded that the risks from occupational exposure to dinoseb outweighed its benefits, not only in the long-term, but also in the interval of time required to conduct usual cancellation proceedings.

On October 14, 1986, EPA issued a Decision and Emergency Order Suspending the Registrations of All Pesticide Products Containing Dinoseb, 51 Fed. Reg. 36,634 (1986). The emergency order prohibited, as of that date, any sale, distribution, or use of a pesticide product containing dinoseb. The United States Court of Appeals for the Ninth Circuit affirmed the final order of the EPA Administrator cancelling registrations of products containing dinoseb and permitting the use of existing stocks, subject to the restrictions and conditions of such order. *Northwest Food Processors Association v. Reilly*, 869 F.2d 542 (9th Cir.), decision at 886 F.2d 1075 (9th Cir. 1989), *cert. denied*, 110 S. Ct. 3239 (1990).

Lindane. In February 1977, EPA initiated a Special Review of the insecticide lindane, which is 99 percent pure gamma-hexachlorocyclohexane. 42 Fed. Reg. 9816 (1977). EPA took this action because it considered that the following three risk criteria had been met or exceeded: (1) acute toxicity to aquatic organisms; (2) oncogenic effects in test animals; and (3) chronic and/or delayed toxicity causing reproductive or fetotoxic effects in test animals. EPA consequently requested registrants and other interested parties to submit additional data.

After its review of the requested comments and additional studies, EPA issued Lindane Position Document 2/3, in which the Agency proposed to cancel most of the uses of lindane, based upon a finding that the risks associated with lindane far outweighed the benefits. 45 Fed. Reg. 45,362 (1980). In particular, EPA, at the time, asserted that the benefits of lindane use were outweighed by the risks of oncogenicity and reproductive and fetotoxic effects. In addition, EPA expressed concern (although criteria were not exceeded) about acute central nervous system effects and about childhood sensitivity to the toxic effects of lindane.

EPA then received comments from the FIFRA Scientific Advisory Panel, the United States Department of Agriculture, and interested members of the public, which caused the Agency to reconsider and re-evaluate much of the available data on lindane and to redo its risk/benefit analysis. As a result of this reconsideration, EPA reversed its initial position on lindane, concluding in Lindane Position Document 4 (PD 4) that lindane was a relatively safe chlorinated hydrocarbon, and that the continued registration of virtually all uses of lindane should be allowed. 48 Fed. Reg. 48,512, 48,514 (1983).

EPA's PD 4 provides a detailed analysis of the risks and benefits of lindane uses. EPA set forth several reasons for reversing its earlier position that most lindane uses should be cancelled: (1) EPA significantly improved its risk estimates through its review of additional use information and data on lindane; (2) upon review, the evidence of potential human cancer risk was limited, and measures less stringent than cancellation could reduce to reasonable levels the risks associated with continued use of lindane; (3) EPA's risk reduction measures also sufficiently reduced the perceived potential risks of fetotoxicity; and (4) in forming its previous position, EPA had not adequately considered the significant benefits of lindane use. Lindane Position Document 4, at ii-iii.

With regard to cancer risk, EPA stated that "[t]here is positive evidence [from two chronic studies] that lindane causes liver tumors in mice," as well as "supportive evidence" from two subchronic studies. 48 Fed. Reg. at 48,514. The Agency acknowledged the controversy over "whether or not increased mouse liver tumors are predictive of human carcinogenicity." Lindane Position Document 4, at 53.

Indeed, EPA subsequently has noted that "[f]or a number of reasons, there are widely diverging scientific views . . . about the validity of mouse liver tumors as an indication of potential carcinogencity in humans when such tumors occur in strains with high spontaneous background incidence and when they constitute the only tumor response to an agent." 51 Fed. Reg. 33,992, 33,995 (1986). Notwithstanding the limited utility of mouse liver tumor data as an indicator of human risk, EPA assumed lindane's carcinogenic potential and applied the most conservative cancer risk estimates to assess the risks to applicators and the general public, and concluded that the risks were not unreasonable. 48 Fed. Reg. at 48,519.

In addressing non-cancer risks, EPA concluded that reproductive and selective fetotoxic effects could be controlled adequately by protecting mothers from general toxic effects. Lindane Position Document 4, at 22-24. EPA also found it unlikely that lindane would cause permanent neurological damage at the levels and under the conditions of foreseeable human exposure. Lindane Position Document 4, at 55. Furthermore, EPA concluded that: (1) the general use applications of lindane pose no unreasonable adverse effects when sold with childproof packaging and used with various precautions and, thus, did not require specific label warnings with respect to cancer or reproductive effects; (2) recent studies demonstrated no statistically significant association between lindane exposure and blood dyscrasias, including aplastic anemia; and (3) lindane was not causing reductions in populations of non-target avian species. 48 Fed. Reg. at 48,515. Accordingly, EPA reversed its earlier position and determined not to cancel the registration of lindane.

Ecological Concerns

Dicofol. In 1984, EPA initiated a Special Review of dicofol to determine the environmental and human health risks posed by dicofol's DDTr content. In conjunction with the Special Review, EPA requested the Agency Office of Endangered Species (OES) to initiate a consultation on dicofol pursuant to § 7 of the Endangered Species Act (ESA), which provides, in relevant part, that "[e]ach Federal Agency shall, in consultation with and with the assistance of the Secretary [of Interior], insure that any action [it takes] . . . is not likely to jeopardize the continued existence of any endangered species." ESA § 7(a)(2). OES opined that the use of dicofol is likely to jeopardize the existence of two falcon species.

Engaging in a risk/benefit analysis, EPA first quantified the risks posed by the DDTr contaminants in dicofol, and determined that DDTr concentrations of 0.6 ppm to 3.0 ppm could result in eggshell thinning, which can reduce avian populations. Next, EPA determined that while the concentration of DDTr has steadily declined since the cancellation of DDT, the current DDTr concentration in dicofol was high enough to make significant contributions to the overall level of DDTr in major use areas. "This contribution would likely result in a slowing of the decline of DDTr residues in areas of dicofol use and would likely result in a stabilization of DDTr residues at levels which would continue to cause adverse effects in wildlife." Of special concern was the risk to endangered species inhabiting areas of dicofol use. 51 Fed. Reg. at 19,512. In assessing the benefits, EPA noted that dicofol: (1) was recommended in integrated pest management programs; (2) was effective where other pesticides were not; and (3) sometimes has no alternatives. The Agency also calculated the economic impact due to the loss of use of dicofol.

Both EPA and the SAP considered OES's opinion, but concluded that dicofol would not threaten the environment, including endangered species, if its DDTr content were reduced to 2.5 percent for a period of time, and reduced thereafter to 0.1 percent. 50 Fed. Reg. 33,008, 33,010, 33,014 (1985). EPA requested that OES reinitiate its dicofol consultation to take into account the EPA and FIFRA SAP conclusion. As a result, OES rendered a second opinion in November 1986, stating that use of dicofol with no more than 0.1 percent DDTr would not jeopardize the peregrine falcon, but that use of dicofol with any higher DDTr content appeared likely to jeopardize the species. In December 1985, EPA again reinitiated consultation with OES to consider the effects of dicofol in the intervening period before registrants could meet the 0.1 percent DDTr standard. In March 1986, OES issued a third opinion which concluded that payment of $325,000 over four years to the Peregrine Nesting Program in California would prevent any jeopardy from the use of dicofol with DDTr greater than 0.1 percent between October 1985 and December 31, 1988.

On May 29, 1986, EPA concluded its Special Review of dicofol and found that dicofol products "can meet the statutory standard for continued registration under FIFRA," 51 Fed. Reg. at 19,517, if registrants (1) reduce the DDTr in dicofol initially to 2.5 percent and, after December 31, 1988, to 0.1 percent, and (2) add to all labeling a requirement that handlers wear gloves. Thus, through use of the foregoing techniques, EPA was able to maintain the registration of the product.

Diazinon. While the dicofol decision indicates EPA's movement toward risk management, the diazinon case indicates that EPA is also cautious, especially when addressing acutely toxic effects. In 1986, EPA initiated cancellation proceedings for golf course and sod uses of diazinon which, for the first time, concerned a pesticide that posed a risk solely to birds (and not to man). 51 Fed. Reg. 35,034 (1986).

In assessing the hazard of diazinon, EPA considered laboratory data indicating that diazinon was acutely toxic to a wide variety of avian species. 53 Fed. Reg. 11,119, 11,122 (1988). EPA also had evidence of reported bird kills "associated with diazinon use on golf courses and other large grassy areas." 53 Fed. Reg. at 11,124. Using residue studies and modeling, EPA estimated exposure. EPA's risk assessment model indicated that after grazing for only a short period of time, several avian species could reach their LD_{50} levels. 53 Fed. Reg. at 11,123. Against this risk, EPA weighed the benefits of continued use, and found that alternative pesticides were less toxic, and that increased costs to golf course operators were negligible. 53 Fed. Reg. at 11,126. Accordingly, the Administrator determined that EPA had made its *prima facie* case, and that the burden had shifted to the registrants to rebut the agency's case.

The registrant proposed to mitigate the risk by reducing the application rate and imposing geographic and seasonal restrictions on use. Acknowledging that label changes to this effect were of some help, EPA determined that they would not eliminate exposure and that, without eliminating exposure, diazinon use "would generally pose a serious risk of regularly repeated bird kills." 55 Fed. Reg. 31,138, 31,141 (1990).

Through simulated and actual field studies, the registrant attempted to show that EPA's exposure levels were inflated. Yet, EPA dismissed these studies as preliminary and insufficient "to quantify the adverse effects." 53 Fed. Reg. at 11,125. Finding its exposure estimates irrebutted, EPA cancelled sod and golf course uses of diazinon. 55 Fed. Reg. at 31,144-45. The diazinon case indicates the heavy burden that EPA places on a registrant whose pesticide is acutely toxic to wildlife.

RECENT AND FUTURE TRENDS

Mesocosm Studies

Although laboratory testing is useful to begin a determination of risks, often the complexities of nature cannot be measured in small-scale microcosm studies. "[E]cologists and aquatic toxicologists have recognized the weaknesses of using single-species tests alone for assessing potential ecosystem impacts." EPA, Hazard Evaluation Division Technical Guidance Document — Aquatic Mesocosm Test to Support Pesticide Registrations 1 (Feb. 1988). Further, field tests may not be suitable because while in field tests a "catastrophic event" can be detected, subtle effects which change or alter the system are not easily identified. Thus, some "[e]cologists who have recognized this deficiency have developed physical models . . . for aquatic ecosystems which allow the necessary control and replicability to detect ecosystem-level effects." Aquatic Mesocosm Test at 3.

Presently, EPA favors mesocosm studies in which experimental ponds with *in situ* enclosures simulate the toxic effects of a pesticide on an aquatic ecosystem. EPA only requires mesocosm studies when single-species laboratory data indicate a serious hazard or impact to aquatic organisms. So far, EPA has required mesocosm studies only to assess the aquatic risks posed by synthetic pyrethoids used on cotton. Because the protocol for mesocosm studies is still rather experimental and the data based on such studies limited, industry has been reluctant to venture into this experimental testing, despite EPA's pressure to do so. Further, aquatic toxicologists still disagree on whether the studies are useful or the resulting data can be meaningfully interpreted.

Criticism of the Maximum Theoretical Dose

In August 1990, the Bush administration issued a guidance document entitled Regulatory Program of the United States Government, which discusses regulatory risk assessment. The report criticized the methods that regulatory agencies use to assess risks. It suggests that "conservative (worst-case) assumptions distort[ ] risk assessment, yielding estimates that may overstate likely risks by several orders of magnitude." Any such excessive risk assessment necessarily "distorts the regulatory priorities of the Federal Government, directing societal resources to reduce what are often trivial carcinogenic risks while failing to address more substantial threats to life and health." Regulatory Program at 14. The report further notes, moreover, that "[b]y combining very high doses with highly sensitive test subjects, some animal bioassays are predisposed to discover apparent carcinoginic effects." Regulatory Program at 18.

In September 1990, the National Academy of Sciences convened a panel on risk assessment to examine the way scientists calculate human health risks from chemicals, focusing specifically on the use of Maximum Theoretical Dose (MTD) testing. MTD is the dose "an animal species can tolerate for a major portion of its lifetime without significant impairment of growth or observable toxic effect other than carcinogenicity." J. Cohrssen & V. Covello, Risk Analysis: A Guide to Principle and Methods for Analyzing Health and Environmental Risks 53 (1989). There is growing concern that the sheer volume of the test chemical, and not its inherent toxicity, may cause the observed effect. If so, a useful compound may be unfairly indicted as a carcinogen when the real cause is the MTD overwhelming the test animal's detoxification mechanisms. The NAS panel hopes to develop alternatives to the MTD method. Wash. Post, Sept. 8, 1990, at A5.

Chapter 10
FEDERAL GROUND WATER REGULATION

In response to mounting evidence of ground water contamination by pesticides,[24] EPA is developing a generic policy for agricultural chemicals found in ground water. Because EPA has taken several years to develop a final policy, the Agency has, in effect, produced two different policies: the generic policy that has yet to be finalized and the Agency's policy in practice while the final rule is pending.

Under EPA's existing practice, ground water issues are addressed on a case-by-case basis. The Agency focuses its efforts on specific chemicals that commonly leach into ground water and have demonstrated adverse effects. EPA's case-by-case approach has led some states, such as California, to promulgate comprehensive ground water regulations to fill the regulatory gaps left by EPA.

While the Agency's current approach is largely piecemeal, in 1988 EPA proposed a generic policy on pesticides in ground water and then followed it up with the first real-world application of the policy in the proposed Special Review for the pesticide aldicarb. 53 Fed. Reg. 24,630 (1988). Under the policy, varying degrees of protection would be accorded ground water depending on its use. Maximum Contaminant Levels under the Safe Drinking Water Act would be used as primary reference points for determining specific goals for prevention and response. The policy would be implemented through a combination of uniform national measures, geographically variable use restrictions imposed by EPA, and

---

[24] Initial EPA survey results indicate that normal agricultural usage has contributed to contamination of ground water by at least forty-six pesticides in twenty-six states. EPA, Pesticides in Groundwater Data Base: 1988 Interim Report (1988). The first phase of EPA's National Survey of Pesticides in Drinking Water Wells, which covered 1,350 community and domestic drinking water wells in fifty states, concluded that about 10.4 percent of community wells and 4.2 percent of rural domestic wells contain pesticides. EPA, National Pesticide Survey: Phase I Report (1990). EPA estimates, however, that less than 1 percent of all wells contain pesticides at concentration above the Maximum Contaminant Levels of Health Advisory Levels.

"state management plans." Neither the policy nor its application in the aldicarb Special Review have yet been finalized.

At this crossroads, several controversial issues remain unresolved. Initially, an important empirical question remains unanswered, namely the extent of ground water contamination from agricultural chemicals. Although initial survey results indicate the presence of some pesticides in certain geographical regions, EPA is awaiting final survey results to determine the severity of the problem nationwide.

Second, the Agency must address the role of ground water contamination in its risk assessment matrix. Under its existing policy, EPA appears to have concluded that the mere presence of a pesticide in ground water — without any evidence of adverse toxicological effects — is sufficient to trigger regulatory requirements. It is questionable, however, if the mere presence of a chemical in ground water constitutes an "unreasonable risk" sufficient for FIFRA regulation.

Finally, and most importantly, EPA is faced with the problem of adapting the generic, nationwide regulatory scheme associated with FIFRA to ground water problems that are by their nature site- and use-specific. EPA's generic policy and its application in the aldicarb Special Review remain in proposed form because of the myriad of practical and legal problems associated with implementing a ground water policy under FIFRA.

EPA'S CURRENT POLICY

EPA's approach to ground water regulation to date has been largely ad hoc. With the exception of an environmental data call-in for 135 active ingredients and the inclusion of pesticides with ground water concerns in Class B for reregistration purposes, regulatory action has been based on the specific characteristics of a product and its use pattern. The Agency has utilized its existing regulatory requirements, including data submissions, labels, tolerances, restricted-use classification and cancellation, and simply applied them to ground water concerns.

Regulatory action has focused on products which: (1) have been detected in ground water in a reasonably large number of instances; (2) are believed to have leached into ground water from normal agricultural use; and (3) demonstrate serious toxicological effects or have significant data gaps. Controversy has arisen surrounding the relative importance of item (3) — that is, whether regulation can be based on the existence of a pesticide in ground water without evidence of adverse effects. The Agency's response to this issue has varied depending on the burden of the regulatory action involved.

In its 1985 Criteria and Procedures for Special Review the Agency concluded that "the mere detection of a pesticide in ground water is not a sufficient basis for a Special Review." 50 Fed. Reg. 49,003, 49,008 (1985). The Agency went on to state that

> [i]f a pesticide is found in ground water, the Agency will evaluate known contamination, the likelihood of more extensive contamination in the future, and available data concerning environmental fate and toxicity before determining whether the potential hazard is sufficient to initiate a Special Review.

50 Fed. Reg. at 49,008. See also 40 C.F.R. § 154.7 (1989). In practice, pesticides that have been cancelled partially due to ground water concerns have also had associated toxicity problems. 48 Fed. Reg. 46,228, 46,234 (1983) (ethylene dibromide); 50 Fed. Reg. 41,938, 41,939 (1985) (carbofuran); 44 Fed. Reg. 65,135, 65,161 (1979), 50 Fed. Reg. 1119, 1121 (1985) (DBCP); 52 Fed. Reg. 49,480, 49,481 (1987) (alachlor); 52 Fed. Reg. 589, 593 (1987) (cyanazine).

EPA has attempted, however, to impose less stringent requirements, e.g., label warnings and restricted use classification, in the absence of a demonstrated toxic effect. EPA has taken the position that the existence of data gaps for chronic studies, coupled with a demonstrated potential for leaching into ground water, is sufficient basis to require label warnings and restricted use classification. The registration standard for simazine is one example. The Agency justifies its position on the grounds that once contaminated, ground water may be difficult or impossible to clean up because of the slow recharge of aquifers, the low level of biological activity and degradation, and the persistence of some pesticides.

Despite EPA's position, it is questionable if the mere presence of a pesticide in ground water constitutes "unreasonable risk" justifying regulatory action.

USDA GROUND WATER INITIATIVES AND THE FOOD, AGRICULTURE, CONSERVATION, AND TRADE ACT OF 1990 (THE 1990 FARM BILL)

The USDA has supplemented EPA's regulatory efforts in the ground water arena with research and development initiatives designed to decrease ground water contamination. In particular, USDA has invested substantial resources in improving management practices in the Midwest where concentrated corn and soybean production involves widespread pesticide use. Other major research has included: (1) determining the extent of agricultural contributions to non-point source pollution; (2) developing improved farm and ranch management practices; (3) improving the understanding of agricultural chemical movement in soil and water; and (4) developing less costly ground water sampling techniques.

The Food, Agriculture, Conservation, and Trade Act of 1990, Pub. L. No. 101-624, 104 Stat. 3359 (the 1990 Farm Bill) expanded USDA's role in ground water protection. The Act established a new water quality protection program which requires USDA to enter into agreements to provide financial assistance for agricultural water quality protection practices upon the request of owners and operators of eligible farms. An "agricultural water quality protection practice" is defined as a "farm-level practice or a system of practices designed to protect water quality by mitigating or reducing the release of agricultural pollutants, including nutrients, [and] pesticides." 136 Cong. Rec. H11,095 (daily ed. Oct. 22, 1990). The program has additional incentives for wetlands preservation and wildlife habitat improvement. The water quality program is, however, restricted to lands that either currently pose ground water concerns or are likely to do so in the future. Eligible lands include:

(1) areas that are not more than a thousand feet from a public well unless a larger well head area is established by the Secretary into consultation with EPA and the state;

(2) areas that are in shallow Karst topography areas where sink holes convey runoff water directly into ground water;

(3) areas that are critical crop land areas within hydrological units identified in a plan submitted by the state under section 319 of the Water Pollution Control Act;

(4) areas where agricultural non-point sources have been determined to pose a significant threat to habitat utilized by threatened and endangered species;

(5) areas recommended by state lead agencies for protection;

(6) areas recommended by EPA or the Secretary of Interior for protection;

(7) areas that, if permitted to continue to operate under existing management practices, would defeat the purpose of the program; and

(8) areas contributing to identified water quality problems in areas designated by the Secretary.

136 Cong. Rec. at H11,097. Under the program, the Secretary must give priority to lands where agricultural production contributes to the potential for failure to meet applicable water quality standards. In addition to the funding provided under the program, USDA provides technical assistance to participating farmers. In 1994, the Secretary is required to provide a summary of the program to Congress.

The Act created three committees to aid in the administration of the water quality protection plan and other aspects of USDA's water quality program. First, the Agricultural Council on Environmental Quality will be responsible for coordinating USDA's environmental policy generally. The Council, which will include members from both the Department of the Interior and EPA, is required to develop an environmental quality policy statement that identifies goals and objectives for addressing the effects of agriculture on environmental quality. Second, the Act required USDA to establish a water quality coordination program within each state. Membership in the program will include a variety of state agencies with responsibility over agricultural activities. 136 Cong. Rec. at H11,105.

Third, the Secretary must establish technical committees in each state to develop guidelines for the wetland and wildlife improvement portions of the water quality protection program.

Other sections of the Act address USDA's ongoing efforts in research and development, and technology transfer. Section 1484 of the Act authorizes the Secretary to establish a coordinated water quality and nutrient management research program covering a broad variety of water quality related issues. Alternate methods of pest and disease control and improved methods for pesticide storage, use and disposal are specifically included as areas for further research. 136 Cong. Rec. at H11,105. The Act also requires USDA to establish a repository of agricultural and ground water quality planning information and to determine whether the Natural Resources Inventory can be modified to acquire information on ground and surface water conditions that affect water quality and supply.

Finally, the pesticides subtitle of the Act confirms that USDA is the principal federal agency responsible for the development and delivery of educational programs, technical assistance, and research programs on use, storage and disposal of agricultural chemicals for users and dealers of agricultural chemicals. 136 Cong. Rec. at H11,106.

THE FUTURE OF GROUND WATER REGULATION: THE 1988 PROPOSED STRATEGY

In February 1988, EPA issued its proposed strategy on pesticides in ground water. Agricultural Chemicals in Groundwater: Proposed Pesticide Strategy (1988) (Proposed Strategy). The strategy provides a framework for EPA's approach to the generic issues posed by pesticides in ground water. The Proposed Strategy, and EPA's first practical application of the strategy in the 1988 proposed ground water portion of the Special Review for the pesticide aldicarb, are the most significant published statements on the future of EPA's pesticides in ground water policy. In 1990, EPA expanded on the Proposed Strategy with a Federal/State Relationship Options Paper (EPA Ground Water Task Force, 1990) (Federal/State Options Paper) and an EPA Statement of Groundwater Principles (Jan.

1990). The papers further develop EPA's position in the Proposed Strategy, particularly in the controversial area of state/federal relations.

The policy and its application in the aldicarb Special Review have remained in proposed form because of the practical and legal problems associated with their implementation. FIFRA, for the most part, contemplates broad, national and uniform regulation of pesticides implemented primarily through licensing and use restrictions on product labels, rather than site-specific measures tailored to localized problems. Ground water problems by their nature are site-specific. As EPA moves beyond the general policy stage, the Agency must resolve the issue of how to adapt existing regulatory mechanisms to effectively implement its ground water policies.

Environmental Goals

Differential Protection. The Proposed Strategy adopts EPA's long-standing position on differential protection. Under this approach, EPA would establish three classes of ground water depending on the use, value and vulnerability of the resource and require different levels of protection for each.

The Proposed Strategy identifies Class I waters as "high-priority ground water." These resources are defined to include ground waters in areas with hydrogeologic characteristics that render them highly vulnerable to contamination and which either serve as irreplaceable sources of drinking water or which are ecologically vital. An irreplaceable source of drinking water is one which is located in an area where there is no practical alternative source, or only an insufficient alternative source for a substantial population. Ecologically vital ground waters are those which contribute to maintaining either the base flow or water level for a particularly sensitive ecological system, that would destroy a unique habitat if polluted. An example of an ecologically vital ground water area is one associated with wetlands that are habitats for unique species of flora and fauna or endangered species.

All other ground water currently used or potentially available for drinking water and other beneficial use is included in Class II ground waters. Particular vulnerability to contamination is not a factor in a Class II determination. Most of the usable ground water in the United States will be included in Class II, with Subclass IIA covering current sources and Subclass IIB covering potential sources of drinking water.

Ground water that is not a potential source of drinking water and is of limited beneficial use will be classified as Class III ground waters. This would include waters that are contaminated naturally, i.e., with high saline levels, and those ground waters with totally dissolved solids (TDS) levels over 10,000 mg/l. Also included are ground waters that are so contaminated by human activity that is unrelated to a specific hazardous waste land disposal site, that the water cannot be cleaned up by methods reasonably employed in a public water treatment system. In order to fall into Class III, ground water must not flow into Class I or Class II water, or surface water.

Implementing the policy to consistently apply such terms as "potentially available for drinking water" or "high priority ground water" to a resource that is localized and variable remains a significant obstacle for EPA. Because the Proposed Strategy would vary cleanup levels substantially for different classes of water, class designation is a crucial issue. The issue becomes particularly troubling if EPA delegates broad authority to the states (see discussion below) and various state authorities become responsible for classifying ground water.

Water Quality Standards. EPA plans to adopt MCLs under the Safe Drinking Water Act (SDWA) as reference points for determining unacceptable contamination of ground waters that are drinking water sources, i.e., Class I and II ground waters. Additional measures may be taken to protect Class I ground water.

The SDWA requires EPA to set MCLs as close as possible to Maximum Contaminant Level Goals (MCLGs, which EPA sets at no known or anticipated adverse effects) but taking into account the cost and feasibility of reducing contamination. When no MCL has been established, EPA proposes to use interim drinking water protection criteria, which are significantly more stringent than MCLs, as reference points. For noncarcinogenic pesticides, the interim drinking water criteria level is set equal to the MCLG. For carcinogenic pesticides it is set at a "negligible" risk level, which EPA equates with a one in a million risk of cancer ($1x10^{-6}$). Because MCLs and interim protection measures only address human health concerns and may not be adequate for protecting fragile ecosystems (particularly ones that support endangered species), EPA is also establishing separate water-quality based criteria under the Clean Water Act (CWA) as reference points for these systems.

EPA has not resolved whether MCLs will be used as a floor or a ceiling for allowable contamination. According to the Federal/State Options Paper, however, MCLs will be applied differently for prevention and cleanup purposes. For prevention purposes EPA is considering (1) allowing contamination up to the MCL and then requiring preventive measures or (2) requiring preventive measures prior to reaching the MCL in order to avoid reaching it. For cleanup purposes EPA is proposing (1) using MCLs as maximum cleanup levels but allowing less stringent levels depending on such factors as cost and feasibility; or (2) applying MCLs as reference points and using criteria under the applicable statute to determine the actual cleanup level.

In order to resolve the issue, EPA must address the conflict created by importing cleanup levels from other statutes while implementing the risk-benefit scheme under FIFRA. FIFRA, unlike the Comprehensive Environmental Response, Compensation, and Liability Act (CERCLA), the Resource Conservation and Recovery Act (RCRA) and the SDWA, specifically requires EPA to consider the benefits of pesticide use in deciding whether the presence of a pesticide in ground water presents an unreasonable risk to public health or the environment. Under the other statutes, EPA establishes standards based solely on public health and environmental considerations and does not consider the benefits of the activities which resulted in the presence of the chemical in the ground water.

In importing purely health-based standards from other statutes, EPA runs the risk of shirking its mandate under FIFRA to consider a pesticide's benefits. EPA's solution in the Proposed Strategy is to use the MCLs as a rebuttable presumption to conclude that the benefits of a pesticide's use do not justify a deviation from SDWA/CERCLA/RCRA standards and then to allow registrants to present evidence of the benefits of the pesticide and/or management techniques that could alleviate the risk. In practice, however, EPA's solution is likely to limit deviation from SDWA/RCRA/CERCLA standards to isolated cases where the pesticide has unique benefits.

Prevention

"Prevention of pesticide contamination of ground water is the centerpiece of the Agency's strategy for protecting ground water." Proposed Strategy at 85. The Proposed Strategy calls for preventing contamination through a combination of uniform national measures, geographically variable use restrictions imposed by EPA and "state management plans." State management plans, which provide the flexibility to address local variability, are the linchpin in the Agency's efforts under FIFRA to address the localized problems associated with ground water use.

Federal Regulation. EPA proposes to use its authority at the federal level to require first, uniform action for pesticides that pose widespread national concerns and second, differential management to account for local variations in ground water use, value and vulnerability.

Ground water problems that are not dependent on local variability can be adequately addressed by uniform national standards. For example, EPA has proposed regulations for the application of pesticides through irrigation, commonly known as "chemigation." 51 Fed. Reg. 6798, 6798 (1986). This practice can directly introduce pesticides into ground water if precautions, such as anti-back-siphoning devices, are not used. EPA is also developing a rule to restrict the use of potential leaching pesticides to certified applicators. 55 Fed. Reg. 5861 (1990).

Because of the local variability in pesticide use the most significant efforts, however, will be made in differential management based on local conditions. EPA plans to take a prominent regulatory role in these issues when (1) there is a need to establish regulatory consistency; (2) the scope of the effort requires national resources; (3) state-by-state efforts would create unwarranted and inefficient duplication; and (4) national security is involved. EPA, State Federal Relationship Options Paper: EPA Ground-Water Task Force 2 (1990) (Ground-Water Task Force). In contrast, states will have the lead regulatory role when (1) the areas of concern are numerous or highly localized and present a low to medium potential for ground water contamination; (2) land use management is the principal protection approach; and (3) technologies currently exist or can be easily developed to address the problem. Ground-Water Task Force at 2.

EPA anticipates differential management taking on a variety of forms ranging from a complete ban on a pesticide in some states to controls on the rate, timing or method of application in other states. The pesticide user would be responsible for determining the applicability of the measures. EPA has emphasized that monitoring ground water for pesticide contamination is essential to its program.

EPA's approach leaves unresolved how the user will be able to adequately make a determination regarding use, i.e., how the Agency will provide the user with adequate information on pesticide use and ground water conditions so that the user can determine applicability. Traditionally, all requirements regarding the use of a particular pesticide have been communicated via the product label. Technical, localized label restrictions, however, would be difficult for users to interpret. It remains to be seen whether restrictions can be effectively communicated to users by other means and what will be done to enforce such restrictions if communication problems arise.

State Management Plans. State management plans are the linchpin of the FIFRA ground water program. Combining minimal federal standards with the flexibility necessary to address local concerns, state plans offer EPA the solution to the uneasy fit between FIFRA and ground water problems.

In the Proposed Strategy, the Agency sets forth basic requirements for state plans. The elements of a program would include:

1. Setting Goals and Documenting Progress

    a. Ground water protection goal

    b. Yearly action plan

    c. Mechanism for evaluating progress toward the goal and providing for EPA review

2. Characterizing the Resource and Prioritizing Action

    a. Comprehensive mapping of aquifer systems

    b. Inventory of all potential sources of contamination

    c. Process used for prioritizing action

3. Developing and Implementing a Control Program

    a. Enforceable quality standards that are health based for drinking water and water quality based in areas were ground water affects sensitive surface water ecosystems

    b. Regulatory and non-regulatory authorities to control contaminants

    c. Remediation program which dovetails with RCRA and CERCLA

    d. Monitoring and data analysis to determine risk

    e. Adequate compliance and enforcement authorities

    f. Water well program

    g. Adequate local and federal funding

    h. Public participation

4. State/Federal Roles

    a. Delineation of state authority

    b. Delineation of local authority

    c. Statement of the state's role under federally regulated statutes

    d. Mechanism for dealing with other federal agencies that affect state ground water programs, e.g. the Department of Agriculture and the Department of Interior

    e. Integration of water quantity and quality management

    f. Coordination of ground water programs with other natural resource programs, e.g. surface water management

Proposed Strategy at 5-7.

In implementing state management plans EPA faces a dilemma. Effective EPA enforcement demands significant federal oversight authority. On the other hand, as EPA's oversight authority increases, the ability of the states to effectively address localized problems decreases and the entire purpose of the management plans — to provide flexible regulation based on local conditions — is lost.

As a result of the substantial controversy surrounding this issue, the Agency has been unable to resolve the problem. In the Proposed Strategy EPA is considering two options: a program involving moderate EPA oversight and a program with more substantial oversight. The moderate oversight program would involve flexible approval/disapproval criteria that are generally performance-oriented, and financial and other incentives.[25] Under the substantial oversight program EPA would develop detailed mandatory approval/disapproval

---

25 The program would be similar to EPA's programs under the well-head protection program (CWA) and the underground storage tank program (RCRA).

criteria for both performance and process components. Sanctions for disapproval would include withdrawing funds and/or EPA assumption of inadequate state programs.[26]

Separately, the Agency is proposing to grant significant authority over non-drinking water sources to states with "comprehensive protection programs." In states with comprehensive programs EPA would defer to state use designations in deciding the level of activity to undertake in agency programs. As a result, under this option standards for non-drinking water in states with "comprehensive protection programs" could provide protection below minimal federal standards. EPA would still, however, enforce minimal standards for drinking water.

While EPA remains embroiled in these policy questions, several crucial implementation issues remain unresolved. The same implementation issues that face EPA at the federal level pose problems for the states: most significantly, how to communicate and enforce the laws against local users. The problem is compounded in many states by inadequate funding.

The mechanism EPA will use to enforce minimal standards for state plans remains a troubling and unresolved issue. EPA's enforcement authority could stem from either § 6 which allows cancellation of pesticides that cause "unreasonable adverse effects" or from § 3(d)(1)(C) under which EPA has the authority to conduct rulemaking for restricted use chemicals.

If EPA's decisionmaking process on other chemicals follows aldicarb, see discussion below, there will be an initial Special Review to determine whether a chemical poses a sufficient ground water risk to warrant regulatory restrictions. A second proceeding would follow, during which the specific state management plans would be developed and submitted to EPA for evaluation, and the pesticide cancelled in areas that fail to submit acceptable plans.

---

26 The program would be similar in structure to the permitting program under RCRA and the National Pollutant Discharge Elimination System (NPDES) under the CWA.

It is unclear whether the second proceeding should take the form of a cancellation under § 6 or a rulemaking under § 3(d)(1)(C). The cancellation process under § 6 involves a full adjudicatory hearing and may be too burdensome for EPA to use effectively. Although the rulemaking process would allow EPA to bypass the extended procedures involved in Special Review, if the procedure begins to look too much like Special Review, it would be open to challenge by industry for its failure to provide adequate procedural protection.

Response Actions

Response actions remain a largely unresolved issue. In the Proposed Strategy, EPA discusses the problem in only very general terms, concluding that acceptable state management plans should "consider the development of a valid corrective response scheme." Proposed Strategy at 122.

Despite EPA's position, however, it is questionable if the Agency has the authority to require cleanup of pesticide contamination. FIFRA does not provide EPA with cleanup authority. Although CERCLA gives EPA broad remedial authority, § 107(i) prohibits the recovery of cleanup costs resulting from normal application of a registered pesticide. Therefore, although EPA may require private parties to clean up contamination resulting from pesticide misuse, the Agency is limited to using scarce Superfund money to fund cleanup from normal application.[27]

EPA may, however, have separate authority under § 106 of CERCLA to order private parties to clean up a site. Under § 106 EPA can issue an administrative order or sue in federal court for an injunction ordering private parties to undertake or pay for the abatement of imminent and substantial hazards caused by an actual or threatened release

27 EPA has proposed to list several pesticide contaminated wells in Oahu, Hawaii as Superfund sites. 49 Fed. Reg. 40,320, 40,330 (1984). In order to clean up the sites, however, EPA will be required to expend Superfund monies.

of a hazardous substance. Whether CERCLA § 106 authorizes EPA to order private parties to abate risks from hazardous substance releases for the purpose of cleaning up pesticide contamination, or whether the exception in § 107(i) extends to § 106, remains unresolved. The Proposed Strategy therefore places EPA in the untenable position of requiring states to have remedial authority and funding that EPA itself may lack.

There are several reasons why the § 107(i) pesticide exemption should be viewed to carry over to § 106. Several courts have held that the full scope of liability under CERCLA is defined in § 107 of the Act. *United States v. Ottati & Goss, Inc.*, 630 F. Supp. 1361 (D.N.H., 1985); *United States v. A&F Materials*, 578 F. Supp. 1249 (S.D. Ill. 1984). See also Grumbles, Pesticides in Ground Water and Section 106 CERCLA, 19 Env't Rep. 281 (June 24, 1988) for a discussion of the Conference Report in the 1986 Superfund Amendments and Reauthorization Act where the conferees specifically chose not to alter the state of the law. Thus, to the extent Congress intended a pesticide applicator or manufacturer to be held liable for cleanup, that liability arguably is set forth in § 107, and is subject to the limitations of § 107(i). Second, because EPA has ample authority to respond to and abate substantial threats to public health due to ground water contamination under SDWA § 143(1), there is no reason for EPA to overstep its authority under § 106. Third, Congress intended to exempt pesticide applications from liability altogether by virtue of the § 107(i) exemption and the express exemption from § 103(e) reporting requirements for pesticide releases. Pesticide use is thus exempt from the liability provisions and reporting requirements, but CERCLA does permit EPA expenditures for cleanup of agricultural sites. CERCLA §§ 101(10), 103(e).

THE POLICY IN PRACTICE: THE PROPOSED ALDICARB REGULATION

EPA's first practical application of the Proposed Strategy involves the pesticide aldicarb. EPA began a Special Review of that chemical in 1984 because of its concerns regarding dietary exposure to aldicarb and reported instances of the detection of aldicarb in ground water. In June 1988, EPA published its proposed regulatory decision for the ground water portion of the Special Review. 53 Fed. Reg. 24,630 (1988).

The aldicarb proposal tracks the general outlines of the proposed ground water strategy, containing a mix of uniform nationally applicable label revisions, EPA-imposed geographically limited requirements, and state management plans for "highly vulnerable" areas.

Federal Regulation

EPA is proposing three uniform national requirements: (1) a restriction on use near wells; (2) restricted-use classification; and (3) a monitoring requirement. Under the proposal, aldicarb use would be banned within 300 feet of any drinking water well. The restricted use classification would be in addition to the pesticide's current classification for restricted use based on acute toxicity and would be intended to increase the applicator's awareness of ground water contamination. Monitoring would be required in representative areas having a medium potential to leach. EPA would maintain the option of imposing additional regulatory requirements based on the results of the monitoring.

State Management Plans

EPA is proposing that state management plans be developed for highly vulnerable areas. Under one alternative state management plans would be required based on an evaluation of "Heath Regions" (eleven distinct hydrological ground water regions in the United States, developed by the U.S. Geological Survey), while another alternative would allow EPA to assess vulnerable areas on a county-by-county basis. Under the Heath Region assessment, the following states would be required to submit a plan: Alabama, Florida, Georgia, Maine, Michigan, Minnesota, New York, North Dakota, Pennsylvania and Wisconsin. If EPA chooses the county option, an estimated fifteen to twenty-four states would be required to submit plans for one or more counties.

EPA's state management plan requirements reflect those in the Proposed Strategy. Although EPA has stated that it will take a middle path between simply setting a performance and dictating strict minimal requirements, it is clear that EPA intends to retain the final say over state management plans. What form this oversight will take is unclear. EPA's general requirements for state plans, which largely mirror those in the Proposed Strategy, offer little guidance on this issue. 53 Fed. Reg. 24,630, 24,636-37 (1988).

In proposing a state management approach, however, EPA explicitly rejected regulation based only on labeling changes enforced at the national level. The Agency felt that this option would result in labels that would be too difficult to interpret and enforce. 53 Fed. Reg. 24,629, 24,635-36. The aldicarb decision, if adopted as proposed, would therefore mark a significant departure from pesticide regulation in the past.

Chapter 11
EXPORTS AND IMPORTS UNDER FIFRA

EXPORTS UNDER FIFRA §§ 17(a) AND (b)

Pesticide exports are regulated under § 17 of FIFRA. Pesticides and active ingredients intended solely for export are generally exempt from most FIFRA requirements if packaged or prepared according to the specifications of the foreign purchaser. FIFRA § 17(a)(1). Producers of such pesticides, however, must still comply with the registration establishment requirements of FIFRA § 7 and the recordkeeping requirements of FIFRA § 8. Such producers are also subject to possible civil and criminal enforcement actions under FIFRA § 12 for any labeling or misbranding violations. See Chapter 15.

FIFRA EXPORT POLICY

In 1980, EPA issued a Statement of Policy on the Labeling Requirements for Exported Pesticides, Devices, and Pesticide Active Ingredients and Procedures for Exporting Unregistered Pesticides, 45 Fed. Reg. 50,274 (1980).

Labeling

All pesticide exports, registered and non-registered, must be labeled in accordance with EPA regulations. All exported pesticides are required to have labels which provide:

(a) the name and address of the producer or registrant, and the person for whom the product is produced;

(b) an EPA establishment number;

(c) a statement of net weight or measure;

(d) an ingredients statement;

(e) a warning and caution statement;

(f) where highly toxic, the word "poison" and a skull and crossbones warning signal along with a statement of practical treatment in case of poisoning; and

(g) when the product is not registered in the United States, the statement "Not Registered for Use in the United States of America."

45 Fed. Reg. at 50,275. FIFRA § 2(q)(1)(E) requires that all statements on a pesticide's label be "in such terms as to render it likely to be read and understood by the ordinary individual under customary conditions of purchase and use." For this reason, the Agency's policy requires statements (d) through (g) above to appear on the label of the exported product in English and the language of the importing country. Pesticides classified by EPA for restricted uses must contain the statement "Restricted Use Pesticide" on the label but "[t]he supplementary statement of the terms and conditions of restriction is not required." 45 Fed. Reg. at 50,279. Any exported pesticide product not containing the appropriate labeling is misbranded and the exporter is subject to penalties under FIFRA § 12(a). See Chapter 15.

EPA has recently published a proposed policy statement on exported pesticides which would change the Agency's 1980 policy. 55 Fed. Reg. 4967 (1990). Changes include revision of the labeling requirements. One proposed change would require multilingual labeling if the country of the purchaser is not the country of final destination for the pesticide product. Under these circumstances, label statements would also have to appear on the label in the language of the country of final destination. 55 Fed. Reg. at 4971.

The Agency would also clarify that labels on unregistered products, shipped under the "similar in composition and use" exemption to the foreign purchaser acknowledgment statement requirements (discussed below), must contain ingredient and warning statements like those of registered products which provide the basis for the similar in composition and use exemption. The proposed policy allows deletion from the label statements required by EPA, if such requirements would conflict with label requirements of the country of intended use. 55 Fed. Reg. at 4972. The exporter, however, would be required to provide such information in supplementary labeling. Moreover, the exporter would have the burden of demonstrating that FIFRA's labeling requirements conflict with the foreign country's requirements, failing which the exporter could be liable under FIFRA § 12(a). Finally, the proposed policy states that pesticides shipped from a domestic to a foreign facility of the same company are subject to § 17(a) requirements, be they registered or unregistered products.

Unregistered Pesticides — Foreign Purchaser Acknowledgment Statement

Pesticides intended for export which are not registered for use in the United States are subject to the foreign purchaser acknowledgment statement requirements of FIFRA § 17(a)(2). Before exporting a pesticide not registered in the United States, the exporter must receive from the foreign purchaser a written acknowledgment that the foreign purchaser understands that the pesticide is not registered in and cannot be sold in the United States. The exporter must draft and forward an acknowledgment statement to the foreign purchaser with instructions to provide the following information:

(a) name and address of the exporter;

(b) name and address of the foreign purchaser;

(c) name of the product's active ingredient;

(d) an indication that the purchaser understands the product is not registered for use in the United States;

(e) destination of the export shipment if not the same as the purchaser's address; and

(f) the foreign purchaser's signature and date of signature.

45 Fed. Reg. at 50,277. After receiving the signed acknowledgment from the foreign purchaser, the pesticide producer can export the product. The exporter must provide EPA with the foreign purchaser's acknowledgment statement within seven days of the exporter's

receipt or prior to export, whichever is earlier. For each pesticide product exported, the exporter need only receive acknowledgment from the foreign purchaser for the first shipment of each calendar year to that country. A copy of the foreign purchaser's acknowledgment statement is then transmitted by EPA to an appropriate official of the importing country. Foreign officials also receive notice from EPA should a pesticide product be subject to cancellation or suspension. FIFRA § 17(b).

The Agency's recent export policy proposal seeks to impose additional requirements for submission of foreign purchaser acknowledgment statements. EPA proposes that the foreign purchaser statements be in the language of the purchasing country and in that of any other countries for which the product is ultimately destined. 55 Fed. Reg. at 4970. Finally, EPA solicits comment on requiring an acknowledgment statement for each export shipment rather than on the first annual shipment each year to a foreign country. 55 Fed. Reg. at 4969. Export of a pesticide product without the acknowledgment statement would subject the exporter to civil and/or criminal liability under FIFRA § 12 for selling or distributing an unregistered pesticide, or for failure to maintain records or submission of false data to the Agency. 55 Fed. Reg. at 4974. FIFRA § 12(a)(1)(A), 12(a)(2)(M).

Under EPA's interpretation of § 17(a)(2), a pesticide product is not registered and thus subject to the foreign purchaser acknowledgment statement requirement if the product:

(a) contains an active ingredient not found in a federally registered product;

(b) is the subject of a denial or cancellation of registration proceeding; or

(c) is not similar in composition and use pattern to any federally registered product.

45 Fed. Reg. at 50,274. In order for a pesticide export to be considered similar to a registered pesticide in composition and use pattern, the export must have the same toxicity category and contain the same active ingredient(s) as the registered pesticide. The Agency does not, however, consider minor variations from a registered product's formulation to be

"unregistered" for purposes of its export policy. Under current Agency policy, only where the Agency has made an adverse determination or no determination at all on the use of the pesticide product is the exporter of the pesticide product required to submit a foreign purchaser acknowledgment statement. A change in the use pattern of the registered pesticide from non-food to food, from outdoor to indoor, from terrestrial to aquatic, or from non-domestic to domestic will render the exported pesticide subject to the foreign purchaser acknowledgment requirements.

EPA's proposed export policy statement makes a substantial change in the similar in composition and use exemption. The Agency proposes two options: (1) eliminating the exemption entirely, or (2) retaining the exemption but increasing requirements for verifying the similarity in use and composition claims. 55 Fed. Reg. at 4969-70. If the exemption is retained, it would fall upon the exporter seeking the exemption to substantiate its claim that the unregistered pesticide is similar in composition or use to a registered product. 55 Fed. Reg. at 4973.

In order to support a similar in composition and use exemption, the exporter would be required to document the identity of the registered product, retain copies of labeling from the registered product supporting the exemption, retain records already required under § 8, and maintain any documentation establishing similarity between the registered product and the unregistered product being exported. EPA is considering two recordkeeping options for exporters with such information: (1) records supporting similarity claims must be kept for three years after export, and/or (2) the record support for an exemption must be submitted to EPA prior to each year's first annual shipment of the unregistered product. Under the proposal, if an exporter fails to adequately substantiate the exemption and "EPA determines that the claim of similarity for a particular product is without merit, EPA will consider all exports of the pesticide without a purchaser acknowledgment statement to have been in violation of FIFRA." 55 Fed. Reg. at 4973. If adopted, the proposal would place the onus on exporters to ensure that similarity claims are accurate and verifiable.

EPA is also proposing a research use exemption from the foreign purchaser acknowlegment statement requirements. If an exporter demonstrates that a pesticide is exported solely for research, then an acknowledgment statement would not be required. A pesticide will be considered exported for research only where "the purpose of the shipment is only to determine its value for pesticide purposes or to determine toxicity or other properties, and when the user does not expect to receive any benefit in pest control from its use." 55 Fed. Reg. at 4970. Pesticides used on more than ten acres of land or one acre of water, or used on crops, water, or animals that will later be used for food or feed are presumed not eligible for the research use exemption. As with the similar in composition and use exemption, the exporter bears the burden of showing that the exported pesticide qualifies for the research exemption. Exporters must provide such proof prior to export of the pesticide. EPA proposes that the exporter either maintain records supporting the research claim for three years after export or the exporter submit a report prior to shipment stating that the shipment is being shipped pursuant to the research use exemption. 55 Fed. Reg. at 4970. Failure to comply with these requirements would constitute a violation of FIFRA § 12.

Any pesticide product that can be used in the United States under an experimental use permit pursuant to FIFRA § 5 or an emergency exemption under FIFRA § 18 is *not* considered registered for purposes of EPA's export policy and cannot be exported without the foreign purchaser acknowledgment statement. 45 Fed. Reg. at 50,276. Under EPA's proposed export policy, unregistered manufacturing use and technical grade pesticides are not considered registered pesticides. 55 Fed. Reg. at 4970. A pesticide product registered by a state under FIFRA § 24(c), however, is considered to be registered for purposes of the export policy.

At present, EPA notifies foreign governments of a regulatory action, pursuant to § 17(b), when it takes "a significant regulatory action, such as a cancellation or suspension of a registration based on the finding that the pesticide's risks outweigh its benefits." 55 Fed. Reg. 4956, 4961 (1990). EPA is now proposing clarifications to its notification policy as to when notices will be provided to foreign governments. For certain regulatory actions, notification to foreign governments will be immediate. For other actions, it will be provided annually. EPA proposes immediate § 17(b) notification for:

(a) final and proposed suspensions or cancellations on the basis of unreasonable adverse effects;

(b) denial of a registration application or tolerance on the basis of a formal finding that the risks outweigh the benefits; and

(c) voluntary cancellations or withdrawals for health or environmental reasons.

55 Fed. Reg. at 4964. Other Agency actions which will be sent to foreign governments as part of an annual compilation of important regulatory activities include:

(a) issuance of reregistration documents;

(b) registration of a new active ingredient or registration of the first food use for an already registered active ingredient;

(c) actions eliminating or virtually eliminating all of an active ingredient's registrations; and

(d) any action regarding a tolerance other than a denial because risks outweigh benefits.

55 Fed. Reg. at 4964. Basically the Agency seeks immediate notification where the action taken is on the basis of health or environmental concerns, and annual notification if the action implicates use of the pesticide. 55 Fed. Reg. at 4963.

BOOKS AND RECORDS ON EXPORTS (FIFRA § 8)

Exporters of pesticide products must comply with FIFRA § 8 recordkeeping requirements. The exporter must maintain certain records for two years after shipment of the pesticide product. These records include: (1) copies of the directions from the foreign purchaser for the production of the product, (2) copies of the labels or labeling that were used to comply with FIFRA § 17(a)(1), and (3) copies of the foreign purchaser's acknowledgment statement and the exporter's certification that the product was not shipped to the foreign purchaser prior to receipt of the acknowledgment statement. 40 C.F.R. § 169.2(h).

FIFRA EXPORT ENFORCEMENT ACTIONS

EPA has filed its first enforcement actions for alleged violations of FIFRA § 17(a). The export violation actions seek penalties ranging from less than $20,000 to more than $1.5 million. The focus of the enforcement actions is alleged violations of the bilingual labeling requirements and/or failure to obtain a foreign purchaser acknowledgment statement. The majority of the alleged violations of the foreign purchaser acknowledgment statement requirements have revolved around the exemption from that requirement for pesticides similar in composition and use to a federally registered product. Since the enforcement actions are the first prosecutions for alleged § 17 violations, the final determination in these actions could help define the parameters of the similar in composition and use exemption.

Both the Agency's proposed export policy statement and the export enforcement actions focus on bilingual labeling requirements and the similar in composition and use exemption. Pesticide exporters should be aware that additional § 17 administrative enforcement actions will likely center on these same issues. Exporters should take steps to assure that they are not in violation of either of these provisions when exporting their pesticide products.

IMPORTS

FIFRA § 17(c) grants statutory authority for regulating importation of pesticides and devices. All pesticides imported into the United States must be registered under § 3 of

FIFRA.[28] Importers of pesticides must submit to the EPA region covering the customs port of entry, prior to arrival of the pesticide shipment in the United States, a Notice of Arrival of Pesticides and Devices (EPA Form 3540-1). An EPA Regional official then completes the Notice of Arrival by "indicating the disposition to be made of the shipment of pesticides or devices upon its arrival in the United States." 19 C.F.R. § 12.112. The Notice of Arrival is then returned to the importer, or to a customs broker designated by the importer, for presentation to the U.S. Customs Service at the port of entry.

A pesticide imported without a complete Notice of Arrival is considered a "prohibited importation" and the shipment will be detained at the port of entry and subject to destruction or exportation. 19 C.F.R. § 12.113(c). The U.S. Customs Service may release a pesticide shipment without a Notice of Arrival only if the importer posts a bond and agrees not to use or dispose of the pesticide prior to action taken by the Agency. 19 C.F.R. § 12.114. If the Agency determines importation should not be allowed, Customs may order the redelivery of any pesticide shipment released without a Notice of Arrival. Failure to redeliver, destroy, or export the shipment under Customs supervision, or to produce a proper Notice of Arrival, will subject the importer to penalties equal to the domestic value of the shipment and liquidated damage claims for up to three times the value of the shipment declared to customs. 19 C.F.R. §§ 12.113(c), 113.62(j)(1).

EPA can request from the Customs Service samples of the imported pesticides, along with any literature, such as labels, circulars, and advertising, accompanying the shipment. 19 C.F.R. § 12.116. If the Administrator finds that such products or labeling are in violation of FIFRA, the Agency can elect to refuse entry or to destroy the pesticide product. 19 C.F.R. § 12.117(b). Prior to action by the Administrator, however, the importer is given a period of time not to exceed twenty days, unless an extension is granted, in which to

---

[28] Devices need not be registered but must not contain false or misleading representations. 19 C.F.R. 12.111. Regulations governing importing of pesticides are found at 19 C.F.R. §§ 12.110-12.117.

respond to the EPA decision. The burden is on the importer to "show cause why the shipment should not be destroyed or refused entry." Should the importer fail to convince the Administrator that the product is not in violation of the Act, the Customs Director will either refuse entry or demand redelivery of the pesticide shipment if it has been released under bond to the importer. 19 C.F.R. § 12.117(b).

LEGISLATIVE PROPOSALS

In 1990 Congress undertook to revise FIFRA's export provisions, but was unsuccessful. A primary focus of legislative efforts was to eliminate the so-called "circle of poison" where residues from pesticides not allowed for use in the United States appear on foods imported into this country as the result of foreign use of the non-registered pesticide. The revisions would have banned the export of pesticides that are not registered for use in the United States and do not have established food tolerances under the Federal Food, Drug, and Cosmetic Act, 21 U.S.C. §§ 301-393 (1988). Another focus of legislative efforts was to establish a prior informed consent (PIC) requirement for export of certain designated pesticides. Under the legislative proposal, certain pesticides could not be exported without prior notification to the country of intended destination. The proposals required the exporter of a pesticide to provide EPA with information about the pesticide, the foreign purchaser, the export shipment, the country of intended use, and any intended use on agricultural commodities or food. If the export was the first export of that pesticide to the country of intended use within the last twelve months, the information would be forwarded to the appropriate government official within that country. If a country had informed EPA that it did not wish to import a pesticide, EPA could ban export of that pesticide to that country. Attempts to change the export provisions ultimately failed, but substantially similar amendments are likely to be offered in the 102nd Congress.

Chapter 12
REGULATION OF BIOTECHNOLOGY PRODUCTS

According to EPA, the Agency has four primary goals in regulating biotechnology:

- First, to minimize risks to the public health and safety without imposing crippling costs and restrictions on a fledgling industry;
- Second, to help create a public climate in which the technology can develop safely;
- Third, to establish a *flexible* program that can adjust as the technology evolves and matures; and
- Fourth, to write regulations that take into account all the scientific uncertainties surrounding the behavior of microorganisms in the environment.

Speech by EPA Administrator William K. Reilly to the Industrial Biotechnology Association, Oct. 13, 1989, Washington, D.C. at 5. These four goals have proven elusive.

In the late 1970's, EPA began receiving an increasing number of applications to register new pesticidal ingredients derived from biological rather than chemical sources. Previously, the Agency had dealt with a small number of such materials[29] on an ad hoc basis. When undertaking to codify data requirements for registration of pesticides, EPA recognized the unique nature of biological materials by creating separate data requirements for them (40 C.F.R. § 158.690, biochemical pesticides, and 40 C.F.R. § 158.740, microbial pesticides) and separate Pesticide Assessment Guidelines, Nov. 1982 (Subdivision M).

Today, the biotechnology portion of EPA's pesticide regulatory program controls the registration of natural *biochemical* materials, such as pheromones, insect hormones, plant extracts and products of fermentation, and natural and genetically altered *microbial*

---

[29] Early products of *Bacillus thuringiensis* (B.t.) and herbal extracts.

materials, such as bacteria, fungi, protozoa and viruses. EPA's program will soon claim jurisdiction over plants genetically modified to have pesticidal properties. Because the materials covered by the biotechnology program are so diverse, EPA has of necessity extensively interpreted and enlarged upon the codified data requirements and guidelines. As a result, most biologically derived products now are subject to a host of informal as well as formal requirements within a broad framework established by the regulations and guidelines.

Adding further complexity to the already fluid data requirements for biotechnology products, two particular classes of materials have raised unique regulatory concerns for EPA: nonindigenous naturally occurring microorganisms and genetically altered microorganisms. Because of the additional uncertainties and risks associated with products containing these microorganisms, EPA plans to establish separate regulations for them. The substantial controversy surrounding EPA's drafts of these regulations, however, has stymied their publication. Pending development of regulations, the agency requires companies to notify EPA of all tests of genetically altered microorganisms and, in some cases, to obtain EPA approval prior to testing.

CLASSIFICATION OF BIOCHEMICAL AND MICROBIAL PESTICIDES

Biochemical Pesticides

An initial issue in registering biotechnology products is identifying them as microbial or biochemical pesticides. EPA's regulations describe the general characteristics of biochemical and microbial pesticides, but there is little formal guidance to distinguish biochemical from chemical pesticides. 40 C.F.R. § 158.65.

The regulations describe biochemical pesticides as having "unique" modes of action. In practice, the agency considers that a material must be of natural origin and have a nontoxic mode of action to be included in the biochemical classification.[30] Until recently, EPA had decided that only juvenile hormones and pheromones fit the informal definition of biochemical pesticides. As part of Phase II of reregistration (see Chapter 6), EPA is considering for the first time classification of other materials as biochemicals.[31] The restriction to a non-toxic mode of action has eliminated from the biochemical classification a wide variety of natural materials, including complex botanical extracts like rotenone, pyrethrin and piperonyl butoxide, products obtained from algea like gibberellin and cytokinin, and fermentation products like antibiotics and the β-exotoxin of B.t. (Thuringiensin).

Microbial Pesticides

The agency has included both microorganisms (bacteria, fungi, algae, and protozoa) and viruses in the microbial classification and has little difficulty in distinguishing these materials from conventional chemical pesticides. The agency, however, has never received an application to register a toxic, pathogenic or infective organism. In the event it does, the agency almost certainly would deviate from the tiered testing approach described in Subdivision M and would make its decisions on an ad hoc basis.

EPA has treated a recent product containing the δ-endotoxin of B.t. in a killed, genetically altered bacterium as a microbial pesticide, although the active ingredient is more like a biochemical, being an encapsulated protein. Applications to register products containing entirely cell-free δ-endotoxin are likely in the near future. EPA will not consider such products to be microbial pesticides, and, with a lethal mode of action, the products also

---

30 Interpretation of the biochemical classification in an internal memorandum from John Melone, Director, Hazard Evaluation Division, to Douglas Campt, Director of Registration Division (June 27, 1984.)

31 For example, ethylene gas for fruit and vegetable ripening, and oil of citronella as a repellent.

will not fit the unpublished requirements for biochemical classification. Presumably, cell-free endotoxin will be classified as a chemical for the purpose of determining data requirements, even though it has been registered before as the active ingredient in a microbial pesticide.

An even more complex classification problem will come when EPA assumes authority over plants genetically modified to produce biochemical or microbial toxins. Currently, applications to register genetically modified plants are not submitted to EPA; rather the U.S. Department of Agriculture assumes regulatory oversight. In its semiannual regulatory agenda published on October 29, 1990, EPA announced that in December 1990 it will issue a proposed rule covering transgenic plants. When plant products are brought under FIFRA, they will be subject to their own separate set of regulations governing data requirements and a new classification will be created for them.

It appears from public meetings in advance of the publication of the new regulations that the data requirements for genetically modified plants will be very similar to the requirements for chemicals. The first applications for registration are expected to be for plants containing the δ-endotoxin of B.t., an active ingredient heretofore classified as a microbial pesticide.

FORMAL DATA REQUIREMENTS AND GUIDELINES

EPA developed the data requirements for biochemical and microbial pesticides by modifying the data requirements for chemical pesticides. In some instances, such as for product chemistry, the agency incorporated the existing chemical requirements in toto, and in others it embellished upon existing protocols — for example, adding pathogenicity testing requirements to standard acute toxicity testing protocols. Unfortunately, EPA did not develop new protocols to fit the non-chemical nature of each type of biological material. (For a discussion of data requirements for chemical pesticides, see Chapter 4.)

In developing its separate requirements for biological materials, the agency assumed that they were inherently safer than chemical pesticides. Insect pheromones used in traps were entirely exempted from registration,[32] and EPA set minimum data requirements for other biologically derived products. A tiered testing scheme was proposed with triggers to escalate the data requirements if lower-tier tests indicated the possibility of adverse non-target impacts.

Throughout the public hearings relating to the proposed regulations and guidelines (October 7, 1980, and October 13 and 19, 1983), the Science Advisory Panal (SAP) criticized EPA's basic approach to biological materials on two grounds. First, the SAP considered inadequate EPA's modification of the existing requirements for chemical pesticides to fit biological materials. Second, the SAP considered unwarranted EPA's basic assumption that biological materials were inherently safer than chemical pesticides. The discussions in the hearings were heated and without resolution, and EPA's data requirements today reflect this continuing controversy.

EPA's inability to resolve the issue of safety of biological materials caused inconsistencies between its guidelines and regulations. EPA published its original guidelines for registration of "biorational pesticides"[33] in October 1982 (Pesticide Assessment Guidelines, Subdivision M, Biorational Pesticides) containing an early set of data requirements, but the regulations continued to evolve. Two years later EPA published its regulations containing a significantly different set of data requirements. 49 Fed. Reg. 42,881 (1984) (codified at 40 C.F.R. pt. 158). Where the guidelines and the regulations disagree, the regulations take precedence.

The most serious disagreement is in the area of toxicology testing for biochemical pesticides. The guidelines require only acute toxicity and mutagenicity testing in Tier I.

---

32 As a result of the exemption from registration, insect traps containing over 100 different pheromones are commercially available, and more than 500 pheromones have been identified and synthesized.

33 In its early efforts to develop data requirements, EPA coined the term "biorational pesticides." Pursuant to the SAP's criticism of the term, EPA subsequently adopted the phrase "biochemical and microbial pesticides."

Subchronic studies are not required unless the Tier I test results trigger progession to Tier II. Pursuant to these guidelines, EPA granted the early pheromone registrations based on acute toxicity studies only.

With the October 1984 publication of the final data requirements, EPA moved all subchronic toxicity testing requirements into Tier I. This last minute response to the SAP's criticisms eliminated the tiered testing scheme, and enormously increased the cost of registering nontoxic biochemical pesticides. For several years applications to register such materials virtually disappeared. Today EPA generally waives the subchronic toxicity testing requirements, and there is renewed interest in products like pheromones.

Recently EPA also substantially revised the guidelines for registration of microbial and viral pesticides. The new guidelines (also called "Subdivision M," like the original guidelines) contain no requirements for biochemical pesticides, even though the word "biochemical" is still contained in the title. To find the biochemical requirements, applicants must look to the original version of Subdivision M, where the requirements for biochemical and microbial pesticides are entwined.

PROCEDURES FOR DETERMINING DATA REQUIREMENTS

EPA always advises companies registering a biochemical or microbial pesticide to meet with an agency scientist before preparing an application. Because biologically based products are usually complex natural materials, tests are difficult to standardize and interpret. EPA's published rules and guidelines often poorly describe the efforts necessary to achieve registration. Preapplication conferences to discuss proposed data requirements and testing protocols are mandatory before preparing an application. Once an application is submitted, the review process often involves an ongoing dialogue with EPA over the interpretation of data. Research is often necessary to resolve problems, and the process generally ends in a conditional registration with additional testing requirements.

DATA REQUIREMENTS FOR BIOCHEMICAL PESTICIDES

Pheromones

The data requirements for insect pheromones are the best understood. Pheromones are the easiest of all biological materials to register, and many have been registered.[34]

In addition to acute toxicity testing, EPA officially requests the following long-term studies to register a pheromone, assuming that the pheromone is applied at under twenty grams per acre: (1) a ninety-day rodent feeding study; (2) a teratology study; (3) an immune response study; (4) a neurotoxicity study; (5) a complete battery of three mutagenicity studies; and (6) an avian dietary study. 40 C.F.R. § 158.690. For application at higher rates, EPA also may request crop residue data, analytical methods and more ecological testing. In practice, however, EPA always waives the longer-term data requirements, but the applicant must formally request waivers in writing as part of the registration application. The waiver requests must be justified scientifically on the basis of low toxicity, low exposure potential and low non-target impacts.

Today, an applicant can obtain a registration for a new pheromone active ingredient or reregister an existing pheromone by submitting:

- full product chemistry information;
- a battery of acute toxicity studies,[35] including eye and skin irritation and skin sensitization;
- an Ames mutagenicity test; and

---

34 The insect pheromones registered to data have all been nontoxic, lacking potential for non-target impacts, and, with one exception, requiring application rates below twenty grams per acre.

35 Because pheromones often cost several dollars a gram at early stages in the development of a new product, acute inhalation studies, which require large quantities of test material, can be expensive. In recognition of this fact, EPA has accepted tests where the pheromone is administered by intratracheal instillation, an approach which consumes very little test material. The intratracheal protocol also has the advantage of achieving accurate high doses easily.

- acute toxicity tests in *Daphnia*, one species of fish and one species of birds.

These limited data requirements pertain to those pheromones applied at less than twenty grams per acre that pass the limit test in the acute toxicity studies. In the limit test, a single high dose of the pheromone is administered to the test animals. If all animals survive, no further testing is required.

The data requirements are further simplified because pheromones contained in traps do not require registration under FIFRA. 40 C.F.R. § 152.25(b). Industry has argued that EPA should extend the limited trap exemption to other forms of pheromone products. For example, no registration is required for a plastic dispenser placed in a cardboard container (a trap) and hung from the branch of a tree. Yet it is necessary to have a full registration before attaching the same dispenser to the branch without the trap — even though the identity, formulation and application rate of the pheromone are unchanged. Whether or not EPA should broaden the exemption for pheromones is considered controversial, but most agree that EPA should initiate some policy changes to increase the timely flow of available pheromone-containing products into the market.

To date, EPA has granted exemptions from tolerances for all pheromone products used on feed or food crops. A precedent was set in 1987 that exemptions from tolerances could be granted without the longer term subchronic studies or residue data. An emulsifiable liquid pheromone formulation for control of tetranychid mites was registered as an attractant to be mixed with any registered conventional chemical miticide, the most expansive pheromone use ever registered. An exemption from tolerance was granted for all crops.

Plant Extracts

Many pesticide active ingredients are natural materials derived from plants. Over thirty herbal materials have been registered at one time as chemical pesticides, most before EPA existed. Other plant-derived materials such as rotenone, piperonyl butoxide, and neem extract are treated as chemicals. In connection with reregistration EPA has begun accepting proposals to reclassify natural plant extracts as biochemical pesticides, but the agency appears unwilling to reregister most natural materials under the data requirements for biochemicals. EPA has made no final decisions on classification.

Most plant extracts are difficult to register because of serious technical problems in the regulatory process designed for synthetic chemicals. Oil of citronella, a widely used insect repellent, illustrates the problems. There are two types of oil of citronella, Ceylon and Java, both containing around fifty major chemical compounds.[36] The two mixtures have entirely different compositions, and the compositions of each type may themselves vary dramatically depending upon soil type, climate, altitude, fertilizer, rainfall and other factors such as distillation.

In order to reregister oil of citronella, EPA requires the applicant: (1) to specify the composition following product chemistry requirements developed for single synthetic chemicals of fixed purity and composition; (2) to set narrow certified limits for every chemical component; and (3) to conduct toxicology tests on a sample of the oil representing the entire range of possible compositions. At this time, the only way to satisfy these chemical-type requirements is to reregister a synthetic oil of citronella of fixed composition, not the natural blends of fifty components. EPA has not yet decided how to address the problem of complex variable compositions associated with natural plant extracts such as oil of citronella. The data requirements for registration are uncertain and open to negotiation at this time.

Fermentation Extracts

EPA treats fermentation extracts as conventional chemicals, despite the fact that its position is sometimes illogical and creates testing problems. For example, in registering the

---

[36] Both types are produced by steam distillation of the oil from fresh citronella grass (*Cymbopogon nardus* or *Cymbopogon winterianus*).

δ-endotoxin of B.t., EPA follows the data requirements and guidelines for microbial pesticides, but EPA considers the β-exotoxin of B.t. in beer liquor from the fermentation of B.t. to be a conventional chemical. Biochemical antibiotics used for plant disease control are also treated like chemicals.

The problem with treating fermentation extracts as conventional chemical pesticides is that the data requirements and guidelines for chemicals fit poorly. The experimental difficulties encountered in certain areas are particularly difficult. For example: How are radio-labeled materials synthesized for metabolism studies? What impurities need to be identified and quantified in a fermentation beer liquor to set certified limits? What is the appropriate test material for toxicology studies, the whole beer liquor, a concentrate after evaporation of water, individual biochemical components or all of the above? How are metabolism studies interpreted when the radio label is readily lost in the carbon pool of the test system?[37] How is the active ingredient specified on the label? What should be measured by the enforcement analytical method?

DATA REQUIREMENTS FOR NATURAL, INDIGENOUS MICROORGANISMS

EPA developed many of its registration requirements for non-engineered indigenous microorganisms from its experience with B.t. products. Microorganisms are much more complex than chemicals, and data requirements and test protocols adapted from the requirements for chemicals are often inadequate. Therefore, with microorganisms the true challenge is not identifying the data requirements but rather mastering the testing protocols.

The data requirements for microbial pesticides, found in the recently revised Subdivision M, contain a tiered testing scheme. An orderly progression to higher tiers is unlikely, however, for toxic or pathogenic organisms. At present EPA probably would not

---

37 In testing products containing the β-exotoxin, EPA expects the adenine moiety in the molecule to be labeled with ^{14}C. Adenine is a building block of nucleotides that make up DNA, cell energy systems, and protein synthesis processes in all living matter. Hence, the ^{14}C-adenine is easily incorporated into the carbon pool.

register such organisms. The agency has no experience assessing the risks of pesticides containing living organisms with the potential for significant nontarget impacts and has no conceptual framework for mitigating nontarget impacts through product labeling.

For the purpose of determining data requirements, EPA views each strain of a particular microorganism as a separate active ingredient. EPA has required a completely independent data base for *every* strain of B.t. When developing a new pesticidal microorganism having numerous uncharacterized strains, a company should seek registration without specifying a strain, if possible. This will enable the registrant to make changes in the organism later without having to restart the registration process.

Indigenous microorganisms present several major concerns inapplicable to pesticidal chemicals. To address these issues, EPA has amended study protocols originally designed for chemicals. Subdivision M requires the same battery of acute mammalian toxicity studies and fish and bird studies required for conventional chemicals, but extends the scope of each study to include pathogenicity testing. EPA seldomly requires more studies to register a microbial pesticide than those found in Tier I of the Subivision M Guidelines. EPA may waive the pathogenicity portions of the testing protocols if the applicant submits data showing that the microorganism, for some reason such as temperature sensitivity, will not survive in nontarget systems.

Protocols specify that the registrant must develop and validate analytical methods for detecting and measuring the organism in a variety of test animal tissues. The duration of each test and the number of test animals are increased to allow monitoring for the organism until it clears the test system or is otherwise shown to have no pathogenic effects. Before administration to the test system, the microorganism must be proven viable and tested for microbial contamination.

The addition of pathogenicity testing to conventional toxicity testing protocols causes many problems. Because most microorganisms have very low acute toxicity, very high test doses are generally administered. The high doses clear slowly and the microorganism is generally detectable in viable form many days after administration. EPA may interpret this result to mean that the organism is infective, even if the organism is declining in concentration and causing no apparent harm. Contamination of the testing laboratory with spores is a difficult problem to avoid. Contamination of the test organism or the test system with other living organisms can remain undetected until it is too late to correct the problem, ruining an otherwise well conducted study.

Pathogenicity testing dramatically increases the cost of studies by adding to their complexity and extending their duration. The cost of developing analytical methods for a variety of tissues and of performing the required analyses can be significant. The cost increases dramatically, particularly in the case of fish studies, when expensive equipment normally used for only a few days per test is tied up for a long period of time on a single test (twenty-one to twenty-eight days, or longer).

Deciding what to test also can be problematic. The proposed product may be spore-free but contain an organism which produces spores. Alternately, the product may contain the spores but no microorganism. EPA can request tests on both the organism and the spores, and spores can be troublesome to make in the quantities needed for testing and difficult to keep viable under the conditions of the test. EPA may accept a full set of studies on the microorganism and one or more key studies repeated with the spores.

The data requirements for registration of microorganisms require some specialized tests not needed for chemicals. Because it is difficult to generate respirable aerosols from microorganism preparations, EPA requires an acute pulmonary toxicity test by intranasal or intratracheal administration of the dose. EPA requires a single-dose intravenous toxicity and pathogenicity test in a mouse or rat with the dose consisting of at least 107 organisms per animal. Injecting a large number of particles directly into the blood stream of an animal, however, often causes gross physical damage and death which should not be attributed to the microorganism. EPA requires an avian respiratory pathogenicity test with intranasal or intratracheal administration of the dose daily for five days. The dosing procedure is so traumatic to the test birds that it can cause death unrelated to the test material itself.

EPA worries about the possible impacts of microbial products on nontarget plants and insects, and therefore requires elaborate testing. The requirements may go far beyond the nontarget testing normally encountered for conventional chemical pesticides. EPA requires that the microorganism be characterized and a sample deposited in a nationally recognized culture collection at the time of application for registration. As a production quality control measure, EPA requires that each production batch of microorganisms be tested for toxic material and contamination with pathogenic organisms by subcutaneous injection into five mice prior to formulation into the product.

Identifying the strength and composition of microbial products for labeling purposes is also a serious problem. For years B.t. labels specified concentration in terms of minimum number of organisms per unit volume and in terms of potency as determined by *in vivo* insect assays. Different measures of potency were used by different manufacturers, and purchasers could not compare products in a meaningful manner. To solve this problem, EPA now requires a quantitative chemical analysis for all B.t. products which measures the weight-percentage of the δ-endotoxin in the product, a fallback to chemical-type requirements. Industry spent years developing an analytical method acceptible to EPA, but the method does not measure the potency per unit mass of endotoxin, which still varies from producer to producer and strain to strain.

DATA REQUIREMENTS FOR VIRAL PESTICIDES

The data requirements for viruses and microorganisms are identical, but the experimental difficulties in completing the testing program are more severe for viruses. In general, viruses are nontoxic to nontarget organisms. Hence, the primary focus of each required toxicity test is pathogenicity. Because relatively large quantities of test substance are needed when investigating nontoxic materials, it may be difficult to obtain the candidate virus in sufficient quantity to complete the testing program. The presence of the virus in tissues and its viability cannot be measured using chemical or microbiological techniques. Instead, an *in vivo* bioassay using the target insect pest is generally employed. The insect used as the bioassay system may not normally ingest the tissues to be tested with the analytical method, making successful application of the bioassay problematic.

NONINDIGENOUS AND GENETICALLY MODIFIED MICROORGANISMS

EPA remains embroiled in substantial controversy regarding the scope and extent of regulation for nonindigenous and genetically modified microorganisms. Under pressure from the Administration and from public and private groups with widely diverging interests, EPA has been unable to produce a final FIFRA regulation.

1986 Framework for Regulation of Biotechnology Products

The Agency is currently operating under an initial EPA policy statement contained in the Reagan Administration's June 26, 1986 Federal Framework for Regulation of Biotechnology Products, 51 Fed. Reg. 23,302. The 1986 Framework included policy statements on biotechnology by EPA, FDA, USDA, OSHA, and the National Institutes of Health (NIH). EPA's statement in the 1986 Framework remains the agency's published policy on biotechnology under FIFRA.

Since the publication of the Federal Framework, the agency has attempted to promulgate a final FIFRA rule. To date, EPA has not proposed a regulation but has circulated a draft proposal for informal comment. In 1990, the White House Office of Science and Technology Policy (OSTP) also proposed amendments to the Federal Framework. EPA's Draft Proposed Rule and OSTP's 1990 proposed amendments are briefly discussed at the end of this chapter.

Under the Federal Framework, the Agency requires all companies to notify EPA prior to initiating tests involving genetically altered microbial pesticides and nonindigenous naturally occurring microorganisms, regardless of the test size. The Agency also establishes criteria for test approval. EPA focuses on the requirements for small-scale field tests for genetically engineered microbial pesticides (GEMPs) and nonindigenous naturally occurring microorganisms. The definition of microorganisms includes bacteria, fungui, algae, viruses and protozoa. For larger scale testing requiring an Experimental Use Permit (EUP) and for registration, the data requirements of Subdivision M also apply.

Small-Scale Field Test Exemption Does Not Apply

For conventional chemicals, EPA exempts companies performing small-scale field tests from the requirement of submitting an EUP. The agency presumes that an EUP is not required for:

> **Land Use.** Tests conducted on a cumulative total of not more than 10 acres . . . .
>
> **Aquatic Use.** Tests conducted on a total of not more than one surface-acre of water . . . .
>
> **Animal Treatments.** Tests conducted only on experimental animals. . . .

40 C.F.R. § 172.3(a).

In accordance with the Federal Framework, this small-scale exemption from the EUP requirement does not apply for GEMPs and nonindigenous naturally occurring microorganisms. EPA takes the position that, because of the ability of microorganisms to replicate, a small-scale field test poses as significant a risk as a large scale field test. 49 Fed. Reg. 40,659, 40,660 (1984). Having eliminated the EUP exemption for GEMPs and nonindigenous microorganisms, however, the agency does not automatically require an EUP for these microorganisms. Instead, EPA requires "notification" of all proposed tests, and on the basis of the information provided in the notification, EPA determines whether the submitter must obtain an EUP.

Notification Requirements

For notification purposes, the Federal Framework divides microorganisms into high risk and low risk classifications. The high risk classification consists of (1) intergeneric microorganisms;[38] (2) microorganisms that are derived from source organisms that are pathogens; and (3) nonindigenous naturally occurring organisms. 51 Fed. Reg. at 23,321. EPA subjects these three categories of organisms to a higher level of scrutiny known as "Level II" review. All other microorganisms are subject to "Level I" review.

EPA defines "intergeneric microorganisms" as microorganisms produced from two different genera. By contrast, microorganisms that result from changes in well-characterized noncoding regulatory regions, such as operators, promoters, origins of replication, terminators, and ribosome-binding regions are not considered intergenerics.

EPA defines "pathogen" as "a virus or organism . . . that has the ability to cause diseases in other living organisms." A disease is "an abnormal physiological function." 51 Fed. Reg. at 23,333. The agency has requested further comment on the definition of pathogen.[39]

EPA defines a nonindigenous microorganism simply as a "naturally occurring microorganism[ ] placed in [an] environment[ ] where [it is] not native." 49 Fed. Reg. 50,880, 50,881 (1984).

EPA's Review Process

Level I Review. For microorganisms falling outside of EPA's high risk classification, the notification process is relatively simple. The data requirements include:

(1) Identity of the microorganism, including characteristics, and means and limits of detection.

---

[38] Although EPA is regulating intergenerics on the assumption that they present a substantial risk, this conclusion is questionable. Taxonomic classification is not necessarily an accurate predictor of the potential risk of a microorganism. The phenotypic characteristics of an organism are more accurate predictors of potential risk. In its 1989 Draft Proposed Rule, EPA is considering expanding the reporting requirement to include intrageneric microorganisms as well.

[39] The agency has encountered substantial difficulties in defining the term "pathogen" and as a result is proposing to delete the term in the final regulation.

(2) Description of the natural habitat of the microorganism or its parental strains, including information on natural predators, parasites, and competitors.

(3) Information on the host range of the parental strain(s) or nonindigenous microorganism.

(4) Information on the relative environmental competitiveness of the microorganism, if available.

(5) If the microorganism is genetically engineered, information should be provided on the methods used to genetically engineer the microorganism(s); the identity and location of the rearranged or inserted/deleted gene segment(s) in question; a description of the new trait(s) or characteristic(s) that are expressed; information on potential for genetic transfer and exchange with other organisms, and on genetic stability of any inserted sequence.

(6) A description of the proposed testing program, including site location, crop to be treated, target pest, amount of test material to be applied, and method of application.

51 Fed. Reg. at 23,321. EPA has thirty days to review the Level I notification.

If EPA makes the decision to require an EUP, the applicant has two options: It can either submit an EUP or submit a notification for Level II review. If the applicant chooses to submit a Level II notification, the agency has sixty additional days for review.

Level II Review. Level II notification is significantly more involved than Level I. EPA advises the applicant to schedule a meeting to discuss testing requirements well in advance of preparing a notification. The data requirements include:

1. Background information on the microorganism.

    (a) Identity of the microorganism, including tables of characteristics, and means and limit of detection using the most sensitive and specific methods available.

(b) Description of the natural habitat of the microorganism or its parental strains, including information on natural predators, parasites, and competitors.

(c) Information on host range, especially infectivity and pathogenicity to nontarget organisms.

(d) Information on survival and ability of the microorganism to increase in numbers (biomass) in the environment (e.g., laboratory or containment facility test data).

(e) If the microorganism is genetically altered, the following information should be provided in addition to the information listed in (a) through (d) above:

i. Information on the methods used to genetically alter the microorganism.

ii. The identity and location of the rearranged or inserted/deleted gene segment(s) in question (host source, nature, base sequence data, or restriction enzyme map of the gene(s)).

iii. Information on the control region of the gene(s), and a description of the new trait(s) or characteristic(s) that are expressed.

iv. Information on potential for genetic transfer and exchange with other organisms, and on genetic stability of any inserted sequence.

v. Information on relative environmental competitiveness compared to the parental strains.

2. Description of proposed field test.

(a) The purpose or objectives of the proposed testing.

(b) A detailed description of the proposed testing program, including test parameters.

(c) A designation of the pest organism(s) involved (common and scientific names).

(d) A statement of composition for the formulation to be tested, giving the name and percentage by weight of each ingredient, active and inert, production methods, contamination with extraneous microorganisms, potency and amount of any toxins present, and where applicable the number of viable microorganisms per unit weight or volume of the product (or other appropriate system for designating the quantity of active ingredient).

(e) The amount of pesticide product proposed for use and the method of application.

(f) The State(s) in which the proposed program will be conducted, and specific identification of the exact location of the test site(s) (including proximity to residences and human activities, surface water, etc.).

(g) The crops, fauna, flora, geographical description of sites, modes, dosage rates, frequency, and situation of application on or in which the pesticide is to be used.

(h) A comparison of the natural habitat of the microorganism with the proposed test site.

(i) The number of acres, number of structural sites, or number of animals/plants, by State, to be treated or included in the area of experimental use, and the procedures to be used to protect the test area from intrusion by unauthorized individuals.

(j) The proposed dates or period(s) during which the testing program is to be conducted, and the manner in which supervision of the program will be accomplished.

(k) A description of procedures for monitoring the microorganism within and adjacent to the test site during the field test.

(l) The method of disposal or sanitation of plants, animals, soils, etc., that were exposed during or after the field test.

(m) Means of evaluating potential adverse effects and methods of controlling the microorganism if detected beyond the test area.

51 Fed. Reg. at 23,321-22. The agency has ninety days to make a decision if a prior Level I was involved. The ninety days includes the thirty days initially allotted for the Level I Review.

Pursuant to its review of a notification, EPA develops its initial written scientific position document identifying the problems with the microorganism and the overall likelihood of significant risk. EPA circulates the draft document to other federal agencies and to the relevant state pesticide regulatory authorities for comment. The agency may also submit its conclusions to a subpanel of the SAP for independent scientific review.

At the completion of the review process, the agency writes a decision document that sets out the potential risks associated with the microorganism, identifies the questions that remain, specifies if an EUP is required, and recommends test restrictions or modifications to address areas of concern. When EPA makes the decision to require an EUP, the applicant should consider redefining the test in order to alleviate EPA's concerns and thereby obviate the need for an EUP.

Special Consideration for Nonindigenous Organisms

An applicant wishing to register a nonindigenous organism under the current Federal Framework will face enough additional testing requirements and other difficulties to justify making a substantial research effort to show that the organism is in fact indigenous to the United States. The applicant should conduct such research before agreeing to the much expanded scientific and procedural requirements for registration of nonindigenous and genetically modified microorganisms.

NEW DIRECTIONS IN BIOTECHNOLOGY REGULATION: EPA'S DRAFT PROPOSAL

Since the publication of the 1986 Federal Framework, EPA has reconsidered several aspects of its policy and is planning to amend sections of the Framework. To date, EPA has not formally proposed these amendments, but has circulated an unpublished Draft Proposed Rule. The following discussion is largely based on EPA's draft proposal. The agency's rulemaking process, of course, is ongoing and the agency may introduce significant changes to the proposal prior to publication.

Changes in the Scope of Regulation

The most significant change in EPA policy signalled in the Draft Proposed Rule is in the scope of regulation. EPA circumvents the earlier problems in defining pathogenic organism by defining two new categories of genetically modified microbial organisms without ever using the term "pathogenic." EPA subjects these categories to notification and data requirements much like those found in the present Level II review. EPA further excludes from the notification requirement all genetically modified organisms not in Category I or II, all nonindigenous organisms and any organisms to be tested in contained facility experiments. An experiment is not contained if it involves a release outside a laboratory, greenhouse, fermentor or other conditions not specifically determined by EPA to be adequately contained.[40] The level of containment required depends upon the characteristics of the microbial pesticide and its use.

Category I includes microorganisms that have been altered by the addition of genetic material intentionally manipulated to modify, enhance, or impart pesticidal properties. "Intentionally manipulated" means the direct addition, rearrangement or removal of nucleotides from genetic material. The standard definition of pesticide is applied. FIFRA § 2(u), 40 C.F.R. § 152.3(s). Category II includes intergeneric microorganisms as defined in the 1986 Policy Statement.

Nonindigenous microobial pesticides are excluded from the notification requirement for small-scale field testing because EPA believes that the U.S. Department of Agriculture, Animal and Plant Health Inspection Service (APHIS) and the U.S. Public Health Service (PHS) already exercise adequate regulatory oversight regarding the release of such organisms. Genetically modified microorganisms not falling into Categories I or II may include organisms derived solely by conjugation, transduction or transformation.

---

[40] The agency defines contained use as the opposite of a release. In contrast, in the proposed TSCA biotechnology rule, EPA defines "contained" as

> a building or other structure that has a roof, floor, and walls composed of materials, other than soil, which are typically impervious to water. The structure may have doors, windows, and a ventilation system, but the structure, including those features, shall be designed to provide pest control. An appropriate pest management program must be in place:
>
> (1) If a microorganism is to be used, applied, or disposed of in the facility in the form of an airborne suspension of the microorganism, airborne suspensions of the microorganism shall be controlled to minimize release of the suspension outside the facility. . . .
>
> (2) To qualify as a contained facility, any waste which may contain a viable microorganism, including liquid wastes and water runoff, must undergo inactiviation by a validated inactivation procedure prior to release from the facility.

Draft proposed TSCA Rule at 163 (Dec. 1, 1988). Under TSCA, EPA has also stated that unintended releases to the envirionment, such as mistakenly washing a small amount of a microorganism down the drain, would not be considered a release, as long as an adequate containment policy is in effect. Draft proposed rule at 39 - 40.

Environmental Biosafety Committees

In response to the escalating number of biotechnology submissions and the limits of EPA resources, the agency contemplates establishing private committees, known as Environmental Biosafety Committees (EBCs), to review notifications and EUPs. Modeled on the NIH Biosafety Committees, the EBCs would consist of at least five members, with a minimum of three technical experts and two public representatives. Draft Proposed Rule at 44. EPA proposes to use the EBC to perform an initial review prior to EPA review of notifications. EBCs would conduct site visits, act as a forum for receiving and evaluating public comments on proposed experiments, review test modifications and serve in a scientific advisory role to experimentors on issues of containment. Draft Proposed Rule at 14, 50.

The agency also proposes to use EBCs to perform significant independent review under TSCA, and as a result EBCs have generated substantial controversy within the agency. Although there is less controversy surrounding EBCs under FIFRA, it is presently unclear whether the agency will go forward with its EBC proposal under either statute.

THE BIOTECHNOLOGY SCIENCE COORDINATING COMMITTEE'S PROPOSAL

Scope of the Proposal

Separate from EPA's contemplated proposal, the Biotechnology Science Coordinating Committee (BSCC),[41] an interagency committee responsible for coordination of science policy, also published an alternate definition for the scope of biotechnology regulation. 55 Fed. Reg. 31,118, 31,120 (1990). The BSCC proposal would extend EPA's oversight to include all organisms with deliberately modified hereditary traits.

[41] The BSCC is a branch of the Office of Science and Technology Policy (OSTP). Although not directly responsible for regulation under any of the specific statutes involving biotechnology (including FIFRA), OSTP took a lead role in developing the 1986 Federal Framework.

The most significant aspect of the BSCC proposal, however, is not its broad scope but its equally broad exemptions. Oversite is not required "unless information concerning the risk posed by the introduction indicates that oversite is necessary." The BSCC emphasizes that "[t]o the extent permitted by law, planned introductions into the environment . . . should not be subject to oversite . . . unless information concerning the risk posed by the introduction indicates that oversite is necessary." 55 Fed. Reg. at 31,120.

Proposed Exemptions

The BSCC proposes to identify several specific exemptions. The proposed exemptions are based on similarity to traditional modification techniques and include:

> [1.] Plants and animals that result from natural reproduction or the use of traditional breeding techniques. These include, for example, for vascular plants, mutagenesis and hand pollution, and for vertebrate animals, artificial insemination, superovulation and transfer of embryos.
>
> [2.] Microorganisms modified solely: (a) Through chemical or physical mutagenesis; (b) by the movement of nucleic acids using physiological processes including, but not limited to, transduction, transformation, or conjugation; or (c) by plasmid loss or spontaneous deletion.
>
> [3.] Vascular plants regenerated from tissue culture, including those produced through selection of somaclonal variants, embryo rescue, protoplast fusion, or treatments that cause changes in chromosome number.
>
> [4.] Organisms that have been modified by the introduction of non-coding, non-expressed nucleotide sequences that cause no phenotypic or physiological changes in the parental organism.
>
> [5.] Organisms resulting from deletions, rearrangements and amplifications, within a single genome, including its extrachromosomal elements.
>
> [6.] Organisms with a new phenotypic trait(s) conferring no greater risk to the target environment than the parental strain, which is considered to be safe.

55 Fed. Reg. at 31,121.

The BSCC also proposes to exempt all organisms that have risks equal to or less than a "similar organism" which was previously introduced into a similar environment. A definition of "similar organism" is not restricted to organisms that have been modified by the same techniques:

> An introduction is considered similar to a previous introduction when the level of risk of the introduction is the same as or less than a previous introduction. An organism can be similar to a previously used organism regardless of the process by which the organism has been modified. Thus, regardless of how the organism has been modified, an introduction should not be subject to oversite used in past safe introduction in a similar target/test environment.

55 Fed. Reg. at 31,120. In determining risk, emphasis will be placed on the nature of the organism, confinement techniques, and relevant factors in the environment.

Chapter 13
STATE REGULATION OF PESTICIDES

While EPA is viewed as the principal regulator of pesticides, FIFRA provides that states also may regulate the sale and use of pesticides. FIFRA § 24(a). Political subdivisions of states such as counties and towns also have sought to regulate pesticides, but their authority to do so is the subject of an ongoing legal controversy.

This chapter first reviews the authority of states to regulate pesticides and then reviews the registration of pesticides in California, which has the most comprehensive and aggressive state pesticide regulatory program.

STATE AUTHORITY TO REGULATE PESTICIDES

State Authority Under FIFRA

The 1972 FIFRA amendments extended the federal regulation of pesticides to include products sold and used intrastate. As a result, any pesticide must be registered with EPA regardless of whether it is sold solely within a state. FIFRA § 3. Despite this extension of federal authority, considerable power to regulate pesticides was reserved to the states. Section 24(a) of FIFRA provides:

> A State may regulate the sale or use of any federally registered pesticide or device in the State, but only if and to the extent the regulation does not permit any sale or use prohibited by this [Act].

State regulation of pesticide sales is accomplished through state registration programs. Most states will grant a state registration on the strength of the registrant's federal registration; however, some states, such as California, may require additional data.

A state may prohibit or restrict a federally registered product. States will contend that since they have the authority to restrict or prohibit a pesticide, they may impose any reasonable requirements as a condition to permitting the sale or use. A state may require a registrant to submit safety and environmental data in addition to the data previously submitted to EPA. States do *not* have authority, however, to impose requirements for labeling or packaging "in addition to or different from" those required under FIFRA. FIFRA § 24(b) (see below).

While a state may not allow the sale or use of a pesticide prohibited by EPA, a state may grant additional uses of federally registered pesticides to meet special local needs (§ 24(c) registrations). 40 C.F.R. §§ 162.150-.156. States also may grant emergency exemptions (§ 18 exemptions) under certain conditions. 40 C.F.R. §§ 166.1-.53.

Judicial Examination of § 24(a)

Section 24(a) was reviewed by a federal district court in *National Agricultural Chemicals Association v. Rominger*, 500 F. Supp. 465 (E.D. Cal. 1980). In *Rominger*, federal registrants challenged California's authority to require the submission of additional data beyond that already required by EPA under FIFRA § 3. They argued that Congress intended EPA's § 3 data requirements to serve as a uniform standard and that the state's additional requirements imposed an unreasonable burden on the interstate commerce of pesticides.

The court concluded that § 24(a) provides the states with broad regulatory authority, including the authority to require federal registrants to submit additional data in support of a California registration. The district court concluded that the Congress did not intend to preempt state regulation of pesticides. 500 F. Supp. at 469. Finally, because Congress had authorized state regulation under § 24(a), the court also concluded that the exercise of such power would not represent an "unreasonable" burden on interstate commerce. 500 F. Supp. at 471. The *Rominger* decision has been cited by other courts in support of the states' broad authority to regulate pesticide sale and use. *Ferebee v. Chevron Chemical Co.*, 736 F.2d 1529, 1541 (D.C. Cir.), *cert. denied*, 469 U.S. 1062 (1984); *Florida Citrus Packers v. California*, 549 F. Supp. 213, 216 (N.D. Cal. 1982).

Regulation of Pesticides by Communities

Many local communities have sought to regulate pesticides by imposing "notification requirements" or in some instances by prohibiting the application of certain pesticides. The legislative history of FIFRA, however, suggests that Congress decided not to extend pesticide regulation below the state level.

The House Committee on Agriculture in 1971 considered the question of whether local governments should regulate pesticides and reported that:

> The Committee rejected a proposal which would have permitted political subdivisions to further regulate pesticides on the grounds that the 50 States and the Federal Government should provide an adequate number of regulatory jurisdictions.

H.R. Rep. No. 511, 92d Cong., 1st Sess. 16 (1971). The Senate Committee on Agriculture and Forestry concurred:

> The Senate Committee considered the decision of the House Committee to deprive political subdivisions of States and other local authorities of any authority or jurisdiction over pesticides and concurs with the decision of the House of Representatives. Clearly, the fifty States and the Federal Government provide sufficient jurisdictions to properly regulate pesticides. Moreover, few, if any, local authorities whether towns, counties, villages, or municipalities have the financial wherewithal to provide necessary expert regulation comparable with that provided by the State and Federal Governments. On this basis and on the basis that permitting such regulation would be an extreme burden on interstate commerce, *it is the intent that section 24, by not providing any authority to political subdivisions and other local authorities of or in the States, should be understood as depriving such local authorities and political subdivisions of any and all jurisdiction and authority over pesticides and the regulation of pesticides.*

S. Rep. No. 838, 92d Cong., 2d Sess. 16 (1972), *reprinted in* 1972 U.S. Code Cong. & Admin. News 3993, 4008 (emphasis added). For this reason, the Senate Agriculture and Forestry Committee rejected an amendment proposed by the Senate Commerce Committee which would have specifically given local governments the authority to regulate the sale or use of a pesticide. S. Rep. No. 838 (pt. II), 92d Cong., 2d Sess. 4, 46, *reprinted in* 1972 U.S. Code Cong. & Admin. News 4023, 4026, 4066.

The bill passed by the Senate did not contain any provision explicitly authorizing local regulation. 118 Cong. Rec. S15885 (1972). Senator Allen, Chairman of the Subcommittee on Agricultural Research and General Legislation, inserted into the Congressional Record (with unanimous consent) an excerpt from the Senate Committee on Agriculture and Forestry Report, stating that FIFRA "should be understood as depriving such local authorities and political subdivisions of any and all jurisdiction and authority over pesticides and the regulation of pesticides." 118 Cong. Rec. at S15888. The joint explanatory statement of the Conference Committee did not express a view on the subject of local regulation. S. Conf. Rep. No. 92-1540, 92nd Cong., 2d Sess., *reprinted in* 1972 U.S. Code Cong. & Admin. News 4130.

Judicial Decisions. The absence of statutory language preempting local regulation and expressions of congressional intent opposed to local regulation have resulted in decisions prohibiting local regulation (*Maryland Pest Control Association v. Montgomery County*, 646 F. Supp. 109, 111 (D. Md. 1986), *aff'd*, 822 F.2d 55 (4th Cir. 1987)), and decisions allowing local initiatives such as *People ex rel. Deukmejian v. County of Mendocino*, 36 Cal. 3d 476, 683 P.2d 1150, 204 Cal. Rptr. 897 (1984) (the *Mendocino* decision was effectively overruled by the California legislature which passed A.B. 2025 prohibiting the local regulation of pesticides). Petitions for certiorari in two recent cases request the U.S. Supreme Court to consider the issue. The Court has granted certiorari in one of the cases.

In the first case, *Mortier v. Town of Casey*, 154 Wis. 2d 18, 452 N.W.2d 555 (1990), *petition for cert. granted*, __ U.S.L.W. ____ (U.S. Jan. 14, 1991) (No. 89-1905), the Wisconsin Supreme Court affirmed FIFRA preemption and invalidated a local restriction on pesticide use, finding that FIFRA's legislative history "unmistakably demonstrates the intent of [C]ongress to preempt local ordinances." The Town of Casey's petition to the Supreme Court for review of this decision cites the opposite conclusion reached by the Maine Supreme Judicial Court in *Central Maine Power Co. v. Town of Lebanon*, 571 A.2d 1189 (Me. 1990) that FIFRA does not preempt traditional principles of state sovereignty allowing states to delegate their regulatory authority to political subdivisions. The Maine court relied on the ruling of the California Supreme Court in *People ex rel. Deukmejian*, above, and in *COPARR, Ltd. v. City of Boulder*, 735 F. Supp. 363 (D. Colo. 1989), *appeal filed*, No. 89-1341 (10th Cir. Nov. 1, 1989).

The second petition for review, which remains pending, challenges the August 1990 decision of the Sixth Circuit Court of Appeals that local posting and notification regulations in Michigan were contrary to Congress' intent in drafting § 24(a). *Professional Lawn Care Association v. Village of Milford*, 909 F.2d 929 (6th Cir. 1990), *petition for cert. filed*, 59 U.S.L.W. 3282 (U.S. Aug. 31, 1990) (No. 90-382). The court of appeals reviewed FIFRA's legislative history and concluded "when Congress rewrote the [FIFRA] statute [in 1972], it impliedly preempted the local regulation of pesticides." 909 F.2d at 933. "Nearly every congressional committee . . . deleted [proposed local regulation] provisions in the belief that the state and the federal governments could regulate pesticides adequately without subjecting the pesticide industry to thousands of regulatory jurisdictions." 909 F.2d at 931.

The petition for review of the *Milford* decision argues that even if Congress intended FIFRA to preempt *some* local authority, it did not intend to preempt *all* local authority over pesticides (e.g., notification requirements) which the petition argues are outside the scope of FIFRA and at most only have an "incidental effect" on federal pesticide regulation.

At the Supreme Court's request, the Solicitor General has filed an *amicus curiae* brief in the *Mortier* case. The Government argues that the local regulation issue warrants Supreme Court review, and that congressional intent was not sufficiently clear to establish preemption.

In addition to the FIFRA § 24 preemption issue, the petition for review of *Mortier* raises the issue of whether local regulation of pesticides may be preempted by *state* statutes regulating pesticide sale or use. See also *Pesticide Public Policy Foundation v. Village of Wauconda*, 622 F. Supp. 423 (N.D. Ill. 1985), *aff'd*, 826 F.2d 1068 (7th Cir. 1987); but compare *Mendocino County*, 683 P.2d at 1157, 204 Cal. Rptr. at 904 (1984). A Supreme Court decision in *Mortier* may also clarify the preemptive effect of state statutes on local pesticide requirements.

Section 24(b) — Preemption of Labeling

As noted, a state may require a federal registrant to obtain a state registration. A state, however, may not require the federal registrant to modify a federally approved label as a condition of obtaining the state registration. Although the State of California sometimes engages in *de facto* regulation of federal labels indirectly through its registration process, the command of § 24(b) is clear:

> Such State shall not impose or continue in effect any requirements for labeling or packaging in addition to or different from those required under this Act.

Federal laws preempting state labeling or packaging laws are strictly interpreted. *Chemical Specialties Manufacturers Association v. Lowery*, 452 F.2d 431, 437-38 (2d Cir. 1971); *Chemical Specialties Manufacturers Association v. Clark*, 482 F.2d 325, 328 (5th Cir. 1973); *Cosmetic Toiletry & Fragrance Association v. Minnesota*, 440 F. Supp. 1216 (D. Minn. 1977), *aff'd*, 575 F.2d 1256 (8th Cir. 1978).

At least two courts, however, have been reluctant to find that § 24(b) preempts requirements that can be satisfied with warnings or instructions that are not affixed to a container or package. In *New York State Pesticide Coalition, Inc. v. Jorling*, 874 F.2d 115 (2d Cir. 1989), the Second Circuit held that FIFRA preempts only labeling and that New York state posting and notification requirements could not be construed as labeling. The court found that labeling includes only that which accompanies the product through interstate commerce and that the purpose of label and labeling requirements is providing information to parties selling and applying pesticides. Thus, state posting and notification requirements, the purpose of which is to inform the public at large of pesticide applications, did not conflict with FIFRA and were not preempted.

Similarly, a federal district court in California held that FIFRA § 24(b) does not preempt California pesticide chronic warning requirements (Proposition 65, discussed below) at the point of sale, if they are not on the label. *Chemical Specialties Manufacturers Association v. Allenby*, 744 F. Supp. 934 (N.D. Cal. 1990), *appeal docketed*, No. 90-16485 (9th Cir. Nov. 5, 1990). The court restated its earlier decision in *d-Con v. Allenby*, 728 F. Supp.

605 (N.D. Cal. 1989), noting that "[s]ome of the Prop 65 warning requirements deemed 'safe harbor' methods . . . do not constitute 'labeling' under FIFRA." 744 F. Supp. 935. The court's two opinions authorize point-of-sale, product-specific chronic hazard warnings (e.g., shelf signs) in California even if they conflict with warnings imposed by FIFRA.

SECTION 24(c) — SPECIAL LOCAL NEEDS REGISTRATIONS

Because of the proliferation of minor special local needs registrations within the states, Congress in 1972 authorized states to register pesticides formulated for intrastate distribution to meet "special local needs." Applications for special local needs are submitted to state agencies by both manufacturers and users.

In January 1981, EPA published final § 24(c) regulations. 46 Fed. Reg. 2014 (1981) (codified at 40 C.F.R. §§ 162.150-.156). In general, EPA's § 24(c) registration regulations specify: (1) the types of registrations a state may grant under § 24(c); (2) the circumstances under which a special local needs registration may be granted by a state; and (3) the circumstances under which the Administrator of EPA may disapprove a special local needs registration. Neither § 24(c) nor EPA's implementing regulations thereunder affect a state's basic authority under § 24(a) to require registration of federally registered products. The manufacturer or formulator of pesticide products should note that a product may require a separate state registration in addition to a FIFRA § 24(c) registration.

After consultation with the Agency, a state may register any use of a federally registered product for which other uses have been denied, disapproved, suspended, or cancelled by EPA. Any use voluntarily cancelled by federal registrants, not following a notice of cancellation, may also be registered by a state after consultation with EPA. 40 C.F.R. § 162.152. A state may not grant a § 24(c) registration for an active ingredient but only for additional uses of end-use products.

EPA's § 24(c) regulations define special local needs as:

> [A]n existing or imminent pest problem within a State for which the State lead agency, based upon satisfactory supporting information, has determined that

> an appropriate federally registered pesticide product is not sufficiently available.

40 C.F.R. § 162.151(i). A state has reasonable discretion in determining whether a federally registered product is "appropriate." Economic factors as well as quantity, location and transportation difficulties may be considered by a state in determining whether a federally registered product is sufficiently available within the state.

EPA's regulations describe two examples of situations which are not considered special local needs: (1) pest problems which exist nationwide; and (2) interregional or nationwide § 24(c) registrations. Accordingly, EPA may view a series of nationwide or regional § 24(c) registrations as an attempt to avoid a § 3 registration under FIFRA.

Labeling for a § 24(c) registration must include the name of the state where it is valid and the special local need (SLN) number assigned to the registration by the state. If the registration is for an amendment to a federal label, the § 24(c) labeling also must include the federally registered product name, the § 24(c) directions for use, the name and address of the § 24(c) registrant, the EPA registration number of the federally registered product, and a statement that use of the product in a manner inconsistent with the § 24(c) or federal labeling is prohibited. If the federally registered product is already being sold in the state, the state must ensure that the amended labeling under § 24(c) is made available to purchasers and users within forty-five days after state approval of the final printed supplemental labeling.

If the state approves the § 24(c) registration, it is required to notify EPA within ten days of its approval, forwarding the "Application/Notification" form to EPA. EPA then has ninety days from the date of the registration to disapprove it. Grounds for disapproval may include the existence of a risk of unreasonable adverse effects, the refusal of the state to submit supporting information, insufficiency of the supporting information submitted by the applicant, or the failure of the applicant to obtain residue tolerances. EPA may not disapprove a § 24(c) registration on the ground that another product is available to satisfy the special local need. FIFRA § 24(c)(2)(A). Although EPA has ninety days to disapprove a § 24(c) registration, the registration goes into effect immediately upon approval by the state.

Generally, states authorize § 24(c) registrations for a period of no more than five years. Some states have limited their duration to one year or less.

EPA has expressed its concern that § 24(c) applications not be used as vehicles for circumventing the registration requirements established under § 3. The Agency thus reviews special local needs registrations to ensure that the following criteria have been met:

1. The existence of an established tolerance which will support the registration which is for a food/feed use registration. Assurance that there are no geographical restrictions on the tolerance that may affect the registration.

2. Proper labeling for the registration is present.

3. For food crop products, a determination that all inert ingredients have proper clearance.

4. The registration is not for a product/use that has been denied, disapproved, cancelled or is currently subject to a suspension order.

5. Data, when necessary, have been submitted to support the registration.

6. The product does not contain an active ingredient that is not currently contained in a federally-registered pesticide.

Pesticide Registration (PR) Notice 84-2, at 3 (Apr. 20, 1984).

CALIFORNIA'S PESTICIDE PROGRAM

The trend toward greater state regulation is best exemplified by California. Because California extensively regulates the sale and use of pesticides within the state, the program of the California Department of Food and Agriculture (CDFA) warrants separate examination.

California first regulated the sale of pesticides in 1901, before the first federal regulation of pesticides under the Federal Insecticide Act of 1910, ch. 191, 36 Stat. 331. The

1901 California Act required manufacturers to submit samples of Paris green for analysis. Act of Feb. 28, 1901, ch. 53, 1901 Cal. Stat. 69. As with early federal laws, California's first pesticide laws sought to regulate only the labeling and efficacy of pesticide products (e.g. Act of May 1, 1911, ch. 653, 1911 Cal. Stat. 1248). California adopted a registration statute as early as 1921, Act of June 3, 1921, ch. 729, 1921 Cal. Stat. 1263, and a permit system in 1949. Act of July 20, 1949, ch. 1043, 1949 Cal. Stat. 1939. See *People ex rel. Deukmejian v. County of Mendocino*, 36 Cal. 3d 476, 683 P.2d 1150, 204 Cal. Rptr. 897 (1984).

California adopted its existing registration and cancellation program in 1969 (Cal. Food & Agric. Code §§ 12811-12858 (West 1986 & Supp. 1990)) and in 1971 a three-category pesticide classification and use permit system. At the same time, the legislature granted broad authority to the Director of CDFA to impose time, place, and use restrictions on pesticides and to prohibit specific uses of pesticides entirely. Cal. Food & Agric. Code §§ 12824-12825, 14005-14006.

California's current pesticide regulation program, administered by CDFA, includes the following activities:

1. Pesticide registration, violations and data requirements. Cal. Food & Agric. Code §§ 12751-13141;
2. Licensing of pest control operators. §§ 11401-11897;
3. Licensing of pest control dealers. §§ 12101-12121;
4. Licensing of pest control advisors. §§ 12001-12054;
5. Assessment of pesticide taxes. §§ 12841-12845;
6. Recommendations of pesticide and uses. §§ 12971-12978;
7. Enforcement of worker safety regulations. §§ 12980-12985;
8. Restricted materials and permits. §§ 14001-14063.

In addition to CDFA's regulation of the sale and use of pesticides, the California Department of Industrial Relations and other state agencies regulate pesticides in California.

CDFA's Registration Program

The Pesticide Registration Division of CDFA is responsible for the Department's registration activities. In 1980, registrants and applicants were required to submit summaries of all data submitted to EPA. The Birth Defect Prevention Act of 1984, as amended, now requires companies to submit complete studies and not simply summaries. Cal. Food & Agric. Code §§ 13121-13130 (West 1986 & Supp. 1990). In addition to submitting to CDFA EPA-required data, companies must submit certain "California only" data, including dermal absorption data, mixer/loader, applicator, and reentry data. CDFA's Registration Division has 120 days to review and reach a proposed decision concerning an application for a new active ingredient or a major new use and sixty days for all other products.

CDFA's Medical Toxicology Branch was created in response to the increased data requirements of the Birth Defect Prevention Act of 1984 discussed below. The Environmental Monitoring Branch was created to implement the requirements of the Pesticide Contamination Prevention Act, also discussed below. The Pesticide Enforcement Branch of CDFA enforces compliance with California's pesticide use regulations and enforces the state's various licensing systems, including the licensing of pesticide control operators, advisors, dealers, and pilots.

In 1985, the California legislature increased the penalties for the misuse of a pesticide. Any violation of California's pesticide laws constitutes a misdemeanor. Cal. Food & Agric. Code § 12996. The penalty for a first violation is not less than $500 nor more than $5,000, or six months' imprisonment, or both. For a second violation the penalty is increased to a maximum of $10,000 (and minimum of $1,000). The maximum prison sentence is six months for a second or subsequent violation. An intentional or negligent violation creating a hazard to human health is punishable by one year imprisonment or a minimum fine of $50,000 or both. § 12996. The Director of CDFA, through the Attorney General, may seek to impose civil penalties instead of criminal penalties. §§ 12997-12999.5. In 1989, California increased the maximum civil penalty from $500 to $1,000 for each violation.

Birth Defect Prevention Act

In 1984, the California legislature enacted the Birth Defect Prevention Act (BDPA). This Act required the Department to submit a report to the California legislature by December 31, 1985, which listed "data gaps" for all active ingredients registered in California and specified whether the studies on file with the Department showed adverse effects. Cal. Food & Agric. Code § 13125. "Data gap" is defined by the BDPA as "a full set of valid mandatory health effects studies," including reproductive, chronic toxicity, mutagenicity, neurotoxicity, oncogenicity and teratogenicity studies. § 13123(a). The BDPA has prompted many questions by pesticide companies.

The BDPA required the Department to submit no later than December 31, 1985, a listing of 200 pesticide ingredients which had the most significant "data gaps, widespread use, and which are suspected to be hazardous to people." § 13127(a). Registrants of these products were notified by the Department to fill the data gaps. On January 31, 1986, CDFA sent California Notice 86-1 to all registrants with one or more data gaps. The response required by the Department to this notice is similar to the response required by EPA in response to a Data Call-In Notice under FIFRA § 3(c)(2)(B).

If the notified registrants had not initiated the studies by September 1, 1986, then the Department was required to fill the data gap in accordance with FIFRA § 3(c)(2)(B). Cal. Food & Agric. Code § 13127(c)(1). The Department is directed to "obtain the required test results" if it again determined by March 1, 1987, that the required studies had not been initiated. It is authorized to utilize "assessments charged to those registrants of the active ingredient for which the data gaps exist in amounts necessary to cover the department's expenses in filling the data gaps." § 13127(d)(1). The Department may exempt pesticides with only limited use if there is insignificant exposure to workers or the public. § 13127(e)(2).

The "formulator's exemption" in § 13128 states that applicants for a California registration, or current registrants who propose to purchase a registered pesticide from another producer in order to formulate the pesticide into an end-use product, are not required to submit or cite the mandatory health effects data or to offer to pay compensation for the use of such data if the producer is engaged in fulfilling the requirements of § 13127. Section 13128 was amended in 1989 to limit the formulator's exemption to cases where the producer is engaged in fulfilling the requirements in § 13127. The amendment effectively overrules the decision in *Chemical Specialties Manufacturers Association v. Parnell*, No. C-686703 (L.A. Cal. Super. Ct. writ of mandate issued Aug. 30, 1988), that gave formulators an absolute exemption from the requirements of § 13127, regardless of whether their producers fulfilled these requirements.

The Pesticide Contamination Prevention Act

One year after adopting the Birth Defect Prevention Act of 1984, the California legislature passed a second law, the Pesticide Contamination Prevention Act, Cal. Food & Agric. Code §§ 13141-13152 (West 1986), requiring California registrants of agricultural pesticides to submit data to CDFA concerning a pesticide's potential to pollute ground water. The Act requires that registrants of agricultural chemicals submit the following data or studies to CDFA no later than December 1, 1986:

1. Water solubility.
2. Vapor pressure.
3. Octanol-water coefficient.
4. Soil absorption coefficient.
5. Henry's Law constant.
6. Dissipation studies conducted under California or similar use conditions.

Cal. Food & Agric. Code § 13143(a). By December 1, 1987, the Director of CDFA is required to publish a list of pesticides for which there are "groundwater protection data gaps" or which exceed numerical values for the studies listed above. § 13144(a). These values must be established by December 1, 1986. § 13144(b).

The Director is authorized to fine a registrant $10,000 per day for each day it fails to submit the required ground water data. § 13145(a). A registrant may obtain an extension of two years, and, if necessary, an additional third year to complete the dissipation studies. The data must be submitted, however, no later than December 1, 1990. Failure to submit the information may result in cancellation.

The Act requires the Director to conduct soil and ground water monitoring for pesticides on the Groundwater Protection List. § 13148. This monitoring program will detect the presence of pesticides on the list below the root zone, the active biological zone in the soil, or eight feet below the surface, whichever is deepest. If a pesticide is detected within these statutorily prescribed zones, it is subject to cancellation unless a subcommittee of the Department finds that the pesticide has not polluted the ground water or that the use of the pesticide can be modified to avoid pollution. § 13149.

The Act also provides that in lieu of a registrant's paying a fine for not complying with the deadline to submit data, the registrant may elect to voluntarily surrender its registration. 3 Cal. Admin. Code § 6199(d)(1). A registrant who has surrendered its registration may not sell or distribute covered products in California.

Due to the large number of registrants that expected not to be able to comply with the December 1, 1990 deadline, it is uncertain whether CDFA will aggressively enforce the penalties established for non-compliance. CDFA is contemplating several approaches to dealing with companies that expect not to be able to comply with the deadline.

First, CDFA is contemplating a case-by-case approach. It will look at data that each registrant has submitted to date, and will determine whether the registrant has made a good faith effort to comply with the law's requirements, whether the registrant has been submitting the required data for a lengthy period of time or whether it began compliance efforts only after it received a final notice.

Second, CDFA is considering sending out Notices of Decisions as a form of a "courtesy notice" to warn companies of the possibility of fine or suspension. This will most likely be done by January or February of 1991. At that time, if a registrant can expect that data will be available "sometime soon," CDFA might elect not to fine the registrant. CDFA has indicated that it is more interested in obtaining data than in fining companies.

Safe Drinking Water and Toxic Enforcement Act of 1986

On November 4, 1986, California passed "Proposition 65" — the Safe Drinking Water and Toxic Enforcement Act of 1986, Cal. Health & Safety Code §§ 25180.7, 25249.5-.13 (West 1986 & Supp. 1990). The primary purposes of the Act are to prohibit contamination of drinking water with carcinogens or known reproductive toxins and to require that warnings be given to any persons who may be exposed to such chemicals. The Act calls for the governor to promulgate lists of known carcinogens and reproductive toxins and prohibits businesses from releasing such chemicals into the environment if they are likely to pass into drinking water. §§ 25249.5, .8.

The Act also requires that a warning be given before any person is exposed to any amount of a listed carcinogen or reproductive toxin. § 25249.6. The exposure may be occupational or environmental and even through consumer products.

The prohibition on discharges applies to releases that occur more than twenty months after a chemical is listed. § 25249.9. Similarly, the warning requirement applies to exposures that occur more than twelve months after listing. Warnings are not required where preempted by federal law. § 25249.10.

The Act creates an exemption from the warning requirement if it can be shown that the exposure poses no significant lifetime cancer risk and will have no adverse reproductive effect at exposures 1,000 times the level in question. § 25249.10.

The Act authorizes enforcement by private citizens' actions, as well as government enforcement by the state attorney general, local district attorneys, and city attorneys. A private action may be brought if the government authorities have not commenced an action within sixty days after receiving notice of the alleged violation. § 25249.7.

Discharge of Carcinogens and Reproductive Toxins Prohibited. The Act prohibits "person[s] in the course of doing business" from discharging or releasing any chemical "known to the state to cause cancer or reproductive toxicity" onto land or water if the chemical is likely to pass into any "source of drinking water." § 25249.5. The prohibition would appear to apply if *any* amount of the chemical will enter into drinking water.

Exposure Warnings Required. The Act also requires that a "clear and reasonable warning" be given before any business knowingly and intentionally exposes an individual to such a chemical. § 25249.6.

The term "person in the course of doing business" is defined to include any individual, trust, joint stock corporation, company, partnership, or association, but excludes any governmental department or agency, any entity operating a public water system, or any person employing fewer than ten employees. § 25249.11. The term "source of drinking water" is limited to either present sources of drinking water, or water which has been identified in a water control plan adopted by a regional board as being suitable for domestic or municipal uses. § 25249.11. The term includes ground waters and surface waters, but only those which meet the above requirements.

Lists of Regulated Chemicals. The prohibited releases and required warnings apply only to chemicals known to the state to cause cancer or reproductive toxicity, i.e., chemicals that are included on lists to be published by the governor on or before March 1, 1987 (and revised at least once per year thereafter). § 25249.8. The Act provides that a chemical is known to the state to cause cancer or reproductive toxicity if:

(1) in the opinion of the state's qualified experts, it has been clearly shown through scientifically valid testing according to generally accepted principles to cause cancer or reproductive toxicity; or

(2) if a body considered to be authoritative by such experts has formally identified it as causing cancer or reproductive toxicity; or

(3) if an agency of the state or federal government has formally required it to be labeled or identified as causing cancer or reproductive toxicity.

§ 25249.8.

Import of the Act for Pesticide Producers and Users. Pesticide producers would not normally face liability for a "discharge or release" of a listed chemical, since the act of releasing the chemical into the environment is performed by a pesticide user or applicator. There is a minor potential that a producer could be liable for releases which occur during production, storage, or transportation.

Since the Act provides no exemption for the lawful application of pesticides, users may try to avoid listed products, as well as products which contain a listed chemical as an inert or impurity. Even if a user is willing to buy the product, the commercial applicator may not be willing to apply it. Although EPA may have approved the continued use of a product, notwithstanding the fact that it is an animal carcinogen, this would not necessarily entitle it to an exemption under the Act. Users and applicators simply may not want to accept the risk of using such chemicals and defending themselves in enforcement actions.

Judicial Review of Proposition 65

At least four suits have been filed in federal district courts claiming that Proposition 65 is preempted by one or more federal statutes regulating the marketing or labeling of chemical products. In *Chemical Specialties Manufacturers Association (CSMA) v. Allenby*, 744 F. Supp. 934 (N.D. Cal. 1990), *appeal docketed*, No. 90-16485 (9th Cir. Nov. 5, 1990), CSMA alleged that the Proposition 65 warning requirements are preempted by § 24(b) of FIFRA and § 18(b) of the Federal Hazardous Substances Act (FHSA). The court rejected CSMA's claims, finding no FIFRA preemption of point-of-sale warnings, if they are not placed directly on the label. In an earlier related case, *d-Con v. Allenby*, 728 F. Supp. 605 (N.D. Cal. 1989), the court held that FIFRA expressly preempted the state from implementing the

Proposition 65 warning requirement through pesticide product labeling, but ruled some safe harbor warnings are permissible.

In *Committee for Uniform Regulatory Labeling v. Allenby*, No. C88-0730-EPL (N.D. Cal. filed Feb. 26, 1988), an association of food producers filed a lawsuit challenging Proposition 65 on several grounds, including commerce clause violations and federal preemption of the Act's warning requirements under the Federal Food, Drug, and Cosmetic Act, the Federal Meat Inspection Act, and the Federal Poultry Products Inspection Act. The case is still pending. In *Chemical Manufacturers Association (CMA) v. California Health & Welfare Agency*, No. CIV-F-88-1615 LKK-JFM (E.D. Cal. filed Dec. 16, 1988), the plaintiff filed suit against the Health and Welfare Agency, its secretary, and the California Attorney General seeking a declaration that the hazard communications standard promulgated under OSHA preempts Proposition 65's occupational warnings. Recently, a state court of appeal ruled that the warning provisions of Proposition 65 apply to the workplace (see *AFL-CIO v. Cal-OSHA Board* below). The state ruling will likely result in an adverse decision for CMA.

On July 12, 1990, in *AFL-CIO v. Cal-OSHA Board*, No. A048574 (Cal. Ct. App. 1st Dist. July 12, 1990), the court of appeal concluded that the Board has the duty, under Proposition 97 (the State Occupational Safety and Health Plan Initiative), to include the warning and enforcement provisions of Proposition 65 in the state plan for occupational safety and health. In 1988, Proposition 97 was passed by the voters, requiring the Board to develop and enforce occupational safety and health standards under a state plan that was not inconsistent with the federal Occupational Safety and Health Act. Following passage of the initiative, several labor and environmental organizations petitioned the Board to promulgate a rule that would include Proposition 65's warning requirement in the state plan. The Board denied the petition, maintaining that Proposition 65 was not a "state law governing occupational safety and health." The organizations petitioned for a writ of mandamus, which the court of appeal issued, resulting in the opinion.

In *AFL-CIO v. Deukmejian*, 212 Cal. App. 3d 425, 260 Cal. Rptr. 479 (1989) (Cal. App. 3d Dist. July 20, 1989), the Third District Court of Appeal on July 20, 1989, handed down the first state appellate court decision on Proposition 65. The court ruled unanimously that the initial list of Proposition 65 chemicals should have included animal, as well as human, carcinogens and reproductive toxins.

In October 1988, a coalition of manufacturers and retailers filed *Ingredient Communications Council, Inc. (ICC) v. Van de Kamp*, No. 504601 (Sacramento Super. Ct. filed Oct. 1988), asking the court to declare (1) that the Agency's warning regulations are valid, and (2) that the coalition's use of toll-free telephone messages satisfies the warning requirements of Proposition 65. The state filed a cross-claim alleging that the use of telephone messages was invalid. On August 22, 1989, a Sacramento Superior Court judge ruled that the system did not provide "clear and reasonable" warnings as required by the Proposition and had been ineffective in practice in providing warnings prior to exposure. The decision will affect all businesses which have relied on toll-free messages. If a system is found to be invalid, manufacturers and retailers relying on it may be liable under the Act's bounty-hunter provision for failure-to-warn penalties, Cal. Health & Safety Code § 25192(a)(2).

The Environmental Defense Fund, among other proponents of Proposition 65, targeted the Gillette Company to test the *ICC* ruling. The groups, under the bounty hunter provision, filed a notice of intent to file a lawsuit against Gillette within two weeks of the *ICC* decision. Gillette had been relying on a toll-free number to warn consumers of the trichloroethylene and lead contained in its Liquid Paper product. To avoid potential litigation with the environmental groups and the Attorney General, Gillette agreed to reformulate its product, eliminating the listed substances, and to pay $300,000 in fines and attorneys fees. The environmental groups hailed Gillette's decision because it agreed to do what they wanted most — discontinue the use of suspect chemicals. Under the terms of the settlement, the state will receive $275,000; attorneys for the environmental groups will receive $25,000.

In *AFL-CIO v. Deukmejian*, No. 502541 (Sacramento Super. Ct. filed May 1988), labor and environmental groups filed an action challenging the regulation that exempted food, drugs and cosmetics from Proposition 65's warning requirements. In a major victory for the proponents of Proposition 65, the Sacramento Superior Court ruled on March 1, 1990, that the regulation creates "a categorical exemption from the warning law for selected industries without specific regard to the chemicals involved" and illegally shifts the burden of showing no significant risk from the person responsible for the exposure. The judge granted declaratory relief rather than injunctive relief, resulting in the regulation being left on the books. An appeal was filed by the Grocery Manufacturers Association, an intervenor in the action, on April 16, 1990, the same date the order was issued. A host of associations and the Health and Welfare Agency have joined in the appeal. The Third District Court of Appeal is expected to hear the case before the end of 1990.

In two separate actions filed in October 1989, *California v. Santa Maria Chili, Inc.*, No. SM 64010 (Santa Barbara Super. Ct. filed Oct. 10, 1989), and *California v. McGhan Medical Corp.*, No. 178922 (Santa Barbara Super. Ct. filed Oct. 9, 1989), the Attorney General alleges that the companies emitted ethylene oxide without providing proper warning. Ethylene oxide is used as a sterilizing or fumigating agent for medical supplies, food products and cosmetic and pharmaceutical ingredients. One of the companies, Santa Maria Chili, Inc., has voluntarily discontinued use of the chemical pending resolution of the lawsuit.

Numerous notices of intent to file lawsuits have been filed with the Attorney General's office and local prosecutors. Under provisions of the Act, unless the authorities file an action within sixty days of the filing of the notice, a private enforcement action may be brought. Cal. Health & Safety Code § 25249.7.

Release of Pesticide Data in California

The release of public records in California is governed by the California Public Records Act, Cal. Gov't Code §§ 6250-6268 (West 1980 & Supp. 1990). The Public Records Act requires public disclosure of the same categories of pesticide safety and efficacy information disclosable under § 10(d)(1) of FIFRA. Cal. Gov't Code § 6254.2. Most importantly, this section of FIFRA provides that "data on safety" are available for public disclosure. As with FIFRA § 10(q), a requestor of information from CDFA must execute a Confirmation of Non-multinational Status. A copy of the executed document will be provided to the data submitter *upon request*. The Department makes public Toxicology Evaluation Worksheets prepared by CDFA in response to data submitted under the Birth Defect Prevention Act.

Reporting of Adverse Effects

CDFA's January 1980 regulations specify an applicant's or registrant's reporting requirements concerning adverse effects of its pesticides. Specifically, the California Administrative Code provides:

> Adverse Effect Disclosure. (a) If, during the registration process or at any time after the registration of a pesticide, the registrant has factual or scientific evidence of any adverse effect or risk of the pesticide to human health or the environment (including ambient air quality), the registrant shall immediately submit such evidence to the director. All such information including, but not limited to that required under Section 6(a)(2) of the Federal Insecticide, Fungicide and Rodenticide Act, shall be submitted to the director.

Cal. Admin. Code tit. 3 § 6210 (1990). California's reporting requirement may be stricter (more inclusive) than the federal reporting requirement of FIFRA § 6(a)(2) because it omits the modifier "unreasonable" before "adverse effects." Additionally, CDFA's reporting requirement includes, but is not limited to, any information reportable under § 6(a)(2) of FIFRA.

California registrants' annual renewal application must affirm their compliance with California's reporting requirements during the preceding year.

Chapter 14
REGULATION OF PESTICIDE APPLICATORS

The primary means by which EPA regulates pesticide applicators is through pesticide labels and labeling. Under § 12 of FIFRA, the use of a pesticide "in a manner inconsistent with its labeling" is unlawful. FIFRA § 12(a)(2)(G). As a result, the Agency can regulate pesticide applicators by utilizing labeling to specify the terms and conditions of use. Furthermore, pesticides can be classified either for "general use" or "restricted use." As a policy matter, the Agency only classifies products for restricted use and does not classify products for general use. 53 Fed. Reg. 15,952, 15,962 (1988). A product that is not restricted is unclassified. 40 C.F.R. § 152.160(a). Only certified applicators or persons under their direct supervision can apply a pesticide that is labeled for "restricted use." The Agency has proposed substantial changes to its regulations on the standards for certification of applicators and supervision of non-certified applicators.

PESTICIDE LABELING

Pesticide labeling[42] is EPA's principal method of communicating safety information to pesticide applicators and the public. Labels and labeling provide detailed information regarding proper application methods, protective clothing, warnings and safety precautions, etc. Current FIFRA labeling requirements are set forth at 40 C.F.R. part 156 (see Chapter 3).

---

[42] The term "labeling" encompasses the term "label." FIFRA defines "label" as "the written, printed, or graphic matter on, or attached to, the pesticide or device or any of its containers or wrappers." FIFRA § 2(p)(1). "Labeling" is "all labels and all other written, printed, or graphic matter accompanying the pesticide or device at any time; or to which reference is made on the label or in literature accompanying the pesticide or device." FIFRA § 2(p)(2). The pesticide applicator, therefore, must read and follow any and all labeling, not just the label. Labeling requirements are discussed in greater detail in Chapter 3.

USE CLASSIFICATION

FIFRA also regulates pesticide applicators according to the type of pesticide being applied. As mentioned above, FIFRA § 3(d) requires that each pesticide product be classified either for "general use" or "restricted use." A general-use pesticide is one which "when applied in accordance with its directions for use, warnings and cautions . . . will not generally cause unreasonable adverse effects on the environment." FIFRA § 3(d)(1)(B). A restricted-use pesticide is one which "may generally cause, *without additional regulatory restrictions*, unreasonable adverse effects on the environment, including injury to the applicator." FIFRA § 3(d)(1)(C) (emphasis added).

Restricted-Use Labeling

The Agency believes that the term "general use" is "potentially misleading if a user infers from the statement that the product may be used for 'general' purposes not listed on the label." 49 Fed. Reg. 37,960, 37,966 (1984). As a result, "general-use" pesticides are not identified as such on their labels; instead a pesticide classified for restricted use must bear the statement "Restricted Use Pesticide" on its label. 40 C.F.R. § 156.10(j)(2)(i)(A); 53 Fed. Reg. 15,952, 15,962 (1988). Below this statement must appear the restrictions placed on the use of the product. If the product is restricted to use only by certified applicators, the label must bear the following statement: "For retail sale to and use only by Certified Applicators or persons under their direct supervision and only for those uses covered by the Certified Applicator's certification." 40 C.F.R. § 156.10(j)(2)(i)(B). If the pesticide product is classified restricted-use for some uses and general-use for others, the pesticide must be marketed either as separate products with different labels or marketed under the restricted-use label with both restricted uses and general uses appearing on the label. 40 C.F.R. § 156.10(j).

Application of Restricted-Use Pesticides

A restricted-use pesticide may be applied only "by or under the direct supervision of a certified applicator." FIFRA § 3(d)(1)(C)(i). Certification of applicators ensures that access to restricted-use products is limited to persons who have the knowledge, skills and experience to use such pesticides safely. 40 C.F.R. pt. 171. In most cases, certification is granted by state agencies. FIFRA § 11(a)(2); 40 C.F.R. pt. 171.

An applicator certified pursuant to FIFRA § 11 can use or supervise the use of any pesticide classified for restricted use as long as the applicator has been certified for that particular use. FIFRA § 2(e)(1). FIFRA § 2(e) defines two categories of certified applicators: "private applicators" and "commercial applicators." EPA regulations establish different standards for certification of "commercial" and "private" applicators. Compare 40 C.F.R. § 171.4 with § 171.5. In addition, the distinction between the two is relevant for purposes of imposing civil or criminal penalties for pesticide misuse. Compare FIFRA § 14(a)(1) with § 14(a)(2).

A "private applicator" is a certified applicator who uses or supervises the use of restricted-use pesticides for agricultural purposes on property owned or rented by him or his employer. FIFRA § 2(e)(2); 40 C.F.R. § 171.2(a)(20). Thus, farmers who have fulfilled certification requirements are considered "private" applicators.

A "commercial applicator" is an applicator who uses or supervises the use of any restricted-use pesticide for any purpose or on any property other than as provided in the definition of a private applicator. FIFRA § 2(e)(3); 40 C.F.R. § 171.2(a)(9). Note that a "commercial applicator" need not be certified if he works under the direct supervision of a certified applicator. Commercial pest control operators fall within the FIFRA definition of "commercial applicators" *only* when they apply restricted-use pesticides.

The Food, Agriculture, Conservation, and Trade Act of 1990, Pub. L. No. 101-624, sec. 1491, 104 Stat. 3359 (Nov. 28, 1990), added recordkeeping requirements for all certified applicators applying restricted use pesticides. Where a state requires commercial applicators to maintain records, all certified applicators must maintain similar records; if a state has no recordkeeping requirements for commercial applicators, all certified applicators must maintain records reflecting the type, amount, date and location of restricted use pesticide applications.[43]

Under current regulations (proposed regulations are discussed below) certified applicators and other persons under their "direct supervision" can apply restricted-use pesticides. FIFRA § 3(d)(1)(C)(ii). Unless otherwise prescribed by its labeling, a restricted-use pesticide is considered to be "applied under the direct supervision of a certified applicator if it is applied by a *competent person* acting under the instructions and control of a certified applicator who is available if and when needed, even though such certified applicator is not physically present at the time and place the pesticide is applied." FIFRA § 2(e)(4) (emphasis added); 40 C.F.R. § 171.2(a)(28). The "availability" necessary to qualify as direct supervision is variable and dependent on the hazard presented by each application. If the certified applicator is not present, "direct supervision" includes detailed instructions for proper application and ensuring that the noncertified applicator knows how to contact the supervising certified applicator. Moreover, certified applicators must demonstrate a practical knowledge of federal and state supervisory requirements with respect to the application of a restricted-use pesticide by uncertified applicators. 40 C.F.R. § 171.6(a).

While a number of states have adopted regulations requiring a certified applicator to be physically present, the majority of them have not. Some states, however, require a certified applicator to be present during application of certain restricted-use pesticides which pose a greater hazard to man or the environment (e.g., fumigation). U.S. General Accounting Office, Nonagricultural Pesticides: Risks and Regulation (GAO Report) at 50 (No. 86-97) (Apr. 1986).

Current EPA regulations define "competent" as used in FIFRA § 2(e)(4) to mean "properly qualified to perform functions associated with pesticide application, the degree of

---

43 Although such information must be provided, upon request, to state and federal agencies regulating pesticide use, the records will not be available for public review.

capability required being directly related to the nature of the activity and the associated responsibility." 40 C.F.R. § 171.2(a)(11). Nevertheless, interpretations of the term "competent person" vary among state regulatory agencies. Most pesticide applicator associations have taken the position that the training of service technicians, along with certified supervision, is an effective way to ensure applicator competency. (Whether on-site supervision is necessary depends on the seriousness of risk posed by the application.) Some states have enacted training and registration requirements for service technicians, and/or require that service technicians be tested in pesticide application. A number of states — Kansas, Michigan, Mississippi, Missouri, and Pennsylvania — have adopted mandatory training requirements for service technicians. A number of other states are considering imposing training, registration or testing requirements for service technicians.

Although there are no federal controls over the competency of commercial applicators with respect to general-use pesticides, a majority of states have requirements designed to ensure that such applicators perform competently. GAO Report at 50-51. Generally, states require that commercial applicators either demonstrate competency or work under the direct supervision of someone with demonstrated competency. Some states require firms which apply any pesticide for hire to obtain a business license and to have a certified applicator in their employ at all times. In several states (for example, California and Florida), the controls are more stringent, requiring *each* applicator to demonstrate competency.

EPA categorizes FIFRA § 2(e)(3) "commercial" applicators who use or supervise the use of restricted-use pesticides according to the types of pest control listed at 40 C.F.R. § 171.3. These categories include, for example, "commercial" applicators involved in agricultural pest control, aquatic pest control, forest pest control, and structural pest control. All commercial applicators, regardless of category, must demonstrate competency in label and labeling comprehension, safety considerations, environmental hazards, pest identification and characteristics, pesticide types and effects, equipment usage, application techniques, and applicable state and federal laws and regulations. Not only must applicants demonstrate general competency in safe application of pesticides, but for the specified categories an applicator must meet additional competency requirements. For example, in aquatic applications the applicator "shall demonstrate practical knowledge of various water use situations and the potential of downstream effects." Similarly specific competency requirements apply to the other specialized application categories. 40 C.F.R. § 171.4(c). In a May 1988 regulation, the Agency changed the pesticide labeling language on restricted-use pesticides to reflect that certified applicators who apply restricted-use pesticides must be specially trained and certified to perform the specific uses listed on the label. Thus, the Agency now requires the restricted-use statement to read: "For retail sale to and use only by Certified Applicators or persons under their direct supervision and *only for those uses covered by the Certified Applicator's certification.*" 40 C.F.R. § 156.10(j)(2)(i)(B) (emphasis added). EPA can thereby limit the use of restricted-use pesticides to designated categories of certified applicators.

PROPOSALS REGARDING APPLICATORS OF RESTRICTED-USE PESTICIDES

In November 1990, EPA proposed significant changes to regulations governing the application of restricted-use pesticides. 55 Fed. Reg. 46,890.[44] The major changes embodied in the proposed rule are as follows:

(1) a three-tier level of supervision for application of restricted use pesticides,

(2) general and category-specific minimum federal standards for certification of applicator competency, and

(3) establishment of private applicator categories and additional categories and sub-categories for commercial applicators.

[44] All subsequent references to proposed 40 C.F.R. § 171 are to this proposed rule.

The most significant change in the regulations would be the level of supervision required for application of restricted-use pesticides. The supervision necessary would depend upon the specific pesticide involved. A restricted-use pesticide would fall into one of three categories:

(1) pesticides that only could be applied by a certified applicator,

(2) pesticides that could be applied by noncertified applicators, but only when a certified applicator is on site and capable of reaching the application area within five minutes, and

(3) pesticides that could be applied by a noncertified applicator under the direct supervision of a certified applicator who is not required to be on site, but is available to provide assistance to the noncertified applicator within a "reasonable period of time."

55 Fed. Reg. at 46,892-93 (proposed 40 C.F.R. § 171.35). The proposed rule also provides that any noncertified applicator must, prior to application, be trained by a certified applicator and demonstrate competency. Training includes educating the noncertified applicator on site-specific conditions, appropriate application techniques, and any risks posed to humans or the environment by the pesticide. Additionally, the certified applicator must verify the noncertified applicator's competency to perform the application. Competency is established if the noncertified applicator has completed a state-administered and EPA-approved training program, or if the certified applicator observes the noncertified applicator's first application of the pesticide and determines the application was competently performed. Records of verified competency of noncertified applicators using restricted-use pesticides at particular sites would have to be maintained by commercial applicators.

EPA is also proposing establishing private applicator categories and expanding the categories of "commercial" applicators by adding highly specialized, defined user categories pursuant to its authority under FIFRA § 3(d)(1)(C)(ii). See proposed 40 C.F.R. §§ 171.5, .7. Applicators would have to meet specific standards in order to be certified for each applicator category. By establishing additional and even more specific categories, the

Agency seeks to ensure that applicators have the requisite knowledge and skill necessary for safe pesticide applications in these specialized areas.

In the proposed rule the Agency is also seeking a minimum federal competency standard for certification of all applicators, commercial and private. Proposed 40 C.F.R. §§ 171.25, .27. EPA would no longer differentiate between the competency necessary for certification of private applicators and commercial applicators. Applicators would have to meet both general competency standards for all pesticide applications and specific competency standards for specific applicator categories. Private applicators could be certified based on any program that would assure the applicators competency in pesticide application, and knowledge of labeling requirements and applicable laws and regulations. For commercial applicators, however, initial certification would require passing a written examination demonstrating: (a) knowledge of general "principles and practices of pest control and safety in use of pesticides"; and (b) specific skills and knowledge required for certification in specialized categories or subcategories. Proposed 40 C.F.R. §§ 171.32, .27.

Under FIFRA § 11(a)(2) states can obtain applicator certification authority if EPA approves of a state-submitted plan for the certification of applicators. The plan submitted by the state must: (1) designate a state agency for implementing the plan; (2) identify legal authority and appropriate personnel for effectuating the plan; (3) provide assurance the state will adequately fund the certification program; (4) provide for submission of reports to EPA; and (5) provide assurance that the state certification standards conform to EPA certification standards. Currently, in order to meet EPA certification standards, the state program must require program participants to demonstrate label and labeling comprehension, knowledge of safety requirements, awareness of environmental consequences and principles and practices of safe pesticide application, and knowledge of applicable state laws and regulations. Under the proposed rule the state plan must also require recertification of both commercial and private applicators at least every five years.

EPA's proposed rule would also clarify the definition of what constitutes "use" of a restricted-use pesticide. See proposed 40 C.F.R. § 171.3. If a restricted-use pesticide is being loaded, mixed or applied, then the "user" must either be certified or under the direct supervision of a certified applicator. If storage, transport and disposal routinely take place at the same time as loading, mixing or applying, then the "user" needs to be certified for those uses as well. If, however, the storage, transport or disposal is not conducted by the same person performing the loading, mixing or applying, then storage, transport or disposal will not be considered uses and the person performing those activities need not be certified. 55 Fed. Reg. at 46,891.

Another EPA proposal that could impact pesticide applicators are the agricultural pesticide worker protection standards. 53 Fed. Reg. 25,970 (1988). The proposed worker protection standards would apply to persons who work on or in farms, forests, nurseries or greenhouses at which agricultural pesticides are used. Under the proposal, an applicator has a responsibility to assure that the pesticide application is performed in accordance with, and the supervised workers receive the protections of, the worker protection standards.

The proposal would establish information, notification, and protective equipment requirements, as well as application and reentry restrictions. Additionally, handlers of pesticides (workers engaged in mixing, loading, transporting, applying or disposing pesticides) that are not certified applicators would have to be provided information on label comprehension, safe handling of pesticides, and environmental hazards associated with pesticide handling. EPA has proposed implementing and enforcing the worker protection standards via inclusion of the standards as part of a pesticide's labeling.

Some applicators may also be subject to upcoming proposed regulations governing the storage, transportation, disposal, and recall of pesticides. EPA, Executive Summary for Proposed 40 C.F.R. Part 165 (Oct. 1990). Further, commercial applicators may be required to use containers meeting EPA performance standards, meet minimum standard storage requirements, and install spill containment pads at locations where pesticides are regularly loaded and mixed. See Chapter 16.

USE IN A MANNER INCONSISTENT WITH THE LABELING

FIFRA § 12(a)(2)(G) prohibits use of a pesticide "in a manner inconsistent with its labeling." FIFRA § 2(ee) establishes the following exceptions to the definition of using a pesticide "in a manner inconsistent with its labeling":

> (1) applying a pesticide at any dosage, concentration, or frequency less than that specified on the labeling unless the labeling specifically prohibits deviation from the specified dosage, concentration, or frequency, (2) applying a pesticide against any target pest not specified on the labeling if the application is to the crop, animal, or site specified on the labeling, unless the Administrator has required that the labeling specifically state that the pesticide may be used only for the pests specified on the labeling after the Administrator has determined that the use of the pesticide against other pests would cause an unreasonable adverse effect on the environment, (3) employing any method of application not prohibited by the labeling unless the labeling specifically states that the product may be applied only by the methods specified on the labeling, [or] (4) mixing a pesticide or pesticides with a fertilizer when such mixture is not prohibited by the labeling . . . .

Section 12(a)(2)(G) is the operative provision for § 2(ee). The 1978 amendments added the foregoing exceptions to FIFRA in order to afford applicators more flexibility in determining the manner in which to apply their pesticides. H.R. Rep. No. 663, 95th Cong., 2d Sess. 15, 20, 70, *reprinted in* 1978 U.S. Code Cong. & Admin. News 1988, 1993, 2039. The 1988 FIFRA amendments modified § 2(ee)(1) and (3) to provide that applicators (1) may use dosage concentrations or application frequencies lower than those indicated on a pesticide's labeling, and (2) may use any method of application not prohibited by the labeling, *unless the labeling specifically prohibits such deviations.*

Despite these clarifying amendments, some state agencies have required applicators to use only the dosages, concentrations, application frequencies, or methods of application indicated on the labeling even though the labeling does not expressly prohibit deviations. The validity of such state requirements, which in effect rewrite a pesticide's labeling, is doubtful in view of FIFRA's express prohibition against imposition of state labeling requirements. See FIFRA § 24(b), discussed in Chapter 13.

ENFORCEMENT ACTIONS AGAINST PESTICIDE APPLICATORS

Although FIFRA vests EPA with substantial enforcement authority, allegations involving pesticide misuse are primarily the responsibility of state regulatory agencies. (See Chapter 15 for additional information on EPA enforcement authority.) Pursuant to FIFRA § 26, EPA must determine that a state has adopted adequate pesticide use laws and enforcement procedures before it can acquire primary enforcement responsibility. 40 C.F.R. § 171.7. EPA retains enforcement authority for use violations that occur within a state that has not met EPA's criteria for primary enforcement authority. All states, with the exception of Nebraska and, for private applicators, Colorado, currently have full primary enforcement responsibility. EPA may prosecute a use violation upon the request of a state.

Investigations to determine whether pesticides are being properly used can be divided into two categories — use inspections and misuse investigations. Although EPA derives its authority to conduct such investigations from FIFRA § 12(a)(2)(G), the states normally perform these activities pursuant to their primary enforcement authority under FIFRA, and pursuant to state pesticide statutes. Use inspections include routine investigations or planned observations. Their purpose is to develop data on common practices of applying pesticides, to encourage the proper use of pesticides, and to determine whether pesticides are being used in accordance with their label directions. Misuse investigations include investigations of reported complaints or incidents of pesticide misuse. These investigations are conducted to develop evidence in support of enforcement actions arising from the alleged use of a registered pesticide in a manner inconsistent with its labeling. EPA, FIFRA Compliance/Enforcement Guidance Manual, chs. 1-3 (1983) (Compliance Manual).

The appropriate level of enforcement for pesticide misuse is based on the severity of the violation. There are two types of actions — administrative and judicial. Civil penalties are the preferred enforcement mechanism. Generally, administrative actions are initiated for violations of a lesser nature or against alleged first-time offenders. Judicial actions are reserved for use violations of an especially egregious nature that result in serious harm to human health or the environment, and for willful or repeated violations.

In July 1990, EPA established an Enforcement Response Policy (ERP) for determining the appropriate enforcement response for FIFRA violations, including use violations. The ERP establishes a policy and criteria for assessment and calculation of civil penalties for FIFRA violations. Under the ERP, use of a pesticide "in a manner inconsistent with its labeling," FIFRA § 12(a)(2)(G), is a level 2 violation with civil penalties ranging from $3,000 - $5,000 per violation, depending on the size of the business. Each occasion of inconsistent use is designated a separate violation. Penalties may be adjusted downward for voluntary disclosure or good faith efforts to comply with FIFRA. Additionally, the ERP provides for "settlement with conditions," where in lieu of a portion of the civil penalty the violator can perform environmentally beneficial activities. EPA may also choose to issue a press release on the applicator's violation. The ERP also notes that under FIFRA § 14(b) EPA can seek penalties of up to $25,000 and imprisonment up to one year for "knowing" violations of FIFRA. Finally, the Agency may seek injunctive relief if it is necessary to prevent environmental harm.

Several different administrative enforcement remedies are authorized by FIFRA. For example, EPA, or the states under their primary enforcement authority, may issue a notice of warning pursuant to FIFRA § 9(c)(3) in response to minor offenses rather than bring an administrative action against an applicator for pesticide misuse. Under FIFRA § 14(a)(4), the Agency determines whether a civil violation has occurred "despite the exercise of due care" by the pesticide applicator, or whether it "did not cause significant harm to health or the environment." In such situations, EPA "may issue a warning in lieu of assessing a [civil] penalty." FIFRA § 14(a)(4). State agencies may or may not have state statutory authority to assess civil fines.

Under the regulations relating to certification of pesticide applicators, EPA is authorized to deny, suspend, modify or revoke federally issued applicator certifications if the applicator violates FIFRA or its regulations. 40 C.F.R. § 171.11(f). States have the same authority with respect to state FIFRA certifications. 40 C.F.R. § 171.7(b)(1)(iii)(B). An enforcement action affecting certification status is a strong measure to be taken only after the Agency determines "public health, interest or welfare warrants immediate action." 40 C.F.R. § 171.11(f)(5)(i). Therefore, EPA, as well as the states, will deny, suspend, modify or revoke a certification only in limited circumstances.

Judicial enforcement actions may involve either civil or criminal proceedings. Remedies in civil proceedings include fines under FIFRA § 14(a), injunctions under § 16(c), and seizures under § 13(b). Criminal proceedings are authorized under § 14(b) and could result in fines or even imprisonment of the pesticide applicator. EPA seeks initiation of criminal proceedings infrequently and imposition of criminal sanctions only in the most serious instances of applicator misconduct. The nature of the violation, the history of compliance on the part of the pesticide applicator, and the seriousness of the environmental consequences generally influence EPA's decision on whether to seek criminal prosecution of a pesticide applicator. Compliance Manual at 5-17.

Unlike certain other environmental statutes, FIFRA does not include a provision for citizen suits to enforce the Act. Nevertheless, private individuals sometimes attempt to redress alleged pesticide misuse by filing lawsuits against pest control operators. The agricultural, aviation, and structural pest control industries have been targets of such suits. Although these types of suits are not enforcement actions per se, they focus on alleged pesticide misuse and seek damages from the pesticide applicator (and sometimes, from the pesticide producer). Proof that the application was performed in accordance with the EPA-approved label instructions provides critical evidence to support a pesticide applicator's contention that the pesticide was not improperly used.

STATE REGULATION OF PESTICIDE USE

In addition to their primary enforcement authority under FIFRA, virtually all states regulate the sale and use of pesticides. See Chapter 13. Most states have enacted special laws directed specifically to misuse of pesticides. California, for example, enacted a statute making any site, plant, crop or commodity treated with a pesticide not in accordance with the pesticide's registration a public nuisance and subject to seizure. Moreover, the owner of the treated plant, crop, commodity or site has the burden of providing evidence that the application did not create a hazard to human health or the environment and that use of the pesticide did not create an unfair economic advantage for the owner. Cal. Food & Agric. Code § 12648 (West 1986 & Supp. 1990). A 1989 amendment to the California Code made *negligent* violations of California's pesticide regulations criminal if the "violation . . . created or reasonably could have created a hazard to human health or the environment." Cal. Food & Agric. Code § 12996. By addressing problems of pesticide misuse through legislation, the states provide an additional mechanism by which to regulate pesticide applicators.

One interesting state regulatory trend in some southern states has been state agencies testing soil samples after a termiticide application for purposes of preventing consumer fraud. If the levels of the termiticide in the soil are below a state minimum standard, the states are requesting that the applicator reapply the termiticide. The states' objective is to protect consumers from pesticide applicators who supposedly charge for services not rendered. The scientific and legal validity of such regulatory efforts are, however, subject to debate.

Posting & Notification

In recent years, as an outgrowth of the right-to-know movement, states have enacted legislation or promulgated regulations requiring pesticide applicators to post signs notifying home owners, building occupants or the public that a pesticide application has, is, or will be taking place. The typical posting requirement requires posting of signs near the site of application. Such regulations often require notice to specifically delineated parties, and in some instances distribution of product-specific information about the pesticide.

A number of states, including Colorado, Connecticut, Florida, Illinois, Iowa, Maine, Maryland, Massachusetts, New Hampshire, New Jersey, New York, Ohio, Pennsylvania, Rhode Island, and Wisconsin, have adopted posting requirements for certain pesticide applications, primarily structural pest control and lawn care. Prior notification requirements can entail notification to occupants of the structure, notification to owners of contiguous property, or in some states, notification to parties listed on a statewide "pesticide sensitivity" registry. A statewide registry can list persons who have medical certification that they are unusually sensitive to chemical applications. Some states have further required that the applicator provide additional information about the pesticide product used in an application, such as providing a copy of the labeling. California, for example, requires structural control applicators to provide the contracting party with an information sheet containing the pesticide(s) to be used, its active ingredient(s), and a generic statement about pesticide risks and benefits. California also requires that the information sheet provide notice of symptoms that could indicate an adverse reaction to the pesticide, and telephone numbers of who to contact for further information. See Chapter 13.

Localities (i.e. cities, towns and counties) have also sought to regulate pesticides with local ordinances or regulations. Local regulations have taken various forms, ranging from posting and notification requirements to outright prohibition against use of certain pesticides. Local authority to regulate pesticides remains an ongoing legal controversy. See Chapter 13. Local regulation presents pesticide applicators with the possibility that distinctly nonuniform rules and regulations, adopted by local political bodies which lack the data and scientific and regulatory expertise available to EPA and state agencies, will disrupt pesticide applications.

Chapter 15
CIVIL AND CRIMINAL ENFORCEMENT

EPA ENFORCEMENT RESPONSIBILITIES UNDER FIFRA

Section 12 of FIFRA establishes a long list of unlawful acts, ranging from the sale of a pesticide that is unregistered, adulterated or misbranded to the use of a registered pesticide in a manner inconsistent with its labeling. To enable EPA to enforce these provisions, pesticide producers must register both their products and their establishments with the Agency, maintain extensive records of their production operations, and allow EPA and state officials to inspect these records at reasonable times. EPA officials also may inspect pesticide establishments for purposes of investigating suspected violations of the Act. When the Agency suspects a violation of the Act, EPA has authority to order the suspected violators to stop selling or using the pesticide, and in certain cases to seize the pesticide. EPA may impose civil and criminal penalties against violators under FIFRA § 14, and may seek a court order restraining violations of the Act pursuant to FIFRA § 16(c). EPA shares enforcement authority over pesticide use violations with the Department of Justice and state pesticide authorities.

Enforcement actions typically involve such matters as use of a pesticide in a manner inconsistent with its labeling, sale of adulterated or misbranded products, failure to file required reports, failure to register establishments, violation of cancellation or suspension orders, use of unregistered products, and making claims that differ from the approved labeling. EPA also monitors registrations that have been suspended due to failure to comply with Data Call-In notices.

EPA takes enforcement action through its Office of Compliance Monitoring, which reports directly to the Assistant Administrator for Pesticides and Toxic Substances. Civil penalty actions are prosecuted by the Toxics Litigation Division of the Office of Enforcement, which reports to the Assistant Administrator for Enforcement. When criminal charges are filed, they are prosecuted by the Department of Justice.

AUTHORITY TO PROSECUTE USE VIOLATIONS

Under FIFRA § 26, EPA has authority to grant states primary responsibility for enforcing pesticide use violations. In order to grant a state this responsibility, EPA must determine that the state has adopted adequate pesticide use laws and procedures for enforcing those laws. Nearly all states have qualified for and received primary enforcement responsibility for use violations. See Chapter 14 for a discussion of enforcement actions against pesticide applicators.

It is important to note that EPA does not relinquish its own enforcement authority within a state by granting primary enforcement responsibility to the state. EPA retains such authority and often exercises it in cooperation with state pesticide regulatory agencies. EPA has entered into cooperative enforcement agreements with many states under FIFRA § 23, under which the Agency provides grants-in-aid to implement enforcement programs and to train and certify applicators under FIFRA § 11.

In addition, EPA has authority under FIFRA § 27 to prosecute enforcement actions for use violations upon request by a state, or if the state fails to take timely action. If EPA receives a complaint alleging a pesticide use violation, it may prosecute the matter if the state has not commenced an enforcement action within thirty days. As a matter of practice, however, EPA rarely if ever overrides prosecutorial decisions by a state enforcement authority.

One federal court has ruled that the Justice Department also retains authority to prosecute criminal violations of FIFRA within a state, even if primary enforcement authority for use violations has been granted to the state. *United States v. Orkin Exterminating Co.*, 688 F. Supp. 223 (W.D. Va. 1988). The court held that the grant of primary enforcement authority under FIFRA § 26 is not an exclusive delegation of enforcement jurisdiction and does not detract from the Attorney General's plenary authority to enforce federal statutes. 688 F. Supp. at 225.

UNLAWFUL ACTS

FIFRA § 12 defines acts that are unlawful and punishable by civil and criminal penalties. FIFRA § 14 prescribes the penalties. Under FIFRA § 12, it is unlawful to:

(1) "Distribute or sell": (a) an unregistered, canceled or suspended pesticide; (b) an adulterated or misbranded pesticide; (c) a product whose composition is different from that specified in the registration statement; and (d) a product for which claims are made as part of its distribution or sale which differ substantially from claims made in the registration statement. FIFRA § 12(a)(1).[45]

(2) Alter or deface a label. FIFRA § 12(a)(2)(A).

(3) Fail to prepare, maintain, or submit records or to allow any entry, inspection, copying or sampling of records authorized under the Act. FIFRA § 12(a)(2)(B).

(4) Make a false guarantee. FIFRA § 12(a)(2)(C).

(5) Disclose or use confidential information for one's own advantage. FIFRA § 12(a)(2)(D).

(6) Advertise a restricted-use product without including in the advertisement the classification of the product. FIFRA § 12(a)(2)(E).

(7) Distribute, sell, use or make available for use restricted-use pesticides for uses other than those authorized by their registrations. FIFRA § 12(a)(2)(F).

---

[45] The 1988 FIFRA amendments define the term "to distribute or sell" as

> distribute, sell, offer for sale, hold for distribution, hold for sale, hold for shipment, ship, deliver for shipment, release for shipment, or receive and (having so received) deliver or offer to deliver. The term does not include the holding or application of registered pesticides or use dilutions thereof by any applicator who provides a service of controlling pests without delivering any unapplied pesticide to any person so served.

FIFRA § 2(gg). Thus, distribution of a pesticide does not include application activities by pest control applicators.

(8) Use a registered pesticide in a manner inconsistent with its labeling. FIFRA § 12(a)(2)(G). See Chapter 14.

(9) Use a pesticide contrary to the provisions of an Experimental Use Permit. FIFRA § 12(a)(2)(H).

(10) Violate a "stop sale" order. FIFRA §§ 12(a)(2)(I).

(11) Violate a cancellation or suspension order or fail to notify appropriate officials of the possession of a cancelled or suspended pesticide. FIFRA § 12(a)(2)(J), (K).

(12) Fail to register a pesticide producing establishment. FIFRA § 12(a)(2)(L).

(13) Falsify an application for registration, books or records, or an application to register an establishment. FIFRA § 12(a)(2)(M).

(14) Fail to file reports required of a registrant by the Act. FIFRA § 12(a)(2)(N).

(15) Adulterate a product. FIFRA § 12(a)(2)(O).

(16) Test pesticides on humans without their knowledge and consent. FIFRA § 12(a)(2)(P).

(17) Falsify or submit false data regarding testing of a pesticide, or submit false data in support of a registration. FIFRA § 12(a)(2)(Q), (R).

(18) Violate any registration or storage, disposal, transportation or recall regulation. FIFRA § 12(a)(2)(S).

EPA has interpreted several of these provisions, and has set forth the elements of each violation in the FIFRA Compliance Program Policy Compendium (May 10, 1982), as well as various Policy and Criteria Notices. The Agency's FIFRA Compliance/Enforcement Guidance Manual (1984) and its FIFRA Enforcement Response Policy are useful sources of information regarding FIFRA's enforcement. The Enforcement Response Policy is discussed below.

Misbranding

Section 12(a)(1)(E) of FIFRA makes it unlawful to sell "misbranded" pesticides. To understand what is meant by this provision, one must refer back to the definition of "misbranded" and the evolution of FIFRA. In general, a pesticide is misbranded if its label (or labeling) is false or fails to contain "information required . . . under the authority of this [Act] to appear on the label." FIFRA § 2(q)(1)(E). Among the required information are directions for use and warning statements which, together with any restricted use classification imposed under § 3(d), are "adequate to protect health and the environment." FIFRA § 2(q)(1)(F), (G).

EPA has taken the controversial position that it can order registrants to add warning statements to a registered label and declare that continued use of the previously existing label without the new warning renders the pesticide "misbranded." In the April 1984 Registration Standard for the herbicide simazine, EPA summarily reclassified the product from general to restricted use, and ordered the registrant to change the product label to reflect the reclassification and include a warning against using the product in areas with sandy soil and sensitive ground water conditions. Subsequently, EPA threatened to stop sale of any simazine products that failed to include these label changes, on the grounds that they would be "misbranded" under § 2(q)(1)(F), (G).

The registrant objected to this "constructive suspension" of the simazine registration without affording the registrant the right to a hearing under FIFRA § 6. EPA responded that the FIFRA § 6 procedures are not the sole means by which EPA may require label changes, but that it may take such action under its authority to prohibit sale of a misbranded pesticide. The registrant subsequently filed suit. *Ciba-Geigy Corp. v. EPA*, 801 F.2d 430 (D.C. Cir. 1986). The parties eventually agreed upon a label and settled the suit. Accordingly, the courts have not yet decided whether EPA can summarily reclassify a pesticide through the registration standard process.

"Misbranding" Pesticides Through Advertising. EPA regulates pesticide advertising under FIFRA § 12(a)(1)(B), which prohibits the distribution or sale of a registered pesticide if claims made for the product as part of its distribution or sale differ substantially from any claims made in connection with its registration. This requirement applies to all pesticide advertising, regardless of whether it physically accompanies a product as it is sold or distributed.

EPA also exerts considerable control over pesticide advertising through its authority to prohibit the sale of any product that is "misbranded" under FIFRA § 12(a)(1)(E). Claims made on labeling, including advertisements that constitute labeling, are considered false or misleading under FIFRA § 12(a)(1)(E) if they contain statements indicating, for example, that the product is "safe," "nonpoisonous," "noninjurious," "harmless," "nontoxic," or "among the least toxic chemicals known," regardless of whether the statements are accompanied by a qualifying phrase such as "when used as directed." 40 C.F.R. § 156.10(a)(5)(ix), (x).

A registrant may advocate uses in addition to those appearing on the approved label only to the extent that such uses are permitted under FIFRA § 2(ee) (See Chapter 14). EPA regards recommending uses which are not permitted by the label or by § 2(ee) as a violation of § 12(a)(1)(B). Under § 12(a)(2)(E), an advertisement for a restricted-use pesticide must include the classification of the product, i.e., such advertising must indicate that the product is a restricted-use pesticide.

One issue that arises with frequency is whether registrants may advertise products for which registrations are pending but not yet approved, or for which Experimental Use Permits, emergency exemptions, or special local need registrations have been approved. Under FIFRA § 12(a)(1)(A) and (B), it is unlawful to "sell or distribute" an unregistered pesticide. On January 11, 1989, EPA published a final interpretative rule addressing the question of pesticide advertising for unregistered products and uses. 54 Fed. Reg. 1122 (1989). EPA had originally proposed that unregistered products, which could not be purchased in the U.S., could be advertised in advance of registration approval. In the final rule, however, the Agency rejected this approach as inconsistent with FIFRA requirements that a pesticide's claims be evaluated before it can be commercially marketed. 54 Fed. Reg. at 1123.

EPA has established a narrow exception to this prohibition that permits a registrant to advertise a pesticide or an unregistered pesticide use that is the subject of an emergency exemption or special local need registration — as long as the advertisement is limited to the geographic area in which such use is permitted and contains prominent notice of the limitations on use. It remains unlawful under FIFRA § 12(a)(1)(B) to advertise a pesticide use permitted only under an Experimental Use Permit. 54 Fed. Reg. at 1122-23.

RECORDKEEPING AND REPORTING

Reporting of Adverse Effects Information

Section 6(a)(2) of FIFRA requires all pesticide registrants to submit to EPA" additional factual information regarding unreasonable adverse effects on the environment of the pesticide."[46]

"Unreasonable adverse effects" are defined in FIFRA to mean "any unreasonable risk to man or the environment, taking into account the economic, social, and environmental costs and benefits of the use of any pesticide." FIFRA § 2(bb).

Although the statutory expression of this requirement is simple, EPA's regulatory implementation of the provision is extremely complex. The Agency has published a number of memoranda, policy statements and proposed rules elaborating on the requirement, but has never promulgated effective regulations in final form. None of these documents,

[46] In the absence of registration under FIFRA, all chemicals are subject to similar reporting requirements under § 8(e) of the Toxic Substances Control Act (TSCA), unless they qualify for another TSCA exemption or exclusion. EPA considers candidate pesticides to remain subject to TSCA until an Experimental Use Permit or application for pesticide registration under FIFRA is submitted. Reporting requirements under TSCA § 8(e) are discussed briefly in Chapter 17. For a more detailed discussion of these and other regulatory requirements under TSCA, see McKenna & Cuneo's TSCA Handbook (2d ed. 1990).

moreover, clearly withdraws or supersedes the other. Thus, a well-reasoned reporting decision requires a thorough knowledge of all of these materials.

The more significant principles from these documents are discussed below:

> A **1978 memorandum by EPA's Office of General Counsel,** published at 43 Fed. Reg. 37,611 (1978), sets forth the General Counsel's views regarding what constitutes reportable information, and gives examples of information that is not required to be reported. The memorandum takes a very broad view of the reporting requirements.
>
> A **1979 Enforcement Policy,** published at 44 Fed. Reg. 40,716 (1979), describes a number of circumstances in which EPA will exercise its enforcement discretion and not seek penalties for failure to report certain types of information. Registrants can examine the 1979 policy and evaluate their compliance status by determining whether they fall within any of the circumstances that EPA states it will not charge as violations.
>
> A **1985 "Final Interpretive Rule and Statement of Policy"** published at 50 Fed. Reg. 38,115 (1985) and codified at 40 C.F.R. part 153, subpart D, sets forth an apparently definitive statement of EPA policy and procedural rules for reporting information under § 6(a)(2). However, the rule was never formally adopted, and thus its binding effect on registration standards was never established. Nevertheless, the proposed rule has not formally been withdrawn.
>
> A **draft Proposed Rule and Guidance** for FIFRA § 6(a)(2) reporting entitled Reporting Requirements for Risk/Benefit Information Proposed Rule and Guidance and circulated among registrants by the Agency in November 1989, apparently represents EPA's most current thinking on FIFRA § 6(a)(2) reporting. As of October 1990, however, EPA had not formally proposed, much less adopted, the rule. Thus, it too has no binding effect.

In addition to these policies of general application, the EPA Office of Pesticides and Toxic Substances also has distributed a more particularized document entitled Phase Three Technical Guidance (Dec. 24, 1989) (revised Apr. 10, 1990), setting forth adverse effects information that registrants must submit during the reregistration process. Section 4(e)(1)(E) of FIFRA requires a registrant seeking reregistration to identify any "data that are required to be submitted to the Administrator under [§ 6(a)(2)] indicating an adverse effect of the pesticide." To assist registrants in this regard, EPA included the technical guidance document in its guidance package for registrants proceeding into Phase III of the reregistration process. See Chapter 6.

In three important respects, the technical guidance document expands EPA's interpretation of the scope of information that must be submitted under § 6(a)(2). First, the guidance indicates that registrants must submit any data that show the pesticide is present in surface or ground water at levels of concern to EPA. Second, information on newly discovered metabolites must be reported. Third, the guidelines state that EPA considers any three similar incidents of adverse effects involving a pesticide to constitute a reportable "series" of events which must be reported, if all three incidents occurred within the space of five years (for incidents not involving humans) or ten years (for incidents involving humans).

Whether these requirements are binding upon registrants who have not received this guidance directly, and thus have no notice of them, is questionable. EPA likely will assert, however, that persons who have received the guidance packages have been put on notice of the Agency's interpretations of the FIFRA § 6(a)(2) reporting requirements.

1978 General Counsel's Memorandum. The General Counsel's Memorandum, 43 Fed. Reg. 37,611 (1978) establishes several general criteria for the reporting requirement. First, the requirement applies only to "persons holding registrations." This suggests that individual officers and employees of a registrant are not subject to the reporting requirements, or to civil or criminal penalties for their company's failure to report information. EPA at times has taken the position, however, that officers of a company may be held personally liable for the company's failure to report. Second, the requirement applies only to a registered pesticide. Information submitted in support of an application for registration is not covered. EPA regulations, however, require the submission of any "adverse effects" information in the applicant's possession at the time of application for the original registration. 40 C.F.R. § 162.8. Third, only "additional" information must be reported, i.e., information which has not previously been submitted to the Agency. Finally, the registrant is required to submit only information that it possesses, not information that it knows about but does not possess.

EPA interprets "information regarding unreasonable adverse effects" to mean any information which

> if true, would be relevant to an Agency decision regarding the risks and benefits of the pesticide, i.e., an Agency decision regarding the registrability of the pesticide or regarding the proper terms and conditions of the registration of the pesticide.

43 Fed. Reg. at 37,611.

Under this interpretation, information is reportable if it *pertains* to any adverse effect; EPA does not have to prove that there will be an adverse effect, or that the adverse effect is itself serious or "unreasonable." EPA believes that the determination of whether adverse effects are "unreasonable" must be made by the Agency, not by individual registrants. EPA interprets § 6(a)(2) to include any information on either risks or benefits of the pesticide which would be relevant either to the terms and conditions of registration or to whether the pesticide should be registered at all. This interpretation was upheld in the case of *Chemical Specialties Manufacturers Association v. EPA*, 484 F. Supp. 513 (D.D.C. 1980).

EPA interprets "factual" information to include both fact and opinion. Opinions must be reported, however, only if expressed by:

(1) A person employed or retained by the registrant to express an opinion on the pesticide's effects.

(2) A person requested by the registrant to give an opinion regarding the pesticide.

(3) A person who is considered an "expert" under the Federal Rules of Evidence.

A federal court has stated that opinion information, or at least expert opinion, is not "factual" information and need not be reported under § 6(a)(2). *Chemical Specialties*

Manufacturers Association v. EPA, 484 F. Supp. at 518. EPA has indicated, however, that it considers the court's decision incorrect.

1979 EPA Enforcement Policy. EPA's 1979 Enforcement Policy, 44 Fed. Reg. 40,716 (1979), is a statement of what types of non-reporting EPA will *not* consider violations of § 6(a)(2). By inference, however, registrants can determine from the policy the kinds of information that a registrant must report.

Toxicology Studies. The seriousness of adverse effects shown by toxicological studies will vary widely. EPA requires the submission of a study if it demonstrates any of the following results:

(1) The substance produced a different kind of toxic or adverse effect than previously reported.

(2) The substance produced a more severe toxic effect than previously reported (i.e., an effect at a lower dosage, after a shorter exposure period, after a shorter latency period, or at a higher incidence or frequency).

(3) The substance produced a toxic effect in a different test organism than previously reported (i.e., a different species, strain, sex, or generation of test organism).

(4) A toxic effect was produced through a different route or medium of exposure than previously reported (if humans or non-target organisms conceivably could be exposed through that route or medium).

(5) A toxic effect was produced through a different mechanism (pharmacokinetic, metabolic, or biological) than previously reported.

(6) If the pesticide is subject to a Special Review, suspension or cancellation proceeding, or is the subject of a generic standard or reregistration proceeding, *any* toxic effect is reportable regardless of the above criteria.

44 Fed. Reg. 40,716, 40,717 (1979).

For completed studies, adverse effects information must be submitted within a reasonable period, but no later than thirty days after the study is completed and the registrant receives the information. This gives the registrant a period in which to check for data errors and validate the data before submitting the study.

In limited situations, a registrant will not be required to report the results of a study which is tainted by inaccurate information. The study is not reportable if:

(1) The results of the study were based on erroneous data;

(2) Every author of the study has acknowledged in writing that the conclusions were incorrect because of the erroneous data and have corrected their opinions; and

(3) After the correction, the information no longer triggers reporting obligations.

Information which is essentially corroborative of previously reported information is not required to be submitted. In general, unless the information indicates that a pesticide may present a different or greater hazard than previously identified, or that prior information was inaccurate, the results of a study need not be submitted.

Preliminary or incomplete results of studies must be reported under the following conditions:

(1) The results of an incomplete acute effects study (ninety days or less duration) which would be reportable if completed, must be submitted within ninety days after the testing has been completed and a preliminary data analysis has been conducted. This gives the registrant up to ninety days in which to complete the final analysis.

(2) The results of an incomplete chronic effects study (longer than ninety days), which would be reportable if completed, must be submitted within one year after the testing has been completed and a preliminary data analysis has been conducted. Again, this gives the registrant up to one year in which to complete the final analysis before submission of the study.

(3) Even though a study is incomplete, it must be submitted at any point at which serious adverse effects are observed and there is a reasonable inference that they were caused by the pesticide. This requires submission of preliminary findings showing a serious adverse effect which is sufficiently different from

the effects observed in concurrent; or historical control animals that the effect reasonably appears attributable to the pesticide being tested.

Other Studies. Virtually all epidemiological studies are required to be submitted. EPA believes that there is usually disagreement over the significance of epidemiological studies, yet they provide an important source of information on the risks associated with human exposure. Therefore, any study which involves correlation between exposure to a pesticide and adverse effects in humans is reportable under the 1979 Enforcement Policy.

In general, EPA no longer requires the submission of efficacy data for registration except for products that claim to control organisms which may pose a threat to human health. Therefore, EPA does not require the submission of efficacy studies under § 6(a)(2) after a product is registered, unless the study demonstrates that the product may not perform as claimed in controlling organisms which may pose a risk to human health. Efficacy studies, however, must be submitted on any product subject to a Special Review, suspension or cancellation proceeding, if the study shows a reduction in the claimed efficacy.

Incidents. Complicated rules govern the reporting of specific "incidents" in which a pesticide is alleged to have caused toxic effects. This kind of information may come to a registrant's attention through product liability claims, dealers, growers, users, and a variety of other sources. EPA recognizes that some reports received by registrants are vague or implausible while others may provide useful information on hazards. Incidents involving humans must be reported whenever: (1) the registrant receives a report that a person who may have been exposed to the pesticide or at least one of its ingredients has suffered an adverse effect other than local skin irritation if the label warns of such effect; (2) there is a basis for an inference that the effect resulted from exposure to the pesticide. If facts available to the registrant establish that the person did not suffer an adverse effect or that the person was not exposed to the pesticide, the incident need not be reported to EPA. Also, if the registrant has not received sufficient information to enable it to investigate whether the adverse effect and exposure did occur, the incident need not be reported.

If facts available to the registrant conclusively establish that the adverse effect was not caused by exposure to the pesticide, the incident is not required to be reported. Absent such conclusive evidence, EPA asserts that the incident must be reported if the registrant receives an "expert" or "qualified" opinion that the pesticide caused the injury or if there is *any* basis for an inference that the pesticide caused the injury.

The same reporting rules generally are applicable to exposure of fish, wildlife, domestic animals, and nontarget plants, with two exceptions. Information on fish and wildlife incidents which meets the reporting criteria for human exposures does not have to be reported if: (1) the exposure to the fish or wildlife resulted from the improper use of the product; and (2) the label gives reasonable notice of the risk of the kind of adverse effect that allegedly occurred (e.g., "this product may be toxic to bees if not used as directed").

In some cases, the registrant may become aware of a series or pattern of incidents which, individually, would not be reportable, but, taken together, show the same kind of toxic effect following exposure to a pesticide and provide an inference that the pesticide caused the injury. Except for the causation factor, the reporting rules for a series of incidents are the same as those for human exposures. As to causation, the information must be reported if the pattern of adverse effects would not be "expected" unless they had been caused by the exposures to the pesticide. If evidence conclusively establishes that the pesticide was not the cause of the adverse effect, the series of incidents need not be reported.

This means that where there has been a series of misuses of a pesticide producing a predictable effect on non-target organisms, the misuses must be reported to EPA. This vague requirement has proved difficult for registrants to apply.

Registrants sometimes receive information alleging that a product has failed to perform as claimed. These incidents generally are not reportable, unless they concern a health risk or immediate hazard to life.

A report concerning a single incident in which a pesticide allegedly was not effective is not reportable to EPA under § 6(a)(2) unless the product was used against organisms which may pose an immediate hazard to human life if not controlled. Failures of efficacy against organisms which may pose a risk to public health, but not an immediate hazard to human life, are not reportable unless they show a pattern of comparable failures. A series of efficacy failures during use of the product to control organisms which may pose a risk to human health is required to be reported.

The criteria for reporting other kinds of adverse effects information are vague. Such other information might concern, for example, impurities in products; the fate of a pesticide in the environment; the metabolism of a pesticide in soil, plants, or animals; bioaccumulation of a pesticide; residues; levels of exposure to applicators, users, and consumers; and drift of pesticides to nontarget areas.

EPA asserts that "adverse effects" information of this type must be reported if the registrant "reasonably should know" that EPA would regard the information as raising serious questions about whether the pesticide should be cancelled, suspended, or restricted, or whether the terms of the registration should be modified. However, if the registrant knows of facts which establish that the reports are incorrect, the information is not required to be submitted to EPA.

Because § 6(a)(2) requires the reporting only of "additional" information, any data that were previously reported to EPA are not required to be submitted. Information is considered to have been previously reported if it is contained in: (1) documents previously submitted to EPA by the registrant; (2) a scientific article which appears in certain abstracts; (3) EPA publications or records, or publications cited in EPA *Federal Register* notices; (4) public reports by other federal agencies.

1985 Interpretive Rule. In 1985 EPA issued a Final Interpretive Rule (1985 rule), published at 50 Fed. Reg. 38,115 (1985) and codified at 40 C.F.R. § 153, which was intended as the definitive statement of the Agency's interpretation of the FIFRA § 6(a)(2) reporting requirements, as reflected in the Agency's 1978 and 1979 interpretive documents. The 1985 rule never became effective, however, notwithstanding its codification, and now has been supplanted by another proposed rule. Elements of the 1985 rule which remain are discussed below.

1989 Draft Proposed Rule. In November 1989, EPA circulated a draft proposed rule, entitled "Reporting Requirements for Risk/Benefit Information," that would revise § 6(a)(2) reporting requirements and responds to some of the criticisms directed by industry of the 1985 rule. Although the proposed rule is largely consistent with prior reporting policies, there are important differences of which pesticide registrants should be aware.

The 1985 rule and 1989 draft proposed rule differ from the 1978 and 1979 guidance documents in several respects. Under the 1985 rule and the 1989 draft proposed rule, the failure to submit certain information concerning pesticide residues in food, feed, water or the environment would be regarded as an actionable violations. Moreover, § 6(a)(2) information must be designated as such by the submitter, and must be reported using the format for submission of information under TSCA § 8(e).

Under the 1989 draft proposed rule, EPA has indicated that the § 6(a)(2) requirements apply not only to the registrant of a pesticide, but also to any *former* registrant. Thus, registrants of a pesticide for which the registration has been cancelled, suspended, withdrawn or lapsed must submit any § 6(a)(2) reportable information to EPA even after they cease to hold a pesticide registration.

The draft proposed rule also indicates that EPA deems a registrant to be in receipt of information at the time that the registrant "first comes into possession of, or *knows of*, such information." This interpretation further would expand the reporting requirements to include a communication regarding § 6(a)(2) information (e.g. a phone call), in the absence of its physical receipt by the registrant.

The draft proposed rule would eliminate the reporting exemption for information previously contained in certain federal or scientific publications. However, the draft proposed rule would retain exemptions for: (1) documents that the registrant previously submitted to EPA (not including information to be identified under FIFRA § 4(e)(1)(E)); (2) EPA publications or EPA hearing records, or publications cited in EPA *Federal Register* notices; (3) any other documents in the official files and records of the Office of Pesticide Programs; and (4) to a limited extent, opinions. As in the 1985 rule, EPA rejects, and asserts that it is not bound by, the view expressed by the court in *Chemical Specialties Manufacturers Association v. EPA*, 484 F. Supp. 513 (D.D.C. 1980), that an opinion is not "factual" information within the meaning of § 6(a)(2).

The draft proposed rule would revise the time deadlines for submission of § 6(a)(2) information to the Agency. Incidents would have to be submitted within forty-five calendar days of the registrant's knowledge of the incident, and reportable information from a toxicology study would be required to be submitted within thirty calendar days of the registrant's receipt of the information.

The time period for submitting incident reports commences to run after the registrant has received information sufficient to commence investigation of the incident; the thirty day period for submitting toxicological reports begins after the registrant has completed testing and test validation. Preliminary results from toxicological studies must be submitted within thirty calendar days if: (1) an adverse effect occurs in a period shorter than that of the planned test period *and* the adverse effects are different from or of a greater magnitude than effects previously associated with that substance; or (2) when the information indicates the pesticide "may pose a substantial risk to human health or the environment."

The draft proposed rule would expand reporting requirements to include information relating to pesticide residues found in or on: (1) food or feed; (2) surface water; (3) ground water; or (4) drinking water. The draft proposed rule further would establish guidelines, albeit vague ones, for reporting studies of the possible presence of residues in food, feed or water. For example, a study "which may show"[47] that residues of any active or inert ingredient, impurity, metabolite, or degradate are found on food or feed above established levels[48] must be reported under § 6(a)(2). If a permissible residue level has not been

47 The draft proposed rule does not explain how registrants are to determine whether a study "may show" that pesticide components are found on food or feed or in water at greater than "established levels."

48 "Established levels" are defined as tolerances, action levels or food additives levels which have been set under the Federal Food, Drug, and Cosmetic Act by EPA or FDA.

established, studies suggesting that *any* residue of a pesticide's active or inert ingredients or the pesticide's impurities, metabolites or degradates may appear on food or feed would be reportable.

Information from a study that "may show" that residues in any amount from an active ingredient,[49] metabolite, or degradate appear in surface water or drinking water would be reportable under § 6(a)(2), except where the pesticide is registered for use on surface water or drinking water and the residue levels do not exceed the levels indicated on the registrant's registration application. Any study that "may show" that any amount of an active ingredient, metabolite or degradate of a registrant's pesticide appears in ground water would also be reportable.

In contrast to residue studies for which reporting is required if the study *may* show the presence of residues, incident reports relating to residues *must* be submitted if the incident in fact does show the presence of residues. Studies indicating that residues that are not active ingredients, or residues derived from an active ingredient but which are found in food/feed or ground, surface, or drinking water, would not have to be reported if the substance is "an innocuous component of human diet" and there is no established residue level.

The draft proposed rule would set a new standard for reporting a series or pattern of incidents involving toxic or adverse effects. Under the new rule, a reportable "series" can be any three or more similar alleged incidents that have been reported to the registrant. Such a series must be reported if the registrant cannot disprove the allegation of exposure, even if the reports were not from an employee or an expert. This differs from the 1979 enforcement policy position, where EPA did not set a specific number of incidents that could comprise a reportable series.

---

49 The presence of inert ingredients in surface water, ground water, or drinking water would not have to be submitted since the Agency believes the presence of such chemicals cannot be accurately traced to the use of pesticides.

The draft proposed rule would revise conditions under which "failure to perform" incidents must be reported to EPA. When a pesticide is subject to a Special Review or a suspension or cancellation proceeding, a registrant must submit "failure to perform incidents" if three or more such incidents are reported by a government employee. EPA considers such information sufficiently reliable to consider it in making risk/benefit determinations.

Penalties for Failure to Report. EPA takes the position that the failure to report information under FIFRA § 6(a)(2) is an unlawful act subject to civil and/or criminal penalties under FIFRA § 12(a)(2)(N). 44 Fed. Reg. 40,716, 40,717 (1979). FIFRA § 12(a)(2)(N) makes it unlawful for any person "who is a registrant, wholesaler, dealer, retailer, or other distributor to fail to file reports required by this [Act]."

If EPA establishes at an administrative hearing that a registrant failed to submit information required to be submitted under FIFRA § 6(a)(2), the violator may be assessed a civil penalty of up to $5,000 for each failure to report. In addition, a registrant who "knowingly" violates FIFRA § 6(a)(2) could be charged with a misdemeanor and assessed a fine of up to $25,000 and/or imprisoned up to one year for each failure to report. FIFRA § 14.

Registration of Establishments

Under FIFRA § 7, all establishments which manufacture, process, formulate, or package a pesticide or an active ingredient that is to be used in a pesticide must be registered with EPA. Establishments include every separate site, including domestic establishments that only export and foreign establishments that only import. EPA requires that establishments producing a multi-use chemical that can be used as a pesticide active ingredient register with EPA if the producer has "actual or constructive knowledge that [the] substance . . . is used or intended for use as an active ingredient in the manufacture of a pesticide." 53 Fed. Reg. 35,056, 35,056 (1988). The producer must report the amount of the substance produced at that establishment for which the producer has active or constructive knowledge the substance will be used or is intended to be used as an active ingredient in the manufacture of a pesticide.

An application for registration of an establishment must include the type of ownership and the names and addresses of all production facilities. Each establishment registered with EPA will be assigned a registration number, which must appear on the labels of all pesticide products produced there.

Revision of these regulations removed the registration requirement for custom blending pesticide establishments. 53 Fed. Reg. at 35,057. Custom blenders are those establishments that mix pesticides to a customer's specifications, but who do not hold the end product in inventory, or do not engage in other pesticide production activities. 40 C.F.R. § 167.3. Such establishments, however, remain subject to FIFRA § 8 recordkeeping requirements.

Every establishment must submit a report to EPA within thirty days after registration and thereafter annually on or before March 1. The report must include such information as the types of pesticides manufactured and formulated, the past year's production and sales figures for each product, and an estimate of the amount that will be produced in the current year.

Books and Records

FIFRA § 8 and the regulations promulgated thereunder require that all pesticide producers, registrants and applicants for registration[50] maintain certain records. The regulations require producers to maintain records indicating:

(1) The batches and amounts per batch of all pesticides produced.

(2) The product names and quantities of devices produced.

---

[50] As yet, EPA regulations still refer only to pesticide producers. 40 C.F.R. § 169.2.

(3) The receipt of pesticides and active ingredients used in producing pesticides, including brand name, manufacturer, shipper, carrier, shipping dates, and quantities.

(4) The shipment (but not sales records) of pesticides and active ingredients used in producing pesticides.

(5) The types and amounts of pesticides produced.

(6) For pesticides intended solely for export, specifications, labeling, and an acknowledgment by a foreign purchaser that a pesticide is not registered for use in the United States if the product is not similar in composition and use pattern to a federally registered product.

(7) The method of disposal of pesticides by the producer.

(8) Tests conducted on human beings.

In addition, pesticide producers must maintain:

(1) Copies of domestic advertising of restricted uses of pesticides.

(2) Copies of guarantees.

(3) Records containing research data relating to registered pesticides, including all test reports submitted in support of a registration or tolerance petition, all underlying raw data and interpretations and evaluations thereof (whether in the possession of the producer or in the possession of the independent testing facility or laboratory which performed such tests on behalf of the producer, which records must be retained as long as the registration remains valid).

40 C.F.R. § 169.2.

Upon request of an EPA employee or representative or designated state official (including a showing of appropriate credentials), a pesticide producer, distributor, carrier, dealer, or any other person who sells pesticides must permit such person to have access to, and to copy, all records required to be maintained under FIFRA § 8. Such access does not extend to financial data, sales information other than shipment information, personnel data, or research data (other than data relating to registered pesticides or to pesticides for which an application for registration has been filed).

INSPECTION OF ESTABLISHMENTS

FIFRA § 9 authorizes officers designated by the EPA Administrator to inspect places where pesticides are held for distribution or sale or where a pesticide with a cancelled or suspended registration is being held. It does not authorize inspection of research facilities or places where pesticides are being used, and limits the scope of an authorized inspection to "inspecting and obtaining samples of any pesticides or devices, packaged, labeled, and released for shipment, and samples of any containers or labeling for such pesticides or devices." FIFRA § 9(a)(1)(A). If the inspector takes any samples, he must give to the owner, before leaving the premises, a receipt that describes the samples taken and, if requested, a portion of the sample equivalent in volume or weight to the portion retained by the inspector. The results of any analysis of the sample must be forwarded promptly to the owner, operator, or agent in charge of the establishment.

Prior to conducting the inspection, the inspector must present appropriate credentials and a written statement of the reason for the inspection, including whether a violation of law is suspected. FIFRA § 9(a)(2). EPA's inspection notice most frequently will state either that a violation is suspected or that the inspection is "routine" — i.e., based on random selection. This practice follows the ruling in *Marshall v. Barlow's Inc.*, 436 U.S. 307 (1978), in which the Supreme Court held that warrantless inspection of business establishments under federal statutes could meet constitutional standards if the establishments were chosen for inspection on the basis of a "neutral" or random selection process.

Upon a showing of reasonable belief that a violation has occurred, EPA may obtain from a local court a warrant that authorizes entry, inspection and reproduction of certain records, and administrative seizure of any pesticide or device found to be in violation of FIFRA. The records that may be inspected or reproduced are those showing delivery, movement or holding of the pesticide or device, including the quantity, date of shipment and receipt, and the name of the consignor or consignee. If an establishment cannot produce records required to be retained, "all other records and information relating to such delivery, movement, or holding of the pesticide or device" will be subject to inspection and reproduction. Thus, FIFRA § 9 must be read in conjunction with § 8(b), which authorizes the inspection of records concerning shipment of pesticides.

ENFORCEMENT ACTIONS

Stop Sale, Use, Removal, and Seizure

Under FIFRA § 13(a), EPA may issue a stop sale, use or removal order (SSURO) to manufacturers, distributors, retailers, and even users, thereby freezing product inventory immediately when: (1) on the basis of inspection or tests, there is reason to believe that a product is in violation of the Act; (2) a product has been, or is intended to be, sold or distributed in violation of the Act; (3) a product has its registration suspended or cancelled.

SSUROs are favored by EPA and states as an enforcement mechanism because they are adaptable and procedurally simple. If EPA has "reason to believe" that the product is in violation of FIFRA, a stop sale order may be issued without any prior adjudication of the violation. Stop sale orders may be challenged in federal district court under the general review provision of FIFRA § 16(a). In such a challenge, the issue is whether EPA had a reasonable basis for issuance of the SSURO. However, EPA has indicated that a stop sale order should be supplemented with a civil penalty assessment in order to preserve the recipient's administrative hearing rights. FIFRA Compliance/Enforcement Guidance Manual (Guidance Manual) at 5-5 to 5-10. Stop sale orders may take effect immediately and are limited in scope only by the underlying violation.

EPA regards issuance of a stop sale order to be mandatory if the pesticide or device provides a direct threat to human health or the environment because (a) the pesticide is being sold after the registration has been suspended or cancelled; (b) the pesticide is adulterated, unregistered, improperly labeled, or improperly packaged, so as to make safe or effective use unlikely or impossible; or (c) the pesticide's labeling bears false or misleading claims which are likely to cause a life-endangering health hazard or serious adverse environmental condition.[51] A stop sale order also is mandatory where a civil penalty has already been issued or the pesticide is subject to a voluntary recall and a party does not comply with the administrative action. EPA also has discretion to issue a stop sale order when appropriate after considering the severity of the violation, whether it is part of a series of violations, whether the violation is easily corrected, and the extent of the threat to health and the environment. Guidance Manual at 5-11 to 5-12.

EPA has authority to bring actions in federal court to seize a pesticide which is unregistered, adulterated, or misbranded, or which makes claims differing from those in the registration statement. FIFRA § 13(b). Pesticides or devices which cause unreasonable adverse effects[52] even when used in accordance with their labeling under this Act also may be seized. In addition, EPA may act to seize misbranded pesticidal devices.

EPA may obtain a warrant from a court for the purpose of seizing a pesticide which is in violation of the Act. EPA generally intends to use its seizure authority in circumstances similar to those which would prompt a stop use order. Because stop use orders are preferred, the Agency generally will not turn to a seizure action unless:

(1) The recipient of a stop use order fails to comply;

(2) EPA anticipates that a party will not comply with a stop use order if issued;

(3) A pesticide is so hazardous that it should be removed from the marketplace or place of use;

---

51 Because a stop sale order is ordinarily issued without prior notice, the actual or potential danger to human health or the environment posed by the product must be serious or imminent. Otherwise, EPA guidance suggests that other remedies will generally be appropriate. Moreover, EPA has indicated that there may be circumstances in which it will seek voluntary action, although a stop use order could be justified by the Agency.

52 In the 1988 amendments, Congress revised this provision to provide that "[i]n the case of a plant regulator, defoliant, or desiccant, used in accordance with the label claims and recommendations, physical or physiological effects on plants or parts thereof shall not be deemed to be injury, when such effects are the purpose for which the plant regulator, defoliant, or desiccant was applied." FIFRA § 13(b).

(4) The seizure will be used to support a recall; or

(5) Seizure is necessary to dispose of products which are the subject of a stop use order, and the responsible party has not taken and has stated the intention not to take corrective action.

Guidance Manual at 5-15.

Penalties

Penalties for violation of FIFRA are imposed under FIFRA § 14. As discussed below, FIFRA § 14(a) prescribes civil penalties, which may be imposed by the Agency following a hearing before an EPA administrative law judge (ALJ). The Act lists a number of factors which the Agency must consider in assessing civil penalties. FIFRA § 14(b) prescribes criminal penalties for knowing violations of FIFRA. Such penalties may be levied only by a United States district court, following a criminal trial and conviction.

Civil Penalty Assessment Procedure. EPA may assess civil penalties against any registrant, commercial applicator, wholesaler, dealer, retailer, or other distributor[53] of up to $5,000 for each violation of FIFRA, without regard to the violator's knowledge or intent.[54] FIFRA § 14(a). Before a civil penalty may be assessed, the person charged with a violation must be provided notice and an opportunity for a hearing on the charge. FIFRA § 14(a)(3). A civil penalty hearing has most of the characteristics of a trial in a civil court, although the proceedings are somewhat less formal.

---

53 Lesser penalties may apply to violations by private applicators or other persons outside of these categories. FIFRA § 14(a)(2).

54 While knowledge or intent are not required to assess penalties for most FIFRA violations, the EPA Chief Judicial Officer has determined that a finding of "knowing falsification of records" under FIFRA § 12(a)(2)(M) requires that the person must have specific intent to falsify records required to be maintained pursuant to FIFRA. *Helena Chemical Co.*, No. FIFRA-09-0439-C-85-18 (EPA June 15, 1987), *rev'd*, FIFRA App. No. 87-3 (EPA Nov. 16, 1989).

The procedures for assessing a civil penalty are set forth in EPA's Consolidated Rules of Practice Governing the Administrative Assessment of Civil Penalties and the Revocation of Suspension of Permits. 40 C.F.R. pt. 22. To initiate a civil penalty proceeding, EPA must first issue a complaint setting forth, among other things, the factual basis for the complaint and the amount of the penalty proposed. 40 C.F.R. § 22.14(a). As in a civil trial, the alleged violator (respondent) responds to the complaint by filing an answer admitting or denying the charges, 40 C.F.R. § 22.15, and may file motions regarding questions of law and procedure. 40 C.F.R. § 22.16. If the respondent admits the allegations or the allegations are proved at a hearing, the administrative law judge will assess the penalty. 40 C.F.R. § 22.27(b). The assessment is subject to review by EPA's Chief Judicial Officer, who may lower or raise the penalty, within prescribed guidelines. 40 C.F.R. § 22.31. Of course, the parties may settle the case at any time during the proceedings, and agree on the amount of civil penalty, if any, to be assessed.

In determining the amount of a civil penalty, EPA must consider the appropriateness of the penalty to the size of the business, the effect of the penalty on the person's ability to continue in business, and the gravity of the violation. FIFRA § 14(a)(4). The Agency must explain how these factors were considered when the penalty was calculated. *Katzson Bros., Inc. v. EPA*, 839 F.2d 1396 (10th Cir. 1988). See also the discussion below on the FIFRA Enforcement Response Policy. Instead of assessing a penalty, EPA may issue a warning if the violator exercised due care or did not cause significant harm to health or the environment. See discussion on the FIFRA Enforcement Response Policy below.

In *Helena Chemical Co.*, No. FIFRA-09-0439-C-85-18 (EPA June 15, 1987), *rev'd*, FIFRA App. No. 87-3 (EPA Nov. 16, 1989), EPA clarified the Agency's position regarding: (1) when violations are to be treated as multiple counts; and (2) which party bears the burden of coming forward with evidence on each of the criteria considered in determining the penalty.

Helena was charged with, among other things, twenty counts of unlawful sale of a restricted use pesticide between February and August 1984. The Agency proposed a civil penalty of $135,800. The administrative law judge found that the facts underlying the twenty counts were proven and constituted violations of the Act, but concluded that the Agency erred in treating each sale as a separate violation. The ALJ thus assessed a civil penalty of $12,400.

On appeal, the chief judicial officer (CJO) reversed and held that each sale was a separately prosecutable offense. Thus, in the absence of proof to support Helena's argument that it had engaged in a single sale with twenty deliveries, each delivery was properly treated as a separate violation, and EPA could assess twenty penalties. The CJO thus reviewed the penalty assessed by the ALJ, and increased the penalty to $117,400.

The CJO also examined the burdens of the parties in proving — or disproving — its appropriateness. In reviewing the penalty, FIFRA provides that penalties should be determined taking into account "the size of the business of the person charged, the effect on the person's ability to continue in business, and the gravity of the violation." FIFRA § 14(a)(4). As to the first and third factors, the CJO imposed on the Agency the burden of coming forward with the evidence and the burden of persuasion. Regarding the second criterion — the effect on the person's ability to continue in business — the CJO held that the burden of coming forward and the burden of persuasion must be borne by the respondent. EPA's civil penalty guidelines assume that the penalty will not affect the ability of the person charged to continue in business, and thereby impose on the respondent these burdens.[55]

Civil Penalty Calculation — FIFRA Enforcement Response Policy. The Office of Compliance Monitoring issued a FIFRA Enforcement Policy on July 2, 1990. This supersedes the old FIFRA Civil Penalty Assessment Guidelines published at 39 Fed. Reg.

---

55 The CJO was interpreting EPA's 1974 FIFRA civil penalty guidelines. 39 Fed. Reg. 27,712 (1974). The 1990 FIFRA Enforcement Response Policy, however, makes the same assumption. Enforcement Response Policy at 24.

27,711 (1974) and, in general, continues the trend toward tougher sanctions for environmental violations.

The Enforcement Policy establishes a five-step process for calculating civil penalties. In brief, the Agency must (1) determine the gravity of the violation, in accordance with a standardized table of gravity levels; (2) determine the size of the violator's business according to a list of size categories; (3) determine according to standardized matrices a "gravity-based penalty," based on the calculated gravity level and business size categories; (4) adjust the penalty according to "Gravity Adjustment Criteria" listed in the Enforcement Policy; and (5) consider the effect that payment of the total civil penalty will have on the violator's ability to continue in business. Enforcement Policy at 18.

Under the Enforcement Policy, the violator's ability to pay generally establishes a limit on the penalty amount, i.e., EPA generally will not seek a penalty that might cause a violator to go bankrupt or otherwise be unable to continue in business. Nevertheless, EPA specifically reserves the right to impose penalties of this magnitude under certain circumstances. In such instances, EPA requires additional internal review of the penalty. Enforcement Policy at 24.

The Enforcement Policy encourages violators to disclose violations voluntarily. A civil penalty will be reduced by 40 percent if: (1) the disclosure is made promptly to EPA or to a state with a cooperative enforcement agreement with EPA;[56] (2) the disclosure is made before the violation is discovered by EPA or a state; (3) prior to the disclosure EPA or the state has not announced that it will inspect the facility in question; and (4) the violator immediately takes all steps necessary to comply with FIFRA, and takes steps requested by the Agency to mitigate the violation. Enforcement Policy at 26.

The Enforcement Policy also identifies circumstances which may justify reducing a proposed civil penalty to settle a civil enforcement action. Enforcement Policy at 27. In

---

[56] Generally, "promptly" means "within 30 to 60 days of discovery by the violator," according to the Enforcement Policy. The Policy does not explain what is meant by "within 30 to 60 days."

addition, EPA may enter into a Settlement with Conditions (SWC), under which the Agency will reduce a portion of the proposed penalty in return for performance by the respondent of a specific "environmentally beneficial" activity. Enforcement Policy at 28.

The Enforcement Policy is not a regulation, and the penalties proposed under the Policy are not binding on an ALJ in a civil penalty proceeding. Nevertheless, an ALJ must consider the Enforcement Policy when assessing a penalty. 40 C.F.R. § 22.35. Under the Consolidated Rules of Practice, an ALJ may disregard EPA's proposed penalty calculation but must explain fully the reasons for doing so. 40 C.F.R. § 22.27(b). Notwithstanding the Enforcement Response Policy, EPA has the burden of proving that any proposed civil penalty is appropriate. 40 C.F.R. § 22.24. Mere reliance on the Policy may not be sufficient to discharge this burden.

Criminal Penalties. A criminal penalty may be imposed under FIFRA § 14(b) for a "knowing" violation of the Act. EPA has contended, with one known success to date, that knowledge of the conduct is all that is required to constitute the violation and that the person charged need not have intended to commit a violation of FIFRA. *United States v. Corbin Farm Service*, 444 F. Supp. 510 (E.D. Cal.), *aff'd*, 578 F.2d 259 (9th Cir. 1978). A violation committed by an employee may also in some circumstances be considered a violation by the employer.

The penalty for a criminal violation of FIFRA may include a fine of up to $50,000 or imprisonment for up to one year, or both.[57] The invocation of other federal statutes, however, may result in higher penalties. For example, under the Alternative Fines Act, if any person derives pecuniary gain from the offense, or if the offense causes any person other than the defendant to suffer a pecuniary loss, a court can fine the defendant not more than the greater of twice the gross gain or gross loss. In addition, prosecutors may bring felony

[57] In a criminal prosecution, felony charges also may be brought under numerous other federal statutes, if appropriate, including for example, false statements, mail and wire fraud, obstruction of justice and similar charges.

charges where they apply under the general federal criminal conspiracy laws or statutes that prohibit use of false documents.

Chapter 16
REGULATION OF PESTICIDE CONTAINERS, STORAGE & DISPOSAL

The 1988 amendments to FIFRA § 19 empower EPA to establish a comprehensive regulatory program for the storage, transportation and disposal of pesticides and their containers, rinsates, and other material used to contain or collect excess or spillage of pesticides.[58] In addition, the 1988 amendments authorized EPA to adopt additional requirements for the storage, transportation and disposal of pesticides which have been suspended or canceled and, for the first time, to order the recall of these pesticides.

EVOLUTION OF FIFRA § 19

Even before the 1988 amendments to FIFRA, EPA had authority under FIFRA to regulate unused pesticides, pesticide residues and containers. The Act directed the Administrator to

> establish procedures and regulations for the disposal or storage of packages and containers of pesticides and for disposal or storage of excess amounts of such pesticides, and accept at convenient locations for safe disposal a pesticide the registration which is canceled under section [6(c)] of this title if requested by the owner of the pesticide.

FIFRA § 19(a), *as added by* Pub. L. No. 92-516, § 19(a), 86 Stat. 973, 995 (1972) (prior to 1988 amendment). Under this provision, EPA was solely responsible for receiving, storing and disposing of pesticides that were suspended and canceled.

In 1974, EPA promulgated regulations which set forth the requirements for acceptance of canceled pesticides. 40 C.F.R. part 165. Under these regulations, EPA was required to accept suspended and canceled pesticides upon request, dispose of them at

[58] This chapter does not address EPA's authority to regulate pesticide containers, storage and disposal under other federal statutes such as the Clean Air Act, the Clean Water Act, the Resource Conservation and Recovery Act or the Comprehensive Environmental Response, Compensation, and Liability Act.

government expense, and reimburse the purchase price to owners and end users. 40 C.F.R. § 165.3-.6. In addition, EPA established recommended procedures for the storage and disposal of "excess" pesticides and pesticide containers. 40 C.F.R. § 165.7-.10. Because of uncertainties about EPA's enforcement authority, these storage and disposal procedures were issued as guidelines for the public and made mandatory only for the Agency in carrying out its pesticide and container storage and disposal operations. 40 C.F.R. § 165.2.

These procedures (which basically require triple rinsing) were bolstered by differing storage and disposal instructions on specific pesticide labels. To achieve uniformity, EPA issued a Label Improvement Notice in 1983 requiring all pesticide registrants to amend their labels to set forth uniform storage and disposal requirements for specific types of containers. Pesticide Registration (PR) Notice 83-3 (Mar. 29, 1983). These requirements became effective December 31, 1984, for all pesticide containers except those for household/domestic use products, which became effective December 31, 1985. On September 26, 1984, EPA published proposed revised pesticide labeling regulations which set forth virtually identical requirements as those found in the Label Improvement Notice. 49 Fed. Reg. 37,986 (1984). Although EPA eventually plans to finalize these regulations, the final regulations may differ significantly from the proposed regulations as a result of EPA's activities in response to the 1988 amendments to FIFRA, as discussed below. 54 Fed. Reg. 18,084 (1989).

As a result of the suspensions and cancellations of Silvex, 2,4,5-T, EDB and dinoseb, EPA received large quantities of materials for storage and disposal which resulted in significant costs to the Agency. Thus, in 1988, FIFRA § 19 was amended to empower EPA to establish requirements and procedures for the storage, transportation and disposal of pesticides, containers, rinsates and other material used to contain or collect excess or spillage of pesticides. EPA is authorized to require registrants or applicants for registration to submit data regarding safe methods for storage and disposal of pesticides to support the registration or continued registration of a pesticide pursuant to FIFRA § 3. FIFRA § 19(a)(1)(A). EPA also may require labeling statements regarding procedures for the transportation, storage and disposal of the pesticide, its container, any residues, or any other materials used to contain or collect excess or spilled quantities of the pesticide. FIFRA § 19(a)(1)(B).

In addition, EPA may require a registrant to demonstrate financial resources sufficient to carry out a recall plan and provide for the disposal of a pesticide in the event of its suspension and cancellation. FIFRA § 19(a)(1)(C). EPA may use its authority pursuant to FIFRA § 3 and 6 to require registrants to submit such data for pesticides that are already registered. It is not clear whether these storage and disposal data requirements are applicable to all pesticides, only pesticides undergoing reregistration, or only those on a case-by-case basis.

The 1988 amendments also authorized EPA to shift the responsibility for storage, transportation and disposal of suspended or canceled pesticides to the registrant. EPA may accomplish this shift by establishing requirements — by regulation or through a FIFRA § 6 order — for the storage, transportation and disposal of suspended or canceled pesticides. FIFRA § 19(a)(2). Until these regulations are promulgated, the recommended procedures issued by EPA in 1974 pursuant to FIFRA § 19(a) for disposal of various types of depleted pesticide containers are still in effect.

EPA plans to establish these requirements by providing for the cradle to grave management of pesticides and their containers, rinsates and residues through revisions to the existing regulations found at 40 C.F.R. part 165. 54 Fed. Reg. 18,084 (1989); see figure 16-1 at the end of this chapter. In EPA's Executive Summary for Proposed 40 CFR Part 165 (Oct. 1990) (Executive Summary), EPA states that it plans to issue a "standards" package containing container, storage, disposal and transportation requirements for all pesticides and a "procedural" package that contains additional requirements for recalled pesticides.

STORAGE AND TRANSPORTATION REQUIREMENTS

EPA indicates that the proposed regulations will establish minimum standards for all facilities that store more than 11,000 pounds of pesticide products, or that store any quantity of recalled pesticides. Exceptions will be made for: (1) storage by private applicators for less than sixty days; (2) retail displays; and (3) temporary storage in a vehicle before unloading. EPA is choosing the 11,000 pound trigger in order to avoid application of the requirements to most farmers storing pesticides. Executive Summary at 6.

Pesticide storage areas will be required to meet minimum standards for security, construction, fire protection, spill prevention and cleanup, and container management and inspection. Facilities established after January 1, 1994, will be required to meet minimum siting and construction requirements. Existing facilities immediately will be required to ensure protection of water supplies. The regulations will specify training for employees who handle pesticides, require the posting of warning signs, and require information on worker safety and emergency procedures. Executive Summary at 6.

Finally, EPA plans to work with the Department of Transportation (DOT) to develop transportation requirements in those areas that are not addressed by DOT. These will include requirements for the transportation of bulk and mini-bulk containers. Executive Summary at 7.

PESTICIDE CONTAINER DESIGN

Prior to the 1988 amendments, EPA's authority to regulate the design and manufacture of pesticide containers was limited to setting standards under FIFRA § 25 to protect children and adults from accidental ingestion or contact with pesticides. FIFRA § 25(c)(3); 40 C.F.R. pt. 157. In 1988, however, FIFRA § 19 was amended to require EPA to establish regulations for the design of pesticide containers that will promote safe storage and disposal. FIFRA § 19(e). These regulations must ensure that the container designs facilitate: (1) pesticide removal and rinsing; (2) safe use of the containers, including elimination of splash and leakage of pesticides; (3) safe disposal of the containers; and (4) safe refill and reuse of the containers. EPA is required to promulgate these regulations by December 24, 1991, and they must take effect no later than December 24, 1993.

In its Executive Summary, EPA states that the rules on containers will deal separately with nonrefillable and refillable containers, and will focus on performance standards rather than on specific design requirements. With respect to refillable containers, EPA plans to encourage the use of closed systems in order to reduce mixer/loader exposure by proposing two standard container closure sizes for agricultural pesticides. To promote waste minimization and prevent cross-contamination, EPA intends to require tracking of containers, record keeping for maintenance and cleaning operations, and retirement of damaged or old containers. For nonrefillable containers, EPA plans to encourage the use of recyclable plastics and thereby reduce the currently widely used practice of burning pesticide containers. Executive Summary at 5.

Pesticide Container Residue Removal

By December 24, 1991, EPA also is required to promulgate regulations establishing procedures and standards for the removal of pesticide residues from containers prior to their disposal. The regulations may provide for: (1) triple rinsing or the equivalent degree of pesticide removal; (2) procedures that can be implemented promptly and easily; (3) reuse or disposal of rinse water and residue; and (4) coordination with container rinsing requirements imposed under the Resource Conservation and Recovery Act (RCRA). FIFRA § 19(f)(1).

Effective December 24, 1993, a state may not exercise primary enforcement responsibility under FIFRA § 26, or certify an applicator under FIFRA § 11, unless EPA first determines that the state is carrying out a program to ensure compliance with these regulations. FIFRA § 19(f)(2).

In its Executive Summary, EPA states that the proposed regulations will set a performance standard for all products, including household products, for the active ingredient residue that may remain after a registrant-prescribed rinsing procedure. The standard would be phased in over a three-year period. The Agency believes that registrants of household products should be able to easily attain the proposed standards since these products generally have low active ingredient concentrations. Executive Summary at 5.

Pesticide Container Study

EPA is required to conduct a study of options to encourage or require: (1) the return, refill and reuse of pesticide containers; (2) the development and use of pesticide formulations that facilitate removal of residues from containers; and (3) the use of bulk storage facilities to reduce the number of pesticide containers requiring disposal. In conducting the study, EPA must consult with the heads of other federal and state agencies, industry groups and environmental organizations. In addition, the feasibility, costs and environmental benefits of the alternatives must be assessed. FIFRA § 19(g)(1). A report of the findings must be submitted to Congress by December 24, 1990. FIFRA § 19(g)(2).

The Agency takes a broad interpretation of the scope of the container study mandate, one that is not limited to scrutiny of agricultural and commercial formulations and associated packaging. EPA takes the position that formulations or packaging that promote the introduction of pesticide residues into the environment should be identified and eliminated and that safer designs should be identified and required. 54 Fed. Reg. at 18,084.

Management of Excess Pesticides and Rinsates

FIFRA § 19(a)(1)(B) authorizes EPA to adopt label requirements for the safe storage, transportation and disposal of excess pesticides, rinsates or residues. Thus, EPA plans to establish label requirements regarding the proper and prohibited methods for residue removal, handling and disposal. These may include requirements for reuse of excess pesticides and for rinse pads to prevent soil and ground water contamination. Executive Summary at 6.

ADDITIONAL REQUIREMENTS FOR RECALLED PESTICIDES

Before the 1988 amendments to FIFRA, EPA had no explicit authority to order the recall of a pesticide. Rather, the Agency relied on manufacturers to voluntarily recall such products. Registrants did so knowing that, at their request, EPA was required to accept such products for storage and disposal and to indemnify them for the cost of such products.

For a discussion of FIFRA's indemnification provisions, see Chapter 8. With the prospect of continuing suspensions and cancellations and restrictions on indemnity, the incentives for voluntary recalls were reduced.

As a result, FIFRA § 19(b) was amended to give EPA the express authority to require the recall of a pesticide that has been suspended *and* canceled. Products whose registrations are suspended but not canceled or canceled without first being suspended are not eligible for recall. FIFRA § 19(b)(1) directs EPA to recall a suspended and canceled pesticide when "the Administrator finds that recall of the pesticide is necessary to protect health or the environment." After determining that a recall is necessary, EPA will request that the registrant of the pesticide submit a voluntary recall plan, if EPA determines that a "voluntary recall by the registrant and others in the chain of distribution may be as safe and effective as a mandatory recall." FIFRA § 19(b)(2). The registrant then has sixty days to submit a recall plan. EPA must then approve the plan "unless the Administrator determines, after an informal hearing, that the plan is inadequate to protect health or the environment." FIFRA § 19(b)(2).

If EPA either does not request a voluntary recall plan or determines that such a plan is inadequate, it must issue a mandatory recall plan in the form of a regulation, which "may apply to any person who is or was a registrant, distributor, or seller of the pesticide, or any successor in interest to such a person." FIFRA § 19(b)(3). Both voluntary and mandatory recall plans must include the level in the distribution chain to which the recall will extend, and a method of verifying the effectiveness of the recall. FIFRA § 19(b)(5).

Those subject to a mandatory recall regulation may be required to: (1) make storage facilities available for the recalled pesticide, and inform EPA of their location; (2) accept the recalled pesticide from holders of the pesticide (those further down the chain of distribution), and store the pesticide in the facility; (3) transport the recalled pesticides to the facility when asked to by a holder of the pesticide; and (4) inform holders of the pesticide of the recall terms and how they may arrange for transport of the pesticide to the storage facility. FIFRA § 19(b)(4).

EPA has indicated that its proposed regulations will provide for recalls to be accomplished through generic, rather than product-specific, rulemaking. EPA believes that generic rules will provide the regulated community with a fairly complete picture of what will be required during a recall and avoid assertions that the Agency is inconsistently applying its recall authority. Executive Summary at 7. The regulations will define those criteria that must be included in a storage and disposal plan in order to receive EPA approval.

The recall regulations will impose additional storage requirements on those facilities storing large quantities of recalled pesticides for an extended period of time. These regulations will be modeled after regulations issued pursuant to RCRA in order to avoid inconsistent requirements. Executive Summary at 7. Under these requirements, a facility must: (1) not be located in earthquake zones or flood plains; (2) not use uncovered construction; (3) have temperature and humidity controls; (4) provide for additional safety training and planning; (5) have upgraded inspection procedures; and (6) provide for facility decontamination (if necessary) after disposal is finished. Executive Summary at 8.[59]

Manufacturers that encounter cancellations will find it advantageous to propose voluntary recalls on terms they consider appropriate, rather than await the imposition of more stringent or expensive mandatory recall measures by the Agency. Recall measures are inextricably linked with such issues as indemnification for and the sale and use of existing stocks, and should be discussed with EPA in these contexts. Early submission of a recall plan, moreover, will enhance the recovery from EPA of storage costs for the recalled pesticide, as discussed below.

Under FIFRA § 19(c), EPA must reimburse a registrant for storage costs, but not transportation or disposal costs, incurred as a result of a voluntary or mandatory recall, provided that storage is effected pursuant to a plan submitted by the registrant and

---

[59] EPA also is reviewing whether recalled pesticides may be regulated under both RCRA and FIFRA. The Agency is exploring two alternatives for exempting EPA-recalled pesticides from RCRA requirements: (1) exempt these products under RCRA § 1006(b); or (2) provide a period of time before application of RCRA during which a recall may be completed and a registrant may explore options for material recovery. Executive Summary at 7.

approved by the Agency under criteria established by regulation. Reimbursement is not available for storage costs incurred by the registrant before the date the storage and disposal plan is submitted. Reimbursement is available for:

- 100% of the costs incurred after the date an acceptable plan is submitted or the date the registration is canceled, whichever is later, but before the plan is approved by EPA;
- 50% of the costs incurred during the one-year period beginning on the date the plan is approved or the registration is canceled, whichever is later;
- 25% of the costs incurred during the period beginning on the first day of the fifth year after the plan is approved or the date the registration is canceled, whichever is later, and ending on the date that a disposal plan for the pesticide can be implemented;
- None of the costs incurred during the three-year period beginning on the 366th day after the plan is approved or the registration is canceled, whichever is later; and
- None of the costs incurred before the date of submission of the plan to EPA.

FIFRA § 19(c)(2). Although the provisions for reimbursement of storage costs are self-implementing, EPA plans to issue regulations that will identify the criteria and procedures to be followed in order for a registrant to receive reimbursement. 54 Fed. Reg. 18,085 (1989); Executive Summary at 8.

FIGURE 16-1

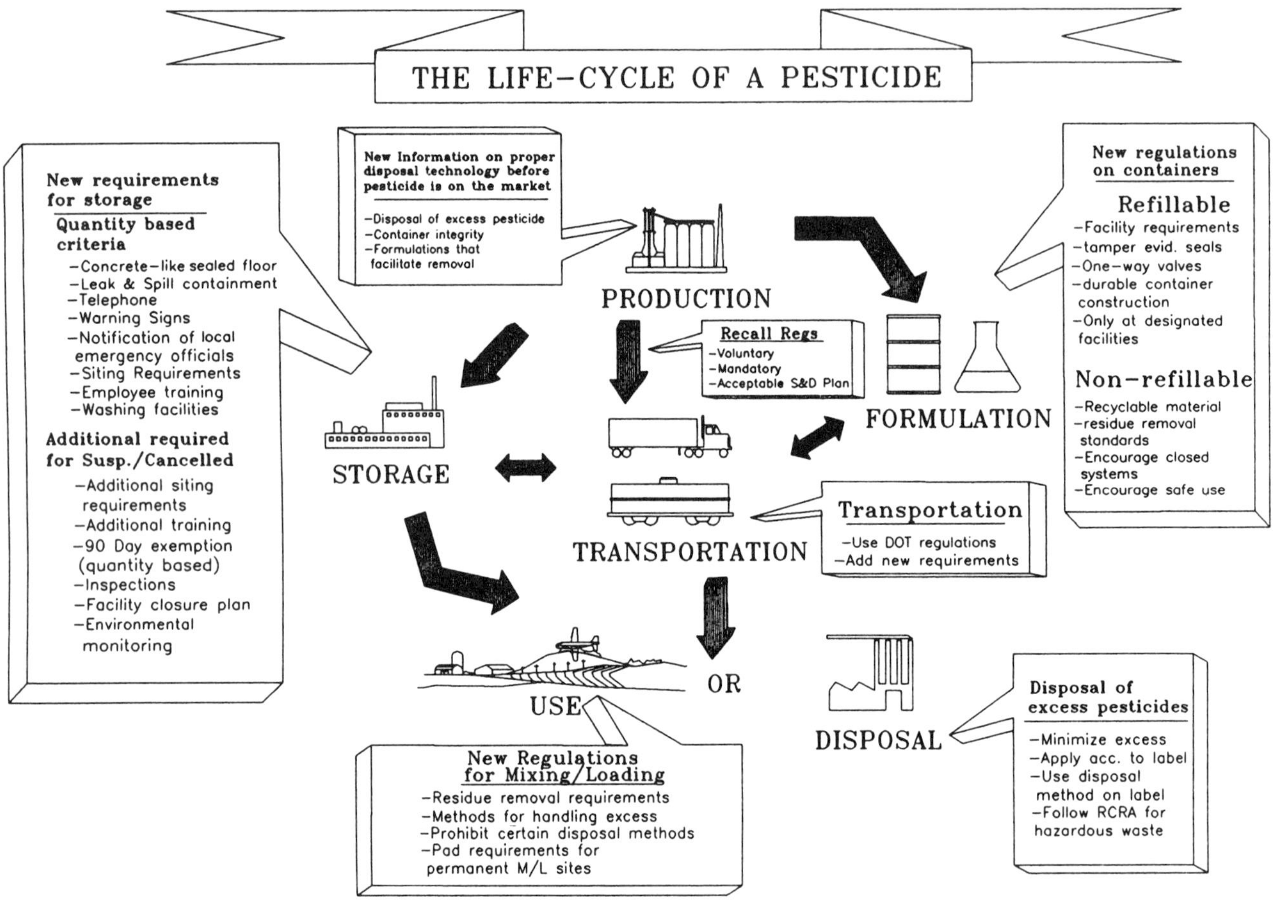
THE LIFE-CYCLE OF A PESTICIDE
New requirements for storage
Quantity based criteria
-Concrete-like sealed floor
-Leak & Spill containment
-Telephone
-Warning Signs
-Notification of local emergency officials
-Siting Requirements
-Employee training
-Washing facilities
Additional required for Susp./Cancelled
-Additional siting requirements
-Additional training
-90 Day exemption (quantity based)
-Inspections
-Facility closure plan
-Environmental monitoring
New Information on proper disposal technology before pesticide is on the market
-Disposal of excess pesticide
-Container integrity
-Formulations that facilitate removal
PRODUCTION
Recall Regs
-Voluntary
-Mandatory
-Acceptable S&D Plan
New regulations on containers
Refillable
-Facility requirements
-tamper evid. seals
-One-way valves
-durable container construction
-Only at designated facilities
Non-refillable
-Recyclable material
-residue removal standards
-Encourage closed systems
-Encourage safe use
FORMULATION
STORAGE
TRANSPORTATION
Transportation
-Use DOT regulations
-Add new requirements
USE
OR
DISPOSAL
New Regulations for Mixing/Loading
-Residue removal requirements
-Methods for handling excess
-Prohibit certain disposal methods
-Pad requirements for permanent M/L sites
Disposal of excess pesticides
-Minimize excess
-Apply acc. to label
-Use disposal method on label
-Follow RCRA for hazardous waste

Chapter 17
OTHER FEDERAL STATUTES REGULATING PESTICIDES

In addition to FIFRA, other federal statutes affect the manufacture, formulation, packaging, and transportation of pesticides. The Federal Food, Drug, and Cosmetic Act (FFDCA) imposes requirements on pesticides intended for use on natural or processed foods or as drugs, new drugs, or animal drugs.[60] The Toxic Substances Control Act (TSCA) regulates raw materials, inert ingredients, and intermediates used in the manufacture and formulation of pesticides. FIFRA § 25(c)(3) requires that EPA standards for child-proof packaging of pesticides be consistent with those established under the authority of the Poison Prevention Packaging Act (PPPA). While FIFRA gives EPA jurisdiction over pesticide use and users, the Occupational Safety and Health (OSH) Act authorizes the Occupational Safety and Health Administration (OSHA) to regulate the workplace where pesticides are manufactured, formulated, stored, or distributed. After leaving the workplace, pesticides that qualify as "hazardous material" must be transported in accordance with the provisions of the Hazardous Materials Transportation Act (HMTA) and its implementing regulations.

FEDERAL FOOD, DRUG, AND COSMETIC ACT (FFDCA), 21 U.S.C. §§ 301-392

Prior to 1938, FFDCA did not require that drugs be proved safe or effective prior to marketing. As amended in 1938, FFDCA defined the term "new drug" and required premarket approval of new drugs for safety. In 1962, Congress further amended FFDCA to require that FDA also approve new drugs for effectiveness. Since 1964, "new drugs" approved for safety between 1938 and 1962 have been reviewed for effectiveness under a program referred to as the Drug Efficacy Study Implementation (DESI). In 1968, Congress again amended FFDCA to require premarket new animal drug approval (NADA).

---

60 Tolerances under FFDCA for pesticide use on crops, food, and feed are discussed in Chapter 5.

Certain pesticide products fall within the statutory definition of "drug" under FFDCA § 201(g)(1), "new drug" under FFDCA § 201(p), or "new animal drug" under FFDCA § 201(w), and as such are subject to the jurisdiction of the FDA. Without special relief, therefore, manufacturers of pesticide products which are "drugs," "new drugs," or "new animal drugs" would also be subject to FDA review and approval prior to marketing. The 1975 FIFRA amendments provided relief for manufacturers of new animal drugs by exempting these drugs from the definition of pesticide under FIFRA § 2(u), leaving them subject only to the FFDCA. As for pesticides which are also drugs or new drugs, however, Congress left EPA and FDA to coordinate their dual jurisdiction.

For those products that fall within the jurisdiction of both FFDCA and FIFRA, EPA and FDA signed a Memorandum of Agreement (MOA) describing the procedures to be followed in the joint review and approval of these products. The MOA specifies: (1) which agency has primary jurisdiction; (2) that the product will be referred to the other agency for review under its law; and (3) that approval for marketing the product will not be granted unless both agencies approve the product under their respective authorities. 36 Fed. Reg. 24,234, 24,235 (1971). This agreement was amended in September 1973. 38 Fed. Reg. 24,233 (1973).

Pesticides that Are New Drugs for Human Use

FFDCA requirements for safety, effectiveness, and other information concerning new drugs are set forth in FFDCA § 505 and 21 C.F.R. part 314. A manufacturer of a pesticide which is also a new drug for human use must obtain FDA approval of a new drug application (NDA) in accordance with FFDCA § 505(b). Like an application for registration under FIFRA, an NDA is comprehensive. It must provide: (1) full reports of investigations to show that the drug is safe and effective in use; (2) the identity of the drug's components; (3) a full statement of the drug's composition: (4) the methods used in, and facilities and controls used for, the manufacture, processing, and packaging of the drug; (5) samples of the drug and drug components; and (6) specimens of the drug's label.

Because the criteria for NDAs are comparable to the registration criteria for pesticides, EPA and FDA concluded that the dual review of pesticides which are new drug products offered solely for human use is wasteful. Hence, the two agencies decided that regulation of these products solely by FDA under the FFDCA would serve the intent of FIFRA adequately. 44 Fed. Reg. 63,749 (1979). Consequently, under authority of FIFRA § 25(b), EPA exempted from the requirements of FIFRA any pesticide that is offered solely for human use which is also a "new drug," as defined by FFDCA § 201(p). 40 C.F.R. § 152.20(b).

The case of Kwell, a lindane head treatment for lice in humans, illustrates the interplay of the procedures which affect pesticides that are also new drugs. Lice are parasitic insects that infest the hair and skin of man and warm-blooded animals. A product which controls human lice is by definition a pesticide under FIFRA and, in the case of Kwell, a "new drug" under FFDCA. In accordance with the agencies' agreement, the registrants of Kwell voluntarily cancelled their FIFRA registrations. 45 Fed. Reg. 40,221 (1980). This product will continue to be marketed as a shampoo, lotion, and cream for human use on the basis that it is adequately regulated for such human use under new drug approval procedures of the FFDCA, specifically 21 C.F.R. parts 310, 312 and 314.

Pesticides that Are Drugs for Human Use (Requiring No NDA)

Pesticides which are drugs for human use, but not "new drugs" within the meaning of FFDCA § 201(p), are not exempt from FIFRA registration.[61] 40 C.F.R. § 152.20(b). Because these "old" human drug products did not require an NDA prior to marketing, EPA will continue to regulate them under FIFRA unless and until FDA issues specific final regulations for these drug products pertaining to the conditions under which they are generally recognized as safe (GRAS) and effective and not misbranded. If FDA

[61] Note that pre-1938 "old" drugs can become "new" if changes (for example, in their intended use) are made.

promulgates these regulations, EPA will exempt these products from FIFRA. 43 Fed. Reg. 47,215, 47,216 (1978).

This contemplated FIFRA exemption for pesticides governed by FDA GRAS regulations will not apply to pesticides/human drug products offered concurrently for use on both humans and inanimate objects (but not on animals). For example, a disinfectant solution would continue to be subject to FIFRA even though it is also offered for use on humans. Because FIFRA provides for premarket control of disinfectants whereas the FFDCA does not (because they are "old" drugs), both agencies agree that these pesticides/human drug products should continue to be subject to EPA regulation under FIFRA. 43 Fed. Reg. at 47,215.

Prior to granting FIFRA registration for a pesticide making human drug claims, however, EPA will refer the application to FDA for review of the labeling and formulation accompanying the EPA submission. If the pesticide is a "new drug" requiring an NDA, EPA will turn jurisdiction over to FDA pursuant to 40 C.F.R. § 152.20(b). If the pesticide is an "old" drug, EPA will regulate it under FIFRA once FDA notifies EPA that it complies with FFDCA. 43 Fed. Reg. at 47,216.

Pesticides that Are Animal Drugs

In 1975, Congress amended FIFRA to exempt new animal drugs from the definition of "pesticide." Specifically, FIFRA § 2(u) provides that the term pesticide shall not include any of the following: (1) a "new animal drug" within the meaning of FFDCA § 201(w), (2) a product determined by FDA not to be a new animal drug by regulation establishing conditions of use, or (3) an animal feed within the meaning of FFDCA § 201(x) which bears or contains a new animal drug. As a result of this amendment, FDA now has sole authority to regulate new animal drugs which are also pesticides. FFDCA requirements for safety, effectiveness and other information concerning new animal drugs are set forth in FFDCA § 512 and 21 C.F.R. part 514.

If a pesticide chemical is used on animal feed, as defined in FFDCA § 201(x), and is not a new animal drug, the animal feed is regulated as a food and the pesticide chemical must be within a tolerance or exempt from a tolerance under FFDCA § 408 or § 409. If the pesticide chemical is a food additive for animal feed, any animal feed containing such pesticide would be subject to the provisions of the Delaney Clause of FFDCA § 409, unless the animal feed is regulated solely under FFDCA § 408 as a raw agricultural commodity (see Chapter 5).

Applications for registration of pesticides with animal drug claims that are submitted to EPA will be referred to FDA before EPA grants registration. Upon referral, FDA will review the labeling and formula accompanying the pesticide registration application and determine whether the animal drug complies with the FFDCA and whether it is or is not "new." If the product is a new animal drug, FDA will so notify EPA; thereafter, FDA will exercise its exclusive jurisdiction under the FFDCA and EPA will exempt it from FIFRA. Conversely, if the pesticide is an "old" animal drug, EPA has jurisdiction and will regulate it under FIFRA once FDA notifies EPA that it complies with the FFDCA. 43 Fed. Reg. at 47,216.

In May 1983, FDA and EPA signed a Memorandum of Understanding (MOU) which is intended to supersede portions of the 1971 MOA with respect to pesticide/drug products intended for use in and/or on animals. 48 Fed. Reg. 22,799 (1983). On August 16, 1983, however, the effective date of this MOU was stayed indefinitely. 48 Fed. Reg. 37,077. The MOU provides that the following products are not drugs, but pesticides that are subject solely to FIFRA: topical treatments to food-producing animals administered only for control of external parasites (such as fleas, ticks, lice, mites, screw-worms, and wool maggots); aquatic treatments solely for control of algae and bacterial slime in ponds or aquariums; and sanitizers for treating aquarium equipment, inanimate surfaces and/or drinking water of animals (if the pesticide makes no claims of disease control). 48 Fed. Reg. 22,799 at 22,800.

Conversely, the MOU specifies that the following products are drugs and regulated solely under the FFDCA: products for oral administration, such as tablets, drinking water preparations (except animal drinking water sanitizers), and medicated feed; parenterally administered products; topically applied treatments for internal parasites; and products applied to poultry eggs used for incubation. 48 Fed. Reg. at 22,800.

For pesticide/animal drug products subject to dual jurisdiction, FDA may require that the product be labeled to indicate that the drug is restricted to use by a licensed veterinarian. If FDA makes such a determination, EPA will relinquish jurisdiction to FDA by rulemaking under FIFRA § 25(b). FDA already has determined that adequate directions for diagnosis by lay persons cannot be written for products intended as treatments for mange in cats, dogs, and horses. This proposed shift of jurisdiction over certain veterinary products from EPA to FDA reflects the basic thrust of the MOU.

TOXIC SUBSTANCES CONTROL ACT (TSCA),[62] 15 U.S.C. §§ 2601-2629

In 1971, the Council on Environmental Quality (CEQ) stated that the legal mechanisms for protecting the environment against chemical hazards were neither cohesive nor adequate and that there was a "high-priority need for a program of testing and control of toxic substances."[63] To rectify this situation, Congress enacted TSCA on October 11, 1976. In so doing, Congress intended that manufacturers assume responsibility for providing data on the health and environmental effects of chemical substances and mixtures, and that the Administrator of EPA have adequate authority to regulate those chemical substances and mixtures that "present an unreasonable risk of injury to health or the environment" or constitute "imminent hazards." TSCA § 2(b). Congress also intended that TSCA fill the gaps among the previously enacted environmental statutes, including FIFRA. S. Rep. No.

---

62 For a comprehensive discussion of requirements under TSCA, see McKenna & Cuneo's TSCA Handbook (2d ed. 1990)

63 Council on Environmental Quality, Toxic Substances (Apr. 1971), *reprinted in* Staff of House Comm. on Interstate and Foreign Commerce, 94th Cong., 2d Sess., Legislative History of the Toxic Substances Control Act (TSCA Legislative History) at 760 (Comm. Print 1976).

698, 94th Cong., 2d Sess. 5, *reprinted in* 1976 U.S. Code Cong. & Admin. News 4491, 4495; H.R. Rep. No. 1341, 94th Cong., 2d Sess. 6 (1976).

To achieve this latter end, TSCA gives the Administrator of EPA broad authority to regulate chemical substances and mixtures, including those used in manufacturing pesticides, unless and until they fall exclusively under the jurisdiction of FIFRA. As Senator Allen of the Senate Committee on Agriculture and Forestry explained in pre-enactment discussion of TSCA and its interrelationship with FIFRA:

> [A]ny chemical or toxic substance would first be subject to the provisions of [TSCA], and yet when it becomes a component of a pesticide, it would be subject to FIFRA. In many instances, the manufacturer of the component is also the manufacturer and registrant of the pesticide.

122 Cong. Rec. 8291 (1976) (statement of Sen. Allen), *reprinted in* TSCA Legislative History at 232.

With a few important exceptions (including pesticides), TSCA § 3(2) defines a "chemical substance" broadly to mean "any organic or inorganic substance of a particular molecular identity." TSCA § 8(b) requires the Administrator to create and keep current an inventory of chemical substances (Inventory) manufactured, imported, or processed in the United States for a commercial purpose. TSCA then divides the chemical universe between existing chemical substances on the TSCA § 8(b) Inventory and "new chemical substances" not on the Inventory. The regulatory options available to EPA depend upon whether the particular chemical substance is classified as "new" or "existing."

TSCA's Pesticide Exclusion

TSCA § 3(2)(B)(ii) defines a "chemical substance" so as to exclude "any pesticide (as defined in the Federal Insecticide, Fungicide, and Rodenticide Act) when manufactured, processed, or distributed in commerce for use as a pesticide." Under FIFRA § 2(u), a pesticide is "(1) any substance or mixture of substances *intended* for preventing, destroying, repelling, or mitigating any pest, and (2) any substance or mixture of substances *intended* for use as a plant regulator, defoliant, or desiccant" (emphasis added). Thus, in order to be a pesticide and thereby exempt from regulation under TSCA, a chemical substance or mixture must satisfy a two-pronged test: (1) it must fall within the FIFRA definition of a pesticide, and (2) it must be intended for use as a pesticide.

The two seemingly distinct parts of the test can become blurred at times because FIFRA defines a pesticide in terms of its intended use. Once intent for use as a pesticide is established, the chemical is excluded from regulation under TSCA and is subject to regulation under FIFRA. Unless and until the requisite intent is established, however, the chemical remains subject to TSCA regulation.

A registered end-use pesticide is a pesticide by definition. A technical or manufacturing-use product is a pesticide under FIFRA if it is intended for use as a pesticide active ingredient. 40 C.F.R. § 152.15. Hence, active ingredients are pesticides under FIFRA even though they require further processing before use as pesticides. By contrast, raw materials and inert ingredients are not pesticides *until* they become components of a pesticide product. 42 Fed. Reg. 64,572, 64,586 (1977). Until then, they are subject to the jurisdiction of TSCA. After becoming a pesticide component, they are pesticides subject to FIFRA.

Intermediate substances, namely those "intended for the production of a pesticide product by chemical reaction with other substances," are not considered pesticides by EPA. 43 Fed. Reg. 9256 (1978); 42 Fed. Reg. 64,572, 64,586 (1977). Therefore, intermediates used in pesticide production generally are regulated under TSCA, not under FIFRA. An interesting jurisdictional question arose when a registered active ingredient was manufactured not just for reformulation into an end-use pesticide, but also for use as an intermediate in the manufacture of another product, in this case another pesticide product. A registered active ingredient is by definition a pesticide under FIFRA. When manufactured for use as a pesticide intermediate, however, it does not satisfy the second prong of the TSCA test for exclusion as a pesticide, namely, that it be manufactured for "use as a pesticide." Therefore, the intermediates, according to EPA, are subject to TSCA. Thus, if the pesticide active ingredient is not on the Inventory (the likely case), the manufacturer should file a premanufacture notification (PMN) before using it as an intermediate.

A jurisdictional status of a candidate pesticide may change as it undergoes research and development. EPA's regulations implementing the Experimental Use Permit (EUP) provisions of FIFRA § 5 provide that certain potential pesticides are not yet pesticides under FIFRA:

> A substance or mixture of substances being put through laboratory or greenhouse tests, or limited replicated field trials to confirm such tests, or other tests, in which the purpose is only to determine its value for pesticide purposes or to determine its toxicity or other properties, and from which the producer, applicator or any other person conducting the test does not expect to receive any benefit in pest control from its use, *is not considered a pesticide within the meaning of the Act* and no experimental use permit will be required.

40 C.F.R. § 172.3(a) (emphasis added). Under this regulatory standard, a product undergoing R&D to determine its pesticidal value is not a pesticide under FIFRA and, therefore, is subject to TSCA including TSCA § 8(e) requirements for notification of substantial risk information. 51 Fed. Reg. 15,096, 15,098 (1986). In other words, EPA considers that adverse test data on a new pesticide undergoing toxicity testing, limited field trials, or other R&D testing are reportable under TSCA § 8(e).

Even after its pesticidal value is determined, the product still is not considered a pesticide subject to FIFRA until the manufacturer intends to receive a "benefit in pest control" from its use. 51 Fed. Reg. at 15,098. Furthermore, EPA has established certain R&D categories within which it will presume that testing is intended solely to determine pesticidal value and not to obtain pesticidal benefit. For example, EPA permits limited field trials on a cumulative total of not more than ten acres of land or one surface acre of water without requiring an EUP. 40 C.F.R. § 172.3(a)(1), (2). These examples, moreover, are not intended to be exclusive or to preclude testing over larger areas if the purpose is only to determine the product's value for pesticidal purposes, its toxicity, or its other properties and not to receive any benefit in pest control. 40 C.F.R. § 172.3(b).

EPA considers submission of an application either for an EUP under FIFRA § 5 or for registration under FIFRA § 3 to be evidence that a chemical substance is "intended for use" as a pesticide within the meaning of FIFRA. 51 Fed. Reg. 15,096, 15,098 (1986); 42 Fed. Reg. 64,572, 64,585 (1977). Once the manufacturer files the EUP or registration application, jurisdiction over the candidate pesticide shifts from TSCA to FIFRA.

Premanufacture Notification Under TSCA § 5

Any company intending to manufacture or import a non-exempt, new chemical substance must give the Administrator ninety days' prior notice under TSCA § 5. This premanufacture notification (PMN) requirement enables the Administrator to evaluate and perhaps regulate a new chemical before its commercial production. If the new chemical presents an unreasonable risk or if there is insufficient information to make a risk assessment, the Administrator can prevent or limit its manufacture, processing, distribution, or disposal. The PMN review, however, is not a registration process. In contrast to the registration of pesticides under FIFRA § 3, TSCA § 5 does not authorize the Administrator to require the PMN submitter to supply premarket data on the new chemical. The PMN submitter instead must submit any existing test data on the new chemical in its "possession or control," but is not required to generate data to support its PMN. TSCA § 5(d)(1)(B).

TSCA § 5 provides a number of important exemptions from the PMN requirements. TSCA § 5(h)(3) exempts small quantities of a new chemical substance from the PMN requirements when used solely for R&D purposes so long as the manufacturer or importer notifies the persons engaged in R&D of any health risks the substance may pose. EPA's regulations implementing the TSCA § 5(h)(3) exemption require the manufacturer and importer of a new R&D chemical to supervise the R&D with a technically qualified individual, evaluate any potential risks associated with the R&D substance and maintain certain records of the R&D activity. Furthermore, if the R&D substance is distributed to other persons, the manufacturer or importer must notify them of the known hazards and of the statutory limitation that it be used solely for R&D purposes. 40 C.F.R. § 720.36.

As a matter of policy, however, EPA has expressly excluded pesticide R&D from even these limited risk assessment, notification, and recordkeeping requirements. The manufacturer or importer of small quantities of a new pesticide chemical intended solely for R&D, therefore, is not subject to these requirements "if the person's exclusive intention is to perform research and development activities solely for the purpose of determining whether the substance can be used as a pesticide." 40 C.F.R. § 720.36(g). Furthermore, a new chemical may qualify for this exclusion even if the potential properties of the chemical are unknown at the time it is first manufactured. 51 Fed. Reg. 15,096, 15,098 (1986).

A manufacturer can show an exclusive intent to develop a pesticide by having a patent for its use as a pesticide or simply by being a company or laboratory devoted entirely to pesticide R&D. When a company or laboratory has a broad range of marketing activities, EPA will look for other positive evidence of the prerequisite exclusive intent to produce a pesticide. Where the R&D activity on a new chemical is *not* solely to determine its pesticidal value, however, the risk assessment, notification, and recordkeeping requirements of the R&D rule apply. Curiously, EPA considers that new chemicals intended for use as intermediates or inerts in the production of a pesticide undergoing R&D remain subject to the risk evaluation, notification, and recordkeeping requirements of the R&D rule, even while the candidate pesticide itself does not. 51 Fed. Reg. at 15,099.

Under TSCA § 5(h)(4), EPA also exempts from PMN requirements any new chemical substance which is a "nonisolated intermediate." 40 C.F.R. § 720.30(h)(8). Intermediates are substances that are consumed in whole or in part in chemical reactions used for the intentional manufacture of other substances or mixtures, or that are intentionally present for the purpose of altering the rate of such reactions. 40 C.F.R. § 720.3(n). An intermediate is "nonisolated" if it is not intentionally removed from the equipment in which it is manufactured. 40 C.F.R. § 720.3(w). EPA chose to exempt nonisolated intermediates from premanufacture notification primarily because they are hard to identify and generally present little exposure and little human or environmental risk. 48 Fed. Reg. 21,722, 21,729 (1983).

Thus, nonisolated intermediates used in the production of pesticides are exempt altogether from PMN requirements.

Under TSCA § 5(h)(4), EPA also exempts new chemical byproducts, such as might result from the manufacture or formulation of pesticides. 40 C.F.R. § 720.30(g). EPA defines a byproduct to mean "a chemical substance produced without a separate commercial intent during the manufacture, processing, use, or disposal of another chemical substance or mixture." 40 C.F.R. § 720.3(d). Furthermore, the byproduct remains exempt from PMN requirements even if it has certain commercial purposes, such as for fuel. 40 C.F.R. § 720.30(g).

The PMN requirements currently apply to companies which manufacture new chemical substances or import them either in bulk or as part of a mixture. 40 C.F.R. § 720.22. Manufacture or importation of a new chemical substance which has not undergone the required PMN review is a violation of TSCA § 15(1). The manufacturer or importer is potentially liable for civil and/or criminal penalties under TSCA § 16, and is susceptible to specific enforcement and seizure under TSCA § 17. Furthermore, TSCA § 15(2) makes it unlawful for a company to "use for commercial purposes" a chemical which it "knew or had reason to know" was manufactured or imported in violation of the PMN requirements. Thus, any pesticide manufacturer or formulator which uses an imported or domestically manufactured chemical which it "knew or had reason to know" did not undergo the required PMN review may be violating TSCA § 15(2) and may be subject to civil and/or criminal penalties under TSCA § 16.

Reporting and Retention of Information Under TSCA § 8

Section 8 of TSCA establishes various reporting and recordkeeping requirements potentially applicable to companies which manufacture or formulate pesticides. These requirements are intended to provide EPA with data on which to base future regulatory and enforcement actions, and to track patterns of adverse reactions to chemicals. TSCA § 8 reporting and recordkeeping requirements are often overlooked. Violations typically are uncovered during routine facility inspections, and are actionable under TSCA §§ 15 and 16.

TSCA § 8(a) — Preliminary Assessment Information

TSCA § 8(a) authorizes the Administrator to promulgate rules on the reporting and retention of categories of information considered necessary to the effective enforcement of the Act. Although EPA has used this authority to impose requirements on specific chemicals, its most significant use has been to promulgate two model rules that apply to multiple chemicals. In 1982, EPA published the Preliminary Assessment Information Rule (PAIR) which requires chemical manufacturers (including certain producers and importers) of each listed chemical to submit information on the quantities manufactured, the amounts directed to certain classes of use, and the potential exposures and environmental releases. 40 C.F.R. part 712, subpart B. In 1985, EPA amended this rule to require automatic reporting of preliminary assessment information on all chemicals designated for testing by the Interagency Testing Committee (ITC) under TSCA § 4(e). 40 C.F.R. § 712.30.

The PAIR exempts manufacturers and importers of chemical substances used solely for purposes of scientific experimentation, analysis, or research, including research and analysis for product development. The rule also exempts "small manufacturers and importers" and manufacturers and importers of fewer than 500 kilograms (1,100 pounds) of a chemical substance at a single plant site during the reporting period. EPA considers a "small manufacturer or importer" to be one who during the reporting period has: (1) total annual parent company sales at all sites below $30 million; and (2) total production of the listed chemical below 45,400 kilograms (100,000 pounds). 40 C.F.R. § 712.25(c)(1), (2).

In 1988, EPA issued a second model recordkeeping and reporting rule, the Comprehensive Assessment Information Rule (CAIR). 53 Fed. Reg. 51,698 (1988) (codified at 40 C.F.R. part 704). The CAIR was published to provide a single form to assist EPA in gathering information on chemical substances. The major differences between the CAIR and PAIR are: (1) the CAIR requires processors as well as manufacturers and importers to provide information; (2) the CAIR reporting form is longer and more detailed than the PAIR form, and the CAIR only requires respondents to answer specifically designated questions; and (3) although both rules can be amended by notice and comment rulemaking, only the PAIR may be amended automatically when the ITC designates or recommends a substance for testing under TSCA § 4. EPA intends to amend the CAIR to allow the same automatic listing and at that time to withdraw the PAIR.

Under the CAIR, companies that manufacture, import or process a listed chemical substance are potentially subject to reporting requirements. 40 C.F.R. § 704.225(a). In the CAIR, EPA expanded the regulatory exemptions to the TSCA § 8(a) reporting rules to include companies that: (1) process or import a CAIR-listed substance solely as part of an article; (2) manufacture a CAIR-listed chemical solely as a non-isolated intermediate; (3) manufacture, import or process small quantities of the substance solely for R&D; (4) are small manufacturers or importers; and (5) only repackage the CAIR-listed substances, so long as they are not otherwise covered. 40 C.F.R. §§ 704.5, .210. In addition, companies which previously reported information to EPA or another federal agency voluntarily using a CAIR form may be partially exempt. 40 C.F.R. § 704.210(c).

TSCA § 8(c) — Significant Adverse Reactions

Section 8(c) of TSCA requires manufacturers, processors, and distributors to keep records of significant adverse reactions to health and the environment alleged to have been caused by the chemical substance or mixture. Such allegations may be made by employees, customers, or surrounding neighbors. Employee allegations must be maintained for thirty years; allegations by others for five years. The allegations are not required to be reported to EPA unless specifically requested by the Agency.

EPA's rule implementing TSCA § 8(c) applies to manufacturers of chemical substances and mixtures covered by TSCA. The rule exempts "processors" except those who combine chemicals to manufacture mixtures. 40 C.F.R. § 717.5(b). Importantly, the TSCA § 8(c) rule exempts firms and sites engaged solely in mining or extraction functions, sole distributors, and retail distributors. 40 C.F.R. § 717.7(a), (c), (d). The TSCA § 8(c) rule notes that multiple, significant adverse reaction reports on a single chemical may trigger TSCA § 8(e) substantial risk reporting requirements.

TSCA § 8(d) — Health and Safety Data Reporting

Section 8(d) of TSCA authorizes EPA to require any person who manufactures, processes, or distributes in commerce any chemical substance or mixture to submit to the Administrator lists and copies of health and safety studies (published or unpublished), whether conducted by or for that person, or known to or ascertainable by that person. In 1982, EPA promulgated the model Health and Safety Data Reporting Rule to implement TSCA § 8(d). The rule requires the submission of unpublished health and safety studies for chemicals or mixtures listed. 40 C.F.R. part 716. EPA uses information from these studies in making decisions under TSCA §§ 4, 5 and 6. In addition, all substances and mixtures either selected or recommended by the Interagency Testing Committee for priority testing consideration are automatically added to the list. Within thirty days after initiating a study of a listed chemical, or within sixty days after EPA adds a chemical to the 40 C.F.R. § 716.120 list, a manufacturer or processor must provide the Administrator with a list of the ongoing studies in its possession or known to it and must submit such studies within thirty days of their completion. 40 C.F.R § 716.60.

Exemptions from this reporting requirement include the following: (a) studies published in the scientific literature; (b) studies previously submitted to the Office of Toxic Substances; (c) studies previously submitted to other federal agencies (and not claimed as confidential); (d) studies conducted or initiated by another person subject to this rule; (e) studies of chemical substances which are not on the TSCA Inventory (i.e., R&D chemicals); (f) certain types of acute and primary studies; and (g) studies of chemical substances or mixtures previously submitted by trade associations. The TSCA § 8(d) rule protects information on confidential processes or quantitative mixture composition if the submitter asserts a claim at the time of submission. The claimant must designate the applicable confidential category under TSCA § 14(b) and also must provide a second, "sanitized" copy of the study in which the confidential information is deleted.

TSCA § 8(e) — Substantial Risk Notification

TSCA § 8(e), much like FIFRA § 6(a)(2), requires disclosure of information on the substantial risks of a chemical. TSCA § 8(e) is self-implementing and provides that a manufacturer, processor, or distributer of a chemical substance or mixture

> who obtains information which reasonably supports the conclusion that such substance or mixture presents a substantial risk of injury to health or the environment shall immediately inform the Administrator of such information unless such person has actual knowledge that the Administrator has been adequately informed of such information.

TSCA § 15 makes it unlawful for any person to fail or refuse to submit substantial risk information required under TSCA § 8(e), and TSCA § 16 provides that civil and criminal penalties may be assessed for violations of TSCA § 15.

Under EPA's enforcement policy and interpretation of TSCA § 8(e), the reporting obligation falls directly upon any "officers and employees of business organizations who are capable of appreciating the significance of pertinent information." Only "by establishing, internally publicizing, and affirmatively implementing procedures for employee submission and corporate processing of pertinent information" can a business relieve its officers and employees of individual responsibility for reporting substantial risk information directly to EPA. 43 Fed. Reg. 11,110, 11,111 (1978). Furthermore, EPA considers that the TSCA § 8(e) obligation applies to virtually all non-exempt chemicals, including catalysts, intermediates, and R&D chemicals, including candidate pesticides.

With the exception of emergency incidents of environmental contamination which must be submitted immediately, EPA permits fifteen working days within which to report substantial risk information under TSCA § 8(e). Under FIFRA § 6(a)(2), by contrast, the Agency allows a reasonable time, which is thirty days, unless a longer period can be demonstrated as reasonable under the circumstances. 44 Fed. Reg. 40,716, 40,717 (1979).

The Agency has proposed to conform the timing and format for reporting under FIFRA § 6(a)(2) with those under TSCA § 8(e). Neither TSCA § 8(e) nor FIFRA § 6(a)(2) requires the reporting of information already known to the Agency or published in scientific literature. Under FIFRA § 6(a)(2), however, EPA requires registrants to submit information showing even modest negative environmental effects, whereas TSCA § 8(e) permits the submitter some discretion in balancing the seriousness of the effect with the likelihood of its occurrence. See also Chapter 15.

Testing Under TSCA § 4

While EPA has no authority under TSCA to require testing of pesticides, a TSCA § 4 test rule could affect intermediates, raw materials, and inerts used in pesticide production. Therefore, some consideration of TSCA § 4 testing authority is warranted. Furthermore, EPA has exercised its TSCA § 4 testing authority over pesticides after their use to assess the environmental hazards associated with their disposal. Solid Waste Chemicals Test Rule, 53 Fed. Reg. 22,300 (1988).

Under TSCA § 4, the Administrator may require testing of existing chemical substances and mixtures if he finds that they "may present" an unreasonable risk of injury to health or the environment, or if they will be produced in substantial quantities with significant or substantial human or environmental exposure. Generally, manufacturers (and not processors) that are subject to a testing rule bear the cost of testing and, much like FIFRA § 3(e)(2)(B) (see Chapter 7), TSCA § 4(c) provides for cost sharing among those manufacturers subject to the testing rule. 40 C.F.R. part 791.

Congress showed a definite bias in favor of testing and developed an ingenious mechanism in TSCA § 4 for implementing testing. TSCA created the Interagency Testing Committee, consisting of one representative from each of eight federal agencies, and charged it to meet regularly and to identify and list those chemicals that deserve EPA's priority consideration for possible testing. The ITC must give special attention to chemicals that are produced in high volume or that pose risks of cancer, gene mutations, or birth defects. After the ITC develops its priority list, EPA has twelve months within which to either initiate rulemaking to require testing, or publish in the *Federal Register* its reasons for not doing so. In deciding whether to require testing, the Agency has two regulatory triggers: a risk trigger and an exposure trigger. TSCA § 4(a)(1).

Under the risk trigger, the Agency determines whether the chemical may present an unreasonable risk to the environment. TSCA § 4(a)(1)(A). (For a discussion of risk assessment under FIFRA, see Chapter 9). Risk takes into consideration both hazard and exposure. Thus, the Agency looks at whether the hazard data are sufficiently reliable to permit the Agency to make a qualitative judgment that the chemical can produce an effect. If there are no reliable data on the hazard, then the Agency cannot make a sound judgment about risk. Even if the data on the hazard are reliable, however, the Agency cannot pull the risk trigger if there is no exposure.

If there are sufficient data on hazard and exposure to conclude that a chemical may present an unreasonable risk, then the Agency must ask, first, whether the existing data are sufficient to evaluate the risk and, second, whether further testing can provide the necessary information. In the latter case, if there is no acceptable test protocol for a particular type of effect, testing would not be statutorily authorized.

The second testing trigger is exposure. If the chemical is produced in substantial quantities and if there is substantial or significant human or environmental exposure, then EPA can pull the exposure trigger. TSCA § 4(a)(1)(B). No evaluation of risk is necessary; exposure alone warrants testing. If exposure is not "substantial" or "significant," however, the Administrator cannot justify a test rule on this basis. The regulations do not define "substantial" or significant, but the Agency follows a pragmatic approach: if there are millions of pounds produced or tens of thousands of people exposed, then exposure is substantial. 46 Fed. Reg. 30,300 (1981). Upon review of the cumene test rule, the U.S. Court of Appeals for the Fifth Circuit held that EPA had inadequately defined the human exposure on which it based the exposure trigger. *Chemical Manufacturers Association v. EPA*, 899 F.2d 344 (5th Cir. 1990).

Having satisfied the requirements of one or both of the triggers, the Agency publishes a test rule. The rule must specifically identify the substance or mixture to be tested, the standards for development of test data and the time period for data to be submitted. TSCA § 4(b)(1).

EPA has used its authority under TSCA § 4 to promulgate its Solid Waste Chemicals Test Rule which requires manufacturers and processors of certain chemicals, including endrin and maleic hydrazide, to test for human health effects and/or chemical fate in support of its regulatory program under the Resource Conservation and Recovery Act (RCRA). Never before had EPA applied TSCA to chemicals like endrin and maleic hydrazide which were manufactured, processed and distributed for use as a pesticide and, thus, appeared to fall within the TSCA § 3(2)(B)(ii) pesticide exclusion. Because all registrations of endrin have been cancelled and the costs associated with testing maleic hydrazide were relatively small, the rule was not challenged. Thus, it is not clear whether EPA's authority to require testing of pesticides will be upheld in the future.

Regulation Under TSCA §§ 6 and 7

If the manufacture, processing, distribution, use, or disposal of the chemical substance or mixture "present[s] an unreasonable risk of injury to health or the environment," the Administrator has broad authority under TSCA § 6 to prohibit or limit production and to impose labeling or other requirements. The determination of unreasonable risk under TSCA § 6, like that of "unreasonable adverse effects" under FIFRA, requires a consideration of both risks and benefits. If, however, a chemical substance or mixture "presents an imminent and unreasonable risk of serious or widespread injury to health or the environment," the Administrator under TSCA § 7 may proceed directly to district court for injunctive and other relief, including seizure and condemnation.

In 1978, EPA used its authority under TSCA § 6 to prohibit virtually all manufacturing, processing, and use of fully halogenated chlorofluoroalkanes (chlorofluorocarbons or CFCs) for aerosol propellant uses. 43 Fed. Reg. 11,318 (1978);

40 C.F.R. part 762. EPA banned CFCs because they are implicated in depleting the earth's ozone layer, thereby increasing the risk of cancer from ultraviolet radiation. EPA extended its ban to include CFC propellants used in pesticide products even though EPA recognized that, once they became components of pesticide products, the CFCs were regulated exclusively under FIFRA. 40 C.F.R part 762; 42 Fed. Reg. 24,542, 24,546 (1977). EPA concluded that it had sufficient authority under TSCA to ban the processing of CFCs for use in pesticides (i.e., their incorporation into aerosol pesticides), despite the exemption for pesticides in TSCA § 3(2)(B)(ii). No one commented upon or protested EPA's use of TSCA § 6 to regulate in effect the manufacture of pesticide products registered under FIFRA. With this as precedent, EPA could invoke TSCA § 6 authority again to regulate hazardous raw materials or inert ingredients before they become components of pesticides.

POISON PREVENTION PACKAGING ACT (PPPA), 15 U.S.C. §§ 1471-1476

To prevent accidental poisonings of children due to inadequate packaging of toxic household products, Congress passed the Poison Prevention Packaging Act of 1970, Pub. L. No. 91-601, 84 Stat. 1670. As first enacted, PPPA authorized the Secretary of Health, Education, and Welfare to promulgate special "child-resistant" packaging (CRP) standards for household substances, including pesticides, if he found that they were required to protect children from "serious personal injury or serious illness resulting from handling, using, or ingesting such substances." 15 U.S.C. § 1472(a)(1) (1988). The required packaging, however, had to be "technically feasible, practicable, and appropriate." 15 U.S.C. § 1472(a)(2).

In 1972, Congress transferred authority to administer PPPA to the Consumer Product Safety Commission (CPSC). Consumer Product Safety Act, Pub. L. No. 92-573, sec. 30, 86 Stat. 1207, 1231 (1972). In the Consumer Product Safety Commission Improvements Act of 1976, Pub. L. No. 94-284, sec. 3(b), 90 Stat. 503, 503, however, Congress exempted pesticides altogether from the definition of household substances and thus eliminated CPSC's jurisdiction over CRP standards for pesticide products.

The elimination of CPSC's jurisdiction reflected the recognition by Congress that EPA had adequate authority under FIFRA § 25(c)(3) to promulgate CRP standards for pesticide products. FIFRA § 25(c)(3) authorizes the Administrator of EPA to:

> [E]stablish standards (which shall be consistent with those established under the authority of the Poison Prevention Packaging Act (Public Law 91-601)) with respect to the package, container, or wrapping in which a pesticide or device is enclosed for use or consumption, in order to protect children and adults from serious injury or illness resulting from accidental ingestion or contact with pesticides or devices regulated by this Act as well as to accomplish the other purposes of this Act.

A pesticide is misbranded under FIFRA § 2(q)(1)(B) if "it is contained in a package or other container or wrapping which does not conform to the standards established by the Administrator pursuant to section 25(c)(3)."

Vested with sole responsibility for pesticide CRP, EPA promulgated CRP regulations. These regulations define CRP as:

> [P]ackaging that is designed and constructed to be significantly difficult for children under 5 years of age to open or obtain a toxic or harmful amount of the substance contained therein within a reasonable time, and that is not difficult for normal adults to use properly.

40 C.F.R. § 157.21(b). CRP is required for all pesticide packages which meet the following criteria: they do not exceed certain size standards, contain pesticides meeting certain toxicity criteria, are intended for residential use, and are not otherwise exempted by the regulations. 40 C.F.R. part 157. For example, a pesticide product not exceeding the size limitations is required to use CRP if it is "highly or moderately toxic by ingestion, dermal absorption, inhalation, or causes severe irritation or corrosion to the skin or eyes." 51 Fed. Reg. 21,276, 21,277 (1986). The criteria correspond for the most part to FIFRA toxicity categories I and II. In addition, if a product's labeling indicates that the product is intended for use at a residential site, such as for use in kitchens, bathrooms, and flower gardens and meets the other criteria stated in part 157, it must use CRP.

Pesticide products in large packages are exempted from the CRP regulations because such large packages are not normally sold for residential use and accordingly do not reach curious children. The actual size cutoffs decided upon by EPA are as follows:

1. **Solid products** (regardless of pesticide type). CRP does not apply to such products packaged in sizes of fifty pounds or greater.

2. **Liquid swimming pool chemicals.** CRP does not apply to liquid swimming pool products in sizes greater than 7.5 gallons.

3. **Liquid non-swimming pool chemicals** (insecticides, rodenticides, herbicides, fungicides, antimicrobials, and mixtures of these). CRP does not apply to products that are packaged in containers of five gallons or greater.

4. **Aerosol products.** CRP does not apply to aerosol pesticide products in sizes of two pounds or greater.

40 C.F.R § 157.24(a)(2)(i)(A)-(D). Products classified for restricted use also are exempt from CRP requirements. 40 C.F.R. § 157.24(a)(1).

Registrants of pesticide products required to meet CRP standards must certify to the Agency that the package meets the applicable standards. They must so certify with each application for new registration or within six months after the Agency determines and notifies the registrant of the need to employ CRP packaging. 40 C.F.R. § 157.34.

OCCUPATIONAL SAFETY AND HEALTH (OSH) ACT OF 1970, 29 U.S.C. §§ 651-678

Congress enacted the OSH Act in order "to assure so far as possible every working man and woman in the Nation safe and healthful working conditions." OSH Act § 2(b). To accomplish this purpose, Congress gave the Secretary of Labor, through OSHA, broad authority under OSH Act § 6 to promulgate occupational safety and health standards to protect workers exposed to hazards in their employment. In addition, OSH Act § 5(a) imposes upon every employer the general duty to provide places of employment "which are free from recognized hazards that are causing or are likely to cause death or serious physical harm to his employees" and also to comply with health and safety standards.

In November 1990, OSHA and EPA signed a Memorandum of Understanding (MOU) intended to improve the enforcement relationship between EPA's Office of Enforcement and OSHA. MOU Between the U.S. Department of Labor, OSHA and the EPA Office of Enforcement (signed Nov. 23, 1990). The MOU establishes a framework for notification, consultation and coordination to assist EPA in identifying environmental problems, such as unsafe handling, storage or use practices involving pesticides, and OSHA in identifying workplace health and safety problems. MOU at 2. This will include cross training of inspectors to promote referrals of possible violations by inspectors of one agency to the other agency. MOU at 12-13.

Occupational Safety and Health Standards

OSH Act § 3(8) defines an occupational safety and health standard as one "reasonably necessary or appropriate to provide safe or healthful employment." Where such standards regulate toxic materials or harmful physical agents, OSH Act § 6(b)(5) requires the Secretary to "set the standard which most adequately assures, to the extent feasible, on the basis of the best available evidence, that no employee will suffer material impairment of health or functional capacity even if such employee has regular exposure to the hazard dealt with by such standard for the period of his working life."

Workplace Standards for Farmworkers

While OSHA has broad authority to set workplace standards, the OSH Act does not apply "to working conditions of employees with respect to which other Federal agencies . . . exercise statutory authority to prescribe or enforce standards or regulations affecting occupational safety or health." OSH Act § 4(b)(1). In 1974, EPA promulgated regulations which address occupational safety and health of agricultural workers performing hand labor operations in fields during or after application of pesticides. 40 C.F.R. part 170. EPA has proposed to revise these regulations to expand the scope of the standards to also include workers in forests, nurseries and greenhouses, and workers who handle pesticides in those locations. 53 Fed. Reg. 25,970 (1988). In *Organized Migrants in Community Action, Inc. v. Brennan*, 520 F.2d 1161 (D.C. Cir. 1975), the court upheld EPA's authority under FIFRA to issue these regulations. Because EPA had set reentry standards, the court held that OSH Act § 4(b)(1) barred OSHA from setting similar standards.

Once EPA has exercised its authority under FIFRA by promulgating health and safety standards or regulations affecting working conditions, therefore, OSHA cannot duplicate this effort. In the absence of EPA's formal exercise of authority under FIFRA, however, OSHA has jurisdiction over workers exposed to pesticides. *Public Citizen Health Research Group v. Auchter*, 702 F.2d 1150, 1156 n.23 (D.C. Cir. 1983).

With the exception of its labeling provisions, however, FIFRA's reach generally does not extend beyond the site of application. Accordingly, OSHA's jurisdiction over workplace standards in pesticide manufacturing, formulating, and distributing facilities is essentially exclusive.

The Benzene and Cotton-Dust Cases

The U.S. Supreme Court has twice interpreted OSHA's authority to set standards for occupational exposure to toxic agents such as pesticides. In *Industrial Union Department v. American Petroleum Institute*, 448 U.S. 607 (1980), a divided Supreme Court overturned OSHA's occupational standard for airborne concentrations of benzene, which OSHA had reduced from 10 to 1 part per million (ppm) (allegedly the lowest feasible level), and for dermal contact with solutions containing benzene, which OSHA had prohibited altogether. The Court found that OSHA had failed to make a threshold determination under OSH Act § 3(8) that the new standard "is reasonably necessary and appropriate to remedy a significant risk of material health impairment." 448 U.S. at 639.

One year later, in *American Textile Manufacturers Institute, Inc. v. Donovan*, 452 U.S. 490, 506 (1981), the Supreme Court upheld OSHA's cotton dust standard and rejected industry's assertion that OSH Act § 6(b)(5) requires OSHA "to determine that the costs of the standard bear a reasonable relationship to its benefits." In setting occupational exposure standards for toxic substances, the Court said OSHA need consider only "feasibility." Congress had already weighed costs and benefits: worker health was primary and OSHA could qualify a health-based standard only to the limited extent that it was "unachievable."

OSHA's Generic Cancer Policy

In 1980, OSHA published its generic cancer policy with respect to the identification, classification, and regulation of toxic substances in the workplace that may pose a cancer risk to workers. 45 Fed. Reg. 5001 (1980) (codified at 29 C.F.R. part 1990). The cancer policy provides a procedural framework for the regulation of individual chemical substances in future OSHA standard-setting proceedings and also establishes certain scientific principles which the Agency has determined are binding on the parties in those proceedings. These scientific principles may not be rebutted in proceedings on individual chemicals unless the evidence presented on the chemical meets certain detailed criteria. 29 C.F.R. §§ 1990.144-.145; 45 Fed. Reg. 5287-88 (1980).

OSHA's cancer policy was the subject of a court challenge in the Fifth Circuit. Those proceedings, which were held in abeyance for six years pending OSHA's rulemaking to consider modifications to its cancer policy, 47 Fed. Reg. 187 (1982), were dismissed without prejudice by the Fifth Circuit. *American Petroleum Institute v. OSHA*, No. 80-3018 (5th Cir. Jan. 13, 1987). Furthermore, in January 1983, OSHA imposed a partial administrative stay on those provisions of the cancer policy which require the Agency to establish and publish Candidate and Priority Lists. 29 C.F.R. §§ 1990.121(a)-(b), .122, .131, .133. 48 Fed. Reg. 241 (1983). Currently, the Agency's stated position is that it is actively reconsidering its cancer policy, albeit at a very measured pace. 55 Fed. Reg. 16,397 (1990). In the meantime, the cancer policy remains in effect.[64]

---

64 For further discussion of other agencies' policies regarding cancer, see Chapter 9.

Workplace Standards for Pesticides

During the late 1970s, in response to medical evidence in the scientific literature indicating that worker exposure to pesticides results in serious health problems, including damage to the liver, lungs and kidneys, sterility, central nervous system disorders, cancer, genetic changes, and birth defects, OSHA began to consider the development of a generic workplace standard for pesticide producing establishments. 43 Fed. Reg. 54,955 (1978); 44 Fed. Reg. 68,278 (1979). In December 1980, OSHA reconfirmed its plans to develop a generic workplace standard for pesticide manufacturing and formulating plants that would apply to any workplace manufacturing or formulating any substance intended for use as a pesticide and would permit OSHA to establish minimum requirements for the protection of all employees in these industries. After promulgating these general requirements, OSHA would then set substance-specific standards for the most hazardous pesticides. 45 Fed. Reg. 80,078 (1980). In 1981, however, OSHA dropped the foregoing initiative.

Instead, OSHA has focused its attention on exposure levels for individual pesticides. OSHA has identified and established exposure limits for a number of pesticides considered to be air contaminants, such as ethylene dibromide, diazinon and chloropyrifos. 29 C.F.R. § 1910.1000. In other cases, OSHA has issued separate regulations for pesticides of concern, such as ethylene oxide (EtO). For example, on June 22, 1984, OSHA issued a permanent standard for EtO of 1 ppm for an eight-hour time weighted average. 49 Fed. Reg. 25,734 (1984) (codified at 29 C.F.R. § 1910.1047).

OSHA's Access Rule

In 1980, OSHA promulgated its "Access to Employee Exposure and Medical Records" rule, giving employees, designated representatives, and OSHA access to employer-maintained exposure and medical records. 45 Fed. Reg. 35,212 (1980) (codified at 29 C.F.R. § 1910.20). The rule applies to all general industry, maritime, and construction employers whose employees are exposed to toxic substances or harmful physical agents.

Generally, the rule requires employers to maintain and preserve exposure and medical records concerning employees exposed to toxic substances or harmful physical agents for the employee's period of employment plus thirty years. Additionally, employers must ensure that employees, their designated representatives, and OSHA have access to those records and to analyses based on them.

In November 1982, a Louisiana district court upheld the access rule in its entirety. *Louisiana Chemical Association v. Bingham*, 550 F. Supp. 1136 (W.D. La. 1982), *aff'd*, 731 F.2d 280 (5th Cir. 1984). The court held that OSHA had ample statutory authority to promulgate a generic access rule under §§ 8(c)(2) and 8(g)(2) and that the rule did not violate employers' fourth amendment rights, employees' privacy rights, or the Trade Secrets Act (which prohibits disclosure of trade secrets by government employees unless authorized by law).

In September 1988, OSHA revised the rule to: (1) exempt first aid and medical records for short-term employees from the records retention requirements; (2) permit microfilm storage of employee x-rays (except chest x-rays); (3) provide additional protection to employer trade secrets and require comformity with OSHA's Hazard Communication Standard; (4) require union representatives to show a need for records when seeking unconsented access; and (5) provide that no industries are treated separately with respect to trade secret disclosure. 53 Fed. Reg. 38,140 (1988).

OSHA's Hazard Communication Standard

Under OSH Act § 6(b)(7), OSHA is authorized to ensure that workers are informed of the physical and health hazards of the chemical substances to which they are exposed in the workplace. Pursuant to this authority, in 1983 OSHA promulgated its Hazard Communication Standard (HCS), a generic standard for warning employees of hazards. 48 Fed. Reg. 53,280 (1983) (codified at 29 C.F.R. § 1910.1200).

Since November 1985, when the standard became effective, chemical manufacturers and importers have been required to conduct comprehensive analyses of the hazards of their chemical products, and to provide downstream users with chemical hazard labels and highly detailed material safety data sheets (MSDSs) for hazardous substances or materials containing hazardous substances. Originally, only manufacturing sector employers were required to have MSDSs for the hazardous chemicals at their workplaces, and to implement employee training and information programs regarding chemical hazards. 29 C.F.R. § 1910.1200. As a result of judicial challenges to the scope of the standard, however, on August 24, 1987, OSHA issued a final rule extending the hazard communication standard to all employers whose employees handle or may be exposed to hazardous substances. 52 Fed. Reg. 31,852.

HCS applies to all chemicals, including pesticides, determined to be either "health hazards" or "physical hazards." A health hazard is defined as:

> [A] chemical for which there is statistically significant evidence based on at least one study conducted in accordance with established scientific principles that acute or chronic health effects may occur in exposed employees. The term "health hazard" includes chemicals which are carcinogens, toxic or highly toxic agents, reproductive toxins, irritants, corrosives, sensitizers, hepatoxins, nephrotoxins, neurotoxins, agents which act on the hematompoietic system, and agents which damage the lungs, skin, eyes, or mucous membranes.

29 C.F.R. § 1910.1200(c). A physical hazard is defined as:

> [A] chemical for which there is scientifically valid evidence that it is a combustible liquid, a compressed gas, explosive, flammable, an organic peroxide, an oxidizer, pyrophoric, unstable (reactive) or water-reactive.

C.F.R. § 1910.1200(c). HCS, however, does not apply to: hazardous wastes, foods, drugs, cosmetics intended for personal consumption, and wood products. 29 C.F.R. § 1910.1200(b)(6). The "wood products" exemption appears to exempt all wood substances treated with wood preservatives.

All pesticides which meet the foregoing criteria are subject to HCS, with one important exception: pesticides, as defined by FIFRA, have a *limited exemption* from HCS's provision regarding labeling when the pesticide is subject to FIFRA's labeling requirements. 29 C.F.R. § 1910.1200(b)(5). This exemption from the labeling provision, however, *does not* exempt pesticides from other hazard communication provisions, including, according to OSHA, MSDS requirements.

In addition, in proposed modifications regarding the scope and application of the standard, OSHA clarified its position regarding coverage of pesticide applicators. 53 Fed. Reg. 29,822, 29,827 (1988). OSHA indicates that applicators of restricted use pesticides are not covered by HCS because "EPA has clearly exercised statutory authority for protecting them in the area of hazard communication." OSHA's position regarding workers who apply non-restricted use pesticides is not as clear, but should be clarified in the final rule.

Major Elements of HCS

HCS has five principal elements: hazard determination, labeling, material safety data sheets, employee training and information, and the written hazard communication program. Hazard determination is at the core of hazard communication. If, for example, a pesticide ingredient is determined not to be hazardous, a manufacturer, importer, distributor, or employer has absolutely no hazard communication responsibilities with respect to that substance under HCS. Only pesticide *manufacturers* (including formulators) and *importers* are required to evaluate chemicals to determine if they are hazardous. Employers who are not also manufacturers are not required to evaluate the chemicals in their workplaces unless they choose not to rely on the previous evaluations made by the upstream chemical manufacturer or importer. 29 C.F.R. § 1910.1200(d)(1).

The evaluator must determine whether a particular chemical is hazardous on the basis of the physical and health hazard definitions within HCS, although the method for doing so is unspecified and thus discretionary. 29 C.F.R. § 1910.1200(d)(1), (2). HCS, however, does create a "floor" of substances which must be considered health hazards. These are substances regulated by OSHA (Toxic and Hazardous Substances, 29 C.F.R. part 1910, subpart Z), listed by the American Conference of Governmental Industrial Hygienists (ACGIH) in the "Threshold Limit Values for Chemical Substances and Physical Agents in the Work Environment," (latest edition), and listed as carcinogenic by OSHA (Toxic and

Hazardous Substances, 29 C.F.R. part 1910, subpart Z), the National Toxicology Program (NTP) in its "Annual Report on Carcinogens" (latest edition), or the International Agency for Research on Cancer (IARC) in its "monographs" (latest editions). 29 C.F.R. § 1910.1200(d)(3), (4).

Special rules apply to mixtures. If a mixture has been tested as a whole to determine its hazards, the results of such test are dispositive. 29 C.F.R. § 1910.1200(d)(5)(i). If a mixture has not been tested as a whole, the mixture is assumed to pose the same health hazard posed by any component comprising 1 percent (by weight or volume) or more of the mixture. With respect to a carcinogenic hazard, however, the mixture is assumed to pose the same hazard as any carcinogenic component comprising 0.1 percent or more of the mixture. 29 C.F.R. § 1910.1200(d)(5)(ii).

With respect to health hazards, HCS creates what might be called the "one positive study guideline":

> [E]vidence which is statistically significant and which is based on *at least one positive study* conducted in accordance with established scientific principles is considered to be sufficient to establish a hazardous effect *if the results of the study meet the definitions of health hazards.*

29 C.F.R. § 1910.1200(d)(2) (emphasis added). The "one positive study" provision seems to require that a single "positive" study makes a chemical "hazardous." The provision equivocates, however, by including the clause: "if the results of the study meet the definitions of health hazards in this section." The preamble to HCS clarifies the intent of section (d)(2) to some extent:

> First of all, the employer is to evaluate animal data as well as human data. Secondly, if a positive study conducted according to established scientific principles, and resulting in statistically significant findings of an adverse health effect is available, manufacturers or importers must report that effect whether they agree with it or not. They need not state the finding as a conclusion they agree with, but must report it nevertheless. They are also free to report any negative data which they believe is relevant to determining the hazard of the chemical.

48 Fed. Reg. 53,280, 53,299 (1983). The one positive study provision, therefore, requires that all such studies be "reported" to downstream recipients of the chemical at least through the MSDS, if not the labeling.

OSHA considers the MSDS "the primary vehicle for transmitting detailed hazard information to both downstream employers and employees." 48 Fed. Reg. at 53,305. Regarding each hazardous chemical, MSDSs must contain twelve different categories of information listed at 29 C.F.R. § 1910.1200(g)(2). Required information includes the physical and chemical characteristics of the chemical, its physical and/or health hazards, the chemical's "route(s) of entry," exposure limits, whether it is listed as a carcinogen, and any generally applicable precautions for safe handling and use.

Only chemical manufacturers (including formulators) and importers are required to "obtain or develop" MSDSs. Distributors are required to "ensure that material safety data sheets, and updated information, are provided to other distributors and employers." Retail distributors selling hazardous chemicals to commercial customers must provide an MSDS upon request. 29 C.F.R. § 1910.1200(g)(7). Employers are only required to "have" an MSDS for each hazardous chemical they use. 29 C.F.R. § 1910.1200(g)(1). MSDSs must be updated within three months of a preparer's receipt of significant new information regarding the hazards of a chemical. 29 C.F.R. § 1910.1200(g)(5).

With respect to labeling, chemical manufacturers, importers and distributors are required to ensure that every container leaving the workplace displays the identity of the hazardous chemical, appropriate hazard warnings, and the name and address of the chemical manufacturer, importer or other responsible party. 29 C.F.R. § 1910.1200(f)(1)(i)-(iii). Employers must ensure that each container of hazardous chemicals in their facilities displays the identity of the chemical and appropriate hazard warnings. 29 C.F.R. § 1910.1200(f)(5). Employers are not required to label portable containers if the hazardous chemicals were transferred from a labeled container and are intended only for the immediate use of the employee who performed the transfer. 29 C.F.R. § 1910.1200(f)(7). As discussed above, however, these requirements do not apply to pesticides subject to FIFRA's labeling requirements.

To accomplish the required labeling, employers may use "signs, placards, process sheets, batch tickets, operating procedures, or other such written materials in lieu of affixing labels to individual stationary process containers, as long as the alternative method identifies the containers to which it is applicable and conveys the information required." 29 C.F.R. § 1910.1200(f)(6).

The fourth major part of HCS is employee training. Employers, whether they are required to prepare MSDSs or not, must provide training to their employees. Employees must be given information regarding: (1) the provisions of HCS; (2) operations in the work area where hazardous chemicals are present; (3) the location and availability of the employer's written hazard communication program and MSDSs; and (4) ways of detecting and protecting against the presence of hazardous chemicals. 29 C.F.R. § 1910.1200(h)(1). Beyond stating the minimum informational requirements of employee training, HCS leaves the form and approach taken in the training up to the employer. He must conduct training, however, at the time of initial employee assignment and "whenever a new hazard is introduced into [the employee's] work area." 29 C.F.R. § 1910.1200(h).

The last major requirement of HCS is the written hazard communication program. Employers using hazardous chemicals must develop, implement and maintain a program that describes: (1) how the labeling, MSDS and employee information and training requirements are being met; (2) what hazardous chemicals are known to be present; and (3) what methods the employer uses for informing employees of the hazards of "non-routine tasks" and chemicals contained in unlabeled pipes. In the case of multi-employer workplaces, plans must include a description of the employer's methods for providing other employers with relevant MSDSs and for informing other employers of the labeling system used and precautionary measures that need to be taken to protect employees. 29 C.F.R. § 1910.1200(e).

Trade Secrets Under HCS

HCS's trade secrets provision constitutes a qualified exemption from the requirement that MSDSs convey a chemical's *identity*. 29 C.F.R. § 1910.1200(i). A chemical's specific identity may be withheld from the MSDS if: (a) that identity is claimed to be a trade secret; (b) the trade secret claim can be supported; (c) information concerning the properties and effects of the chemical is disclosed; (d) the MSDS indicates that the specific chemical identity is being withheld as a trade secret; and (e) the specific chemical identity is made known in emergency situations to health professionals and in non-emergency situations to health professionals, employees, and their representatives. 29 C.F.R § 1910.1200(i).

In September 1986, OSHA narrowed the scope of the trade secret provision. The rule amended the provision: (1) to eliminate the phrase "chemical name or other unique chemical identifier" from the definition of trade secret in order to preclude the protection of chemical identity information that is readily discoverable through reverse engineering; (2) to extend access to chemical identity information in non-emergency situations to employees and their representatives; and (3) to permit occupational health nurses to have access to trade secret information. 51 Fed. Reg. 34,590 (1986).

The amendment to HCS's trade secret provision increases the risk that confidential information concerning the identities of inert pesticide ingredients may be disclosed. FIFRA's labeling provisions, of course, do not require disclosure of inert pesticide ingredients. The "ingredient statement" is defined to require only "the total percentage of all inert ingredients[] in the pesticide." FIFRA § 2(n)(1). Accordingly, pesticide manufacturers may have to disclose ingredient identity information which heretofore they have protected as confidential.

Preemption Under HCS

One aspect of the Hazard Communication Standard which may actually reduce rather than increase the burdens on the pesticide industry is its preemption provision. 29 C.F.R. § 1910.1200(a)(2). Under this provision, both state and local laws pertaining to occupational hazard communication are preempted by the federal standard, unless a federally approved state plan is in effect. This means that any state or local worker "right-to-know" law requiring the preparation of MSDSs, labeling of chemicals and identification of their hazards, development of written hazard communication programs, and development and implementation of worker chemical hazard training for the *primary* purpose of assuring worker safety would be preempted by HCS. Federally approved plans are the exclusive alternative to preemption.

Several states do not have OSHA-approved state plans, such as Pennsylvania and New Jersey. As a result, portions of New Jersey's Community and Worker Right to Know Law and Pennsylvania's Right to Know Act have been held to be preempted by the federal standard. *Manufacturers Association of Tri-County v. Knepper*, 801 F.2d 130 (3d Cir. 1986), *cert. denied*, 484 U.S. 815 (1987); *New Jersey State Chamber of Commerce v. Hughey*, 774 F.2d 587 (3d Cir. 1985), *cert. denied*, 109 S. Ct. 3246 (1989). The Third Circuit held, however, that these state statutes are "expressly preempted only to the extent that the State act pertains to issues that are addressed by a duly promulgated OSHA standard . . . [A] state right-to-know Act [is] impliedly preempted only to the extent that it is impossible to comply with both the state law and the federal standard, or to the extent the state law serves as an obstacle to the accomplishment of the Congressional purposes served by the OSHA standard."

EMERGENCY PLANNING AND COMMUNITY RIGHT-TO-KNOW

The Emergency Planning and Community Right-to-Know Act of 1986 (EPCRA) (title III of the Superfund Amendments and Reauthorization Act of 1986 (SARA Title III)), Pub. L. No. 99-499, 100 Stat. 1613, requires states and localities to develop mechanisms for handling relevant information on hazardous chemicals and preparing for chemical emergencies. It requires facilities that manufacture, process, or use certain chemicals to routinely provide chemical information to the cognizant state and local agencies. It makes this information publicly available, subject only to a very limited trade secret exemption.

Finally, it requires owners and operators of such facilities to report particular chemical releases to local planning committees and exacts heavy penalties (up to $25,000 per violation per day for first offenders and up to $75,000 per day for second offenders) for failure to comply with the emergency release reporting and other notification obligations.

Many pesticides are subject to EPCRA's provisions, and therefore pesticide producers must implement a compliance program. The following discussion reviews the reporting and other obligations imposed under SARA's emergency planning and community right-to-know provisions.

Emergency Planning Program

EPCRA commands states and localities to create a bureaucracy to plan for responding to emergencies involving the release of hazardous and toxic chemicals. By April 17, 1987, the governor of each state had to appoint a state emergency response commission (SERC). The commission in turn designated local emergency planning districts and appointed local emergency planning committees (LEPCs) responsible for preparing and implementing an emergency plan for their district. EPCRA § 301.

The power of emergency planning committees to gather information is stated in the broadest possible terms. Upon request, the owner or operator of a facility subject to the emergency planning requirements must promptly provide the emergency planning committee with *any* information necessary for the implementation of an emergency plan. EPCRA § 303(d)(3). Undoubtedly, facilities will devote significant amounts of staff time and effort to gather the requested information.

As directed by EPCRA § 302, EPA has promulgated a list of extremely hazardous substances (many of which are pesticides) and established a "threshold planning quantity" for each substance listed. 40 C.F.R. part 355, app. A. The threshold planning quantity is the total amount of a listed substance which, if present at a facility at any time, subjects the facility to emergency planning requirements. To determine whether a threshold planning quantity has been exceeded by a facility, all quantities of a listed substance, however and wherever present at a facility on a given date, must be aggregated.

By May 17, 1987, each facility which exceeded the threshold quantities of an extremely hazardous substance at any time during the prior year had to notify the SERC of that fact, and stand ready to provide information requested by the LEPC. Any facilities that begin handling an extremely hazardous substance in an amount exceeding the applicable threshold planning quantity must notify the SERC and the LEPC within sixty days. These facilities then also become subject to the emergency planning provisions. Affected facilities must provide similar follow-up notification when EPA adds to its list of extremely hazardous substances.

Information Reporting Requirements

EPCRA builds upon OSHA's HCS and brings right-to-know out of the workplace and into the community at large. It accomplishes this through three distinctive reporting forms: material safety data sheets (MSDSs) (EPCRA § 311), emergency and hazardous chemical inventory forms (EPCRA § 312), and toxic chemical release forms (EPCRA § 313).

Material Safety Data Sheets. Pursuant to EPCRA § 311, facilities required to prepare an MSDS for a "hazardous chemical" under OSHA's HCS must *submit* the MSDS (or a list of MSDSs) to the SERC, the LEPC, and the local fire department. The obligation to submit an MSDS is triggered only if a facility exceeds the reporting threshold established for hazardous chemicals and extremely hazardous substances. Under § 311(e), several categories of hazardous chemicals, such as pesticides used for "routine agricultural operations" and substances used for "household purposes," are not subject to the requirement to submit MSDSs. Agricultural chemicals, such as pesticides, however, are subject to reporting by retailers and suppliers of such chemicals.

Emergency and Hazardous Chemical Inventory Forms. Under EPCRA § 312, owners and operators subject to the MSDS requirements also must prepare and submit an Emergency and Hazardous Chemical Inventory Form (Inventory Form) to the SERC, the

LEPC, and the local fire department. The Inventory Form must contain "Tier I" information which includes the following information: (1) an estimated range of the aggregate maximum amount of hazardous chemicals present at the facility at any time during the preceding calendar year; (2) an estimated range of the aggregate average daily amount of hazardous chemicals present; and (3) the general location of hazardous chemicals. In addition, the facility owner must place each chemical into one or more of five hazard categories: "immediate (acute) health hazard," "delayed (chronic) health hazard," "fire hazard," "sudden release of pressure," and "reactive." 40 C.F.R. § 370.2.

If specifically requested by the SERC, the LEPC or the local fire department, owners and operators must also prepare "Tier II" information which further refine Tier I information by specific hazardous chemical. EPCRA § 312(e). Facilities must update both Tier I and Tier II information annually.

Toxic Chemical Release Forms. EPCRA § 313 provides that owners or operators of facilities in Standard Industrial Classification (SIC) codes 20 through 39 that manufacture, process, or otherwise use a "toxic chemical" in excess of a threshold amount must submit to EPA and their SERC a Toxic Chemical Release Form (Form R). For each toxic chemical that is used, manufactured or processed in excess of the threshold amount, a facility owner must estimate annual quantities of the chemical released to air, water, and land, and transfers to off-site facilities. The form must be submitted annually by July 1.

The toxic chemicals subject to the requirements of § 313 are listed in 40 C.F.R. § 372.65. EPA may by rule add or delete a chemical from this list using any one of three criteria set forth under § 313(d). Any person may petition EPA to add or delete a chemical from the list.

Emergency Notification

EPCRA § 304 requires facilities that handle "hazardous chemicals" to report the release of a reportable quantity of either an "extremely hazardous substance" or a substance listed under § 102 of the Comprehensive Environmental Response, Compensation, and

Liability Act (CERCLA or Superfund). Releases which result in exposure to persons solely within the facility site are exempt from the emergency notification requirement. Reportable quantities for extremely hazardous substances are set forth in 40 C.F.R. part 355 appendix A. Reportable quantities applicable to CERCLA-listed hazardous substances are set forth in table 302.4 of 40 C.F.R. part 302.

Note that only facilities which handle "hazardous chemicals" are subject to the emergency release notification requirement. The term "hazardous chemical" has the same meaning under § 304 as under § 311(e), and, therefore, the exemptions under § 311(e) also apply to § 304. EPCRA § 329(5). Thus, facilities which *only* handle exempt hazardous chemicals, such as fertilizers and pesticides used as part of routine agricultural activities, are *not* subject to the emergency release notification requirement.

Immediately following a release, owners and operators of facilities must give notice (by telephone, radio or in person) to the community emergency coordinator for areas likely to be affected and to the SERC. For releases arising out of transportation of extremely hazardous chemicals (or storage incident to such transportation), the facility owner can give notice by dialing 911, or in the absence of a 911 emergency telephone number, by calling the operator. EPCRA § 304(b)(1).

Public Access Versus Trade Secret Protection

The legislative history of EPCRA is replete with remarks that citizens have a right to know about chemicals in their communities — what they are, where they are, and how much is present. EPCRA § 324 implements this congressional intent and provides that virtually every piece of information produced under EPCRA must be made available to the public.

EPCRA § 322 authorizes the withholding of certain trade secret information but significantly limits the scope of trade secret information. Information is entitled to trade secret protection only if four enumerated "trade secret factors" are satisfied. EPCRA

§ 322(b). Trade secret protection, however, does not extend to the emergency release notification requirement of § 304.

HAZARDOUS MATERIALS TRANSPORTATION ACT (HMTA), 49 U.S.C. app. §§ 1801-1812

In HMTA § 102, Congress declared its policy "to improve the regulatory and enforcement authority of the Secretary of Transportation to protect the Nation adequately against the risks to life and property which are inherent in the transportation of hazardous materials in commerce." HMTA § 103(2) defines "hazardous material" to mean any "substance or material in a quantity and form which may pose an unreasonable risk to health and safety or property when transported in commerce."

The Secretary has broad discretionary authority under HMTA § 104 to designate hazardous materials. Hazardous materials may include, but are not limited to, "explosives, radioactive materials, etiologic agents, flammable liquids or solids, combustible liquids or solids, *poisons*, oxidizing or corrosive materials, and compressed gases." HMTA § 104 (emphasis added).

The Department of Transportation (DOT) manages and regulates transportation of hazardous materials through two offices: the Office of Hazardous Materials Transportation, and the Research and Special Programs Administration. 49 C.F.R. §§ 107.1, .3. These two offices oversee the classification, description, marking, labeling, packaging, and condition of hazardous materials transported in the United States. The DOT regulations for the transportation of hazardous materials are set forth at 49 C.F.R. parts 171-180.

Designated Hazardous Materials

A pesticide is subject to DOT's hazardous materials regulations if it or any of its components is either explicitly listed or falls within one of the generic hazard classifications set forth in the Hazardous Materials Table, 49 C.F.R. § 172.101. The table lists those materials designated as hazardous for transportation in commerce. The table also includes a description of the material, the proper shipping name, class, label, and the type of packaging which applies to each designated hazardous material.

There is a regulatory trend toward including specific pesticides on the list of hazardous materials subject to DOT regulation (e.g., chlorpyrifos and diazinon are on the list of hazardous materials subject to DOT regulations). Many of the hazardous materials already listed in the regulations are used as pesticides or chemical ingredients of pesticides.

General Requirements

If a pesticide is in the Hazardous Materials Table, its shipment in commerce is subject to the identification requirements set forth in 49 C.F.R. part 172. These regulations set forth requirements for: (1) shipping papers, including the description and quantity of the hazardous material and the shipper's certification; (2) marking, which identifies the hazardous material (usually by authorized abbreviations) in each package, freight container, and transport vehicle; (3) labeling, including required labeling, prohibited labeling, placement and specification of labels (e.g., NON-FLAMMABLE GAS, CORROSIVE); and (4) placarding, including prohibited placarding, general specification of placards, and the manner in which they are to be affixed.

Shippers must comply with the shipment and packaging requirements, including quantity limitations and restrictions on the use of tank cars, portable tanks, cargo tanks and cylinders. 49 C.F.R. part 173. These include specific requirements for each category of hazardous materials established under the regulations: (1) explosives; (2) flammable, combustible, and pyrophoric liquids; (3) flammable solids, oxidizers, and organic peroxides; (4) corrosive materials; (5) gases; (6) poisonous materials, irritating materials and etiologic agents; (7) radioactive materials; and (8) other regulated material (ORM).

Shippers also must comply with additional operating, handling and loading requirements based upon the mode of transportation: (1) rail, 49 C.F.R. part 174; (2) aircraft, part 175; (3) vessel, part 176; and (4) carriage by public highway, part 177. Finally, the regulations include manufacturing and testing specifications for packaging containers and tank cars. 49 C.F.R. parts 178-179.

Small Quantities Exemptions

To provide some relief from the shipment and packaging regulations, DOT issued standardized packaging requirements for shipments of small quantities of certain hazardous materials, such as experimental chemicals, product samples, or materials contained in diagnostic kits or medical devices, that fall within the following categories: flammable liquids and solids; oxidizing materials; organic peroxides; corrosive materials; poison B; ORM A, B, and C; and radioactive materials. 49 C.F.R. § 173.4.

A package exempted under this section is not required to meet the other requirements of part 173 so long as it complies with the packaging requirements in § 173.4. These packaging requirements specify that (1) the maximum quantity in an inner receptacle be limited to to thirty milliliters of liquid, thirty grams of solids, or one gram for authorized materials classed as poison B; (2) the inner receptacle is packaged in a second container with material sufficient to cushion and, if necessary, absorb the entire contents of each receptacle containing a liquid; (3) the secondary packaging be placed in outside packaging conforming to the requirements of 49 C.F.R. § 173.24; (4) the entire package be capable of withstanding a six-foot free drop; (5) its gross weight not exceed sixty-five pounds; and (6) the shipper mark the outside of the package to indicate compliance.

ORM-D "Consumer Commodity"

Of special interest to those who prepare pesticides in small packages, such as for home lawn and garden use, is the ORM-D classification. An ORM-D material is defined as:

> [A] material such as a consumer commodity which, though otherwise subject to the regulations of this subchapter, presents a limited hazard during transportation due to its form, quantity and packaging. They must be materials for which exceptions are provided in § 172.101 of this subchapter.

49 C.F.R. § 173.500(b)(4). "Consumer commodity" is defined as:

> [A] material that is packaged and distributed in a form intended or suitable for sale through retail sales agencies or instrumentalities for consumption by individuals for purposes of personal care or household use.

49 C.F.R. § 171.8.

The ORM-D consumer commodity classification relieves the shipper of many of the packaging, labeling, placarding, and shipping paper requirements normally required of other classes of hazardous materials. Instead, the shipper is subject to the preparation and shipping requirements found in 49 C.F.R. § 173.1200. These regulations include specific requirements for those consumer commodities which are flammable liquids, corrosive liquids, corrosive solids, flammable solids, oxidizers, organic peroxides, poison B liquids or solids, and compressed gases (such as aerosols).

Even the requirements for strong outside packaging in § 173.1200 are not required for ORM-D materials when "unitized in cages, carts, or similar overpacks and when shipped by a private or contract motor carrier from a distribution center to a retail outlet." 49 C.F.R. § 173.505(b).

Avoiding Conflict with OSHA's Hazard Communication Standard

As discussed above, a key component of OSHA's HCS is the requirement for manufacturers, importers and distributors to ensure that each container of hazardous chemicals "leaving the workplace" is labeled to ensure that hazard information is passed on to employees. 48 Fed. Reg. 53,280 (1983). In contrast, DOT's labeling regulations are directed at transportation in commerce. To avoid potential conflicts and duplicative regulations, OSHA's HCS requires labeling to be done "in a manner which does not conflict with the requirements of the Hazardous Materials Transportation Act . . . and regulations issued under that Act by the Department of Transportation." 29 C.F.R. § 1910.1200(f)(3).

FEDERAL INSECTICIDE, FUNGICIDE, AND RODENTICIDE ACT (7 U.S.C. §§ 136-136y)

FIFRA § 2. DEFINITIONS. [7 U.S.C. § 136.]

For purposes of this subchapter -

(a) **Active ingredient** — The term "active ingredient" means —

(1) in the case of a pesticide other than a plant regulator, defoliant, or desiccant, an ingredient which will prevent, destroy, repel, or mitigate any pest;

(2) in the case of a plant regulator, an ingredient which, through physiological action, will accelerate or retard the rate of growth or rate of maturation or otherwise alter the behavior of ornamental or crop plants or the product thereof;

(3) in the case of a defoliant, an ingredient which will cause the leaves or foliage to drop from a plant; and

(4) in the case of a desiccant, an ingredient which will artificially accelerate the drying of plant tissue.

(b) **Administrator** — The term "Administrator" means the Administrator of the Environmental Protection Agency.

(c) **Adulterated** — The term "adulterated" applies to any pesticide if —

(1) its strength or purity falls below the professed standard of quality as expressed on its labeling under which it is sold;

(2) any substance has been substituted wholly or in part for the pesticide; or

(3) any valuable constituent of the pesticide has been wholly or in part abstracted.

(d) **Animal** — The term "animal" means all vertebrate and invertebrate species, including but not limited to man and other mammals, birds, fish, and shellfish.

(e) **Certified applicator, etc.**

(1) **Certified applicator** — The term "certified applicator" means any individual who is certified under FIFRA § 4 [136a-1][1] of this title as authorized to use or supervise the use of any pesticide which is classified for restricted use. Any applicator who holds or applies registered pesticides, or use[2] dilutions of registered pesticides consistent with subsection (ee) of this section, only to provide a service of controlling pests without delivering any unapplied pesticide to any person so served is not deemed to be a seller or distributor of pesticides under this subchapter.

---

[1] So in original. Should read "FIFRA § 11 [136i]."

[2] So in original. Probably should be "uses".

(2) **Private applicator** — The term "private applicator" means a certified applicator who uses or supervises the use of any pesticide which is classified for restricted use for purposes of producing any agricultural commodity on property owned or rented by him or his employer or (if applied without compensation other than trading of personal services between producers of agricultural commodities) on the property of another person.

(3) **Commercial applicator** — The term "commercial applicator" means an applicator (whether or not he is a private applicator with respect to some uses) who uses or supervises the use of any pesticide which is classified for restricted use for any purpose or on any property other than as provided by paragraph (2).

(4) **Under the direct supervision of a certified applicator** — Unless otherwise prescribed by its labeling, a pesticide shall be considered to be applied under the direct supervision of a certified applicator if it is applied by a competent person acting under the instructions and control of a certified applicator who is available if and when needed, even though such certified applicator is not physically present at the time and place the pesticide is applied

(f) **Defoliant** — The term "defoliant" means any substance or mixture of substances intended for causing the leaves or foliage to drop from a plant, with or without causing abscission.

(g) **Desiccant** — The term "desiccant" means any substance or mixture of substances intended for artificially accelerating the drying of plant tissue.

(h) **Device** — The term "device" means any instrument or contrivance (other than a firearm) which is intended for trapping, destroying, repelling, or mitigating any pest or any other form of plant or animal life (other than man and other than bacteria, virus, or other microorganism on or in living man or other living animals); but not including equipment used for the application of pesticides when sold separately therefrom.

(i) **District court** — The term "district court" means a United States district court, the District Court of Guam, the District Court of the Virgin Islands, and the highest court of American Samoa.

(j) **Environment** — The term "environment" includes water, air, land, and all plants and man and other animals living therein, and the interrelationships which exist among these.

(k) **Fungus** — The term "fungus" means any non-chlorophyll-bearing thallophyte (that is, any non-chlorophyll-bearing plant of a lower order than mosses and liverworts), as for example, rust, smut, mildew, mold, yeast, and bacteria, except those on or in living man or other animals and those on or in processed food, beverages, or pharmaceuticals.

(l) **Imminent hazard** — The term "imminent hazard" means a situation which exists when the continued use of a pesticide during the time required for cancellation proceeding would be likely to result in unreasonable adverse effects on the environment or will involve unreasonable hazard to the survival of a species declared endangered or threatened by the Secretary pursuant to the Endangered Species Act of 1973 [16 U.S.C. § 1531 et seq.].

(m) **Inert ingredient** — The term "inert ingredient" means an ingredient which is not active.

(n) Ingredient statement — The term "ingredient statement" means a statement which contains—

(1) the name and percentage of each active ingredient, and the total percentage of all inert ingredients, in the pesticide; and

(2) if the pesticide contains arsenic in any form, a statement of the percentages of total and water soluble arsenic, calculated as elementary arsenic.

(o) Insect — The term "insect" means any of the numerous small invertebrate animals generally having the body more or less obviously segmented, for the most part belonging to the class insecta, comprising six-legged, usually winged forms, as for example, beetles, bugs, bees, flies, and to other allied classes of anthropods whose members are wingless and usually have more than six legs, as for example, spiders, mites, ticks, centipedes, and wood lice.

(p) Label and labeling

(1) **Label** — The term "label" means the written, printed, or graphic matter on, or attached to, the pesticide or device or any of its containers or wrappers.

(2) **Labeling** — The term "labeling" means all labels and all other written, printed, or graphic matter—
(A) accompanying the pesticide or device at any time; or
(B) to which reference is made on the label or in literature accompanying the pesticide or device, except to current official publications of the Environmental Protection Agency, the United States Departments of Agriculture and Interior, the Department of Health and Human Services, State experiment stations, State agricultural colleges, and other similar Federal or State institutions or agencies authorized by law to conduct research in the field of pesticides

(q) Misbranded

(1) **A pesticide is misbranded if—**
(A) its labeling bears any statement, design, or graphic representation relative thereto or to its ingredients which is false or misleading in any particular;
(B) it is contained in a package or other container or wrapping which does not conform to the standards established by the Administrator pursuant to FIFRA § 25(c)(3) [136w(c)(3)] of this title;
(C) it is an imitation of, or is offered for sale under the name of, another pesticide;
(D) its label does not bear the registration number assigned under FIFRA § 7 [136e] of this title to each establishment in which it was produced;
(E) any word, statement, or other information required by or under authority of this subchapter to appear on the label or labeling is not prominently placed thereon with such conspicuousness (as compared with other words, statements, designs, or graphic matter in the labeling) and in such terms as to render it likely to be read and understood by the ordinary individual under customary conditions of purchase and use;
(F) the labeling accompanying it does not contain directions for use which are necessary for effecting the purpose for which the product is intended and if complied with, together with any requirements imposed under FIFRA § 3(d) [136a(d)] of this title, are adequate to protect health and the environment;

(G) the label does not contain a warning or caution statement which may be necessary and if complied with, together with any requirements imposed under FIFRA § 3(d) [136a(d)] of this title, is adequate to protect health and the environment; or
(H) in the case of a pesticide not registered in accordance with FIFRA § 3 [136a] of this title and intended for export, the label does not contain, in words prominently placed thereon with such conspicuousness (as compared with other words, statements, designs, or graphic matter in the labeling) as to render it likely to be noted by the ordinary individual under customary conditions of purchase and use, the following: "Not Registered for Use in the United States of America".

(2) A pesticide is misbranded if —
(A) the label does not bear an ingredient statement on that part of the immediate container (and on the outside container or wrapper of the retail package, if there be one, through which the ingredient statement on the immediate container cannot be clearly read) which is presented or displayed under customary conditions or purchase, except that a pesticide is not misbranded under this subparagraph if:
(i) the size of[3] form of the immediate container, or the outside container or wrapper of the retail package, makes it impracticable to place the ingredient statement on the part which is presented or displayed under customary conditions of purchase; and
(ii) the ingredient statement appears prominently on another part of the immediate container, or outside container or wrapper, permitted by the Administrator;
(B) the labeling does not contain a statement of the use classification under which the product is registered;
(C) there is not affixed to its container, and to the outside container or wrapper of the retain package, if there be one, through which the required information on the immediate container cannot be clearly read, a label bearing—
(i) the name and address of the producer, registrant, or person for whom produced
(ii) the name, brand, or trademark under which the pesticide is sold
(iii) the net weight or measure of the content, *except that* the Administrator may permit reasonable variations; and
(iv) when required by regulation of the Administrator to effectuate the purposes of this subchapter, the registration number assigned to the pesticide under this subchapter, and the use classification; and
(D) the pesticide contains any substance or substances in quantities highly toxic to man, unless the label shall bear, in addition to any other matter required by this subchapter—
(i) the skull and crossbones;
(ii) the word "poison" prominently in red on a background of distinctly contrasting color; and
(iii) a statement of a practical treatment (first aid or otherwise) in case of poisoning by the pesticide.

(r) Nematode. — The term "nematode" means invertebrate animals of the phylum nemathelminthes and class nematoda, that is, unsegmented round worms with elongated, fusiform, or saclike bodies covered with cuticle, and inhabiting soil, water, plants, or plant parts; may also be called nemas or eelworms.

3 So in original. Probably should be "or".

(s) **Person.** — The term "person" means any individual, partnership, association, corporation, or any organized group of persons whether incorporated or not.

(t) **Pest.** — The term "pest" means (1) any insect, rodent, nematode, fungus, weed, or (2) any other form of terrestrial or aquatic plant or animal life or virus, bacteria, or other micro-organism (except viruses, bacteria, or other micro-organisms on or in living man or other living animals) which the Administrator declares to be a pest under FIFRA § 25(c)(1) [136w(c)(1)] of this title.

(u) **Pesticide** — The term "pesticide" means (1) any substance or mixture of substances intended for preventing, destroying, repelling, or mitigating any pest, and (2) any substance or mixture of substances intended for use as a plant regulator, defoliant, or desiccant, except that the term "pesticide" shall not include any article that is a "new animal drug" within the meaning of § 321(w) of title 21, that has been determined by the Secretary of Health and Human Services not to be a new animal drug by a regulation establishing conditions of use for the article, or that is an animal feed within the meaning of § 321(x) of Title 21 bearing or containing a new animal drug.

(v) **Plant regulator** — The term "plant regulator" means any substance or mixture of substances intended, through physiological action, for accelerating or retarding the rate of growth or rate of maturation, or for otherwise altering the behavior of plants or the produce thereof, but shall not include substances to the extent that they are intended as plant nutrients, trace elements, nutritional chemicals, plant inoculants, and soil amendments. Also, the term "plant regulator" shall not be required to include any of such of those nutrient mixtures or soil amendments as are commonly known as vitamin-hormone horticultural products, intended for improvement, maintenance, survival, health, and propagation of plants, and as are not for pest destruction and are nontoxic, nonpoisonous in the undiluted packaged concentration.

(w) **Producer and produce** — The term "producer" means the person who manufactures, prepares, compounds, propagates, or processes any pesticide or device or active ingredient used in producing a pesticide. The term "produce" means to manufacture, prepare, compound, propagate, or process any pesticide or device or active ingredient used in producing a pesticide. The dilution by individuals of formulated pesticides for their own use and according to the directions on registered labels shall not of itself result in such individuals being included in the definition of "producer" for the purposes of this subchapter.

(x) **Protect health and the environment** — The terms "protect health and the environment" and "protection of health and the environment" mean protection against any unreasonable adverse effects on the environment.

(y) **Registrant** — The term "registrant" means a person who has registered any pesticide pursuant to the provisions of this subchapter.

(z) **Registration** — The term "registration" includes reregistration.

(aa) **State** — The term "State" means a State, the District of Columbia, the Commonwealth of Puerto Rico, the Virgin Islands, Guam, the Trust Territory of the Pacific Islands, and American Samoa.

(bb) **Unreasonable adverse effects on the environment** — The term "unreasonable adverse effects on the environment" means any unreasonable risk to man or the environment, taking into account the economic, social, and environmental costs and benefits of the use of any pesticide.

(cc) **Weed** — The term "weed" means any plant which grows where not wanted.

(dd) Establishment — The term "establishment" means any place where a pesticide or device or active ingredient used in producing a pesticide is produced, or held, for distribution or sale.

(ee) To use any registered pesticide in a manner inconsistent with its labeling — The term "to use any registered pesticide in a manner inconsistent with its labeling" means to use any registered pesticide in a manner not permitted by the labeling, except that the term shall not include (1) applying a pesticide at any dosage, concentration, or frequency less than that specified on the labeling unless the labeling specifically prohibits deviation from the specified dosage, concentration, or frequency, (2) applying a pesticide against any target pest not specified on the labeling if the application is to the crop, animal, or site specified on the labeling, unless the Administrator has required that the labeling specifically state that the pesticide may be used only for the pests specified on the labeling after the Administrator has determined that the use of the pesticide against other pests would cause an unreasonable adverse effect on the environment, (3) employing any method of application not prohibited by the labeling unless the labeling specifically states that the product may be applied only by the methods specified on the labeling, (4) mixing a pesticide or pesticides with a fertilizer when such mixture is not prohibited by the labeling, (5) any use of a pesticide in conformance with FIFRA §§ 5, 18, or 24 [136c, 136p, or 136v] of this title, or (6) any use of a pesticide in a manner that the Administrator determines to be consistent with the purposes of this subchapter. After March 31, 1979, the term shall not include the use of a pesticide for agricultural or forestry purposes at a dilution less than label dosage unless before or after that date the Administrator issues a regulation or advisory opinion consistent with the study provided for in § 27(b) of the Federal Pesticide Act of 1978, which regulation or advisory opinion specifically requires the use of definite amounts of dilution.

(ff) Outstanding Data Requirement —

(1) **In general** — The term "outstanding data requirement" means a requirement for any study, information, or data that is necessary to make a determination under FIFRA § 3(c)(5) [136a(c)(5)] and which study, information, or data —

(A) has not been submitted to the Administrator; or

(B) if submitted to the Administrator, the Administrator has determined must be resubmitted because it is not valid, complete, or adequate to make a determination under FIFRA § 3(c)(5) [136a(c)(5)] and the regulations and guidelines issued under such section.

(2) Factors — In making a determination under paragraph (1)(B) respecting a study, the Administrator shall examine, at a minimum, relevant protocols, documentation of the conduct and analysis of the study, and the results of the study to determine whether the study and the results of the study fulfill the data requirement for which the study was submitted to the Administrator.

(gg) To Distribute or Sell — The term "to distribute or sell" means to distribute, sell, offer for sale, hold for distribution, hold for sale, hold for shipment, ship, deliver for shipment, release for shipment, or receive and (having so received) deliver or offer to deliver. The term does not include the holding or application of registered pesticides or use dilutions thereof by any applicator who provides a service of controlling pests without delivering any unapplied pesticide to any person so served.

FIFRA § 3. REGISTRATION OF PESTICIDES. **[7 U.S.C. § 136a.]**

(a) Requirement of Registration — Except as provided by this Act, no person in any State may distribute or sell to any person any pesticide that is not registered under this Act. To the extent necessary to prevent unreasonable adverse effects on the environment, the Administrator may by regulation limit the distribution, sale, or use in any State of any pesticide that is not registered under this Act and that is not the subject of an experimental use permit under FIFRA § 5 [136c] or an emergency exemption under FIFRA § 18 [136p].

(b) Exemptions — A pesticide which is not registered with the Administrator may be transferred if—

(1) the transfer is from one registered establishment to another registered establishment operated by the same producer solely for packaging at the second establishment or for use as a constituent part of another pesticide produced at the second establishment; or

(2) the transfer is pursuant to and in accordance with the requirements of an experimental use permit.

(c) Procedure for registration —

(1) Statement required — Each applicant for registration of a pesticide shall file with the Administrator a statement which includes—

(A) the name and address of the applicant and of any other person whose name will appear on the labeling;

(B) the name of the pesticide;

(C) a complete copy of the labeling of the pesticide, a statement of all claims to be made for it, and any directions for its use;

(D) except as otherwise provided in paragraph 2(D) of this section, if requested by the Administrator, a full description of the tests made and the results thereof upon which the claims are based, or alternatively a citation to data that appear in the public literature or that previously had been submitted to the Administrator and that the Administrator may consider in accordance with the following provisions:

(i) with respect to pesticides containing active ingredients that are initially registered under this subchapter after September 30, 1978, data submitted to support the application for the original registration of the pesticide, or an application for an amendment adding any new use to the registration and that pertains solely to such new use, shall not, without the written permission of the original data submitter, be considered by the Administrator to support an application by another person during a period of ten years following the date the Administrator first registers the pesticide, except that such permission shall not be required in the case of defensive data;

(ii) except as otherwise provided in clause (i) of this paragraph, with respect to data submitted after December 31, 1969, by an applicant or registrant to support an application for registration, experimental use permit, or amendment adding a new use to an existing registration, to support or maintain in effect an existing registration, or for reregistration, the Administrator may, without the permission of the original data submitter, consider any such item of data in support of an application by any other person (hereinafter in this subparagraph referred to as the "applicant") within the fifteen-year period following the date the data were originally submitted only if the applicant has made an offer to compensate the original data submitter and submitted such offer to the Administrator accompanied by evidence of delivery to the original data submitter of the offer. The terms and amount of compensation may be fixed by agreement between the original data submitter and the applicant, or, failing such agreement, binding arbitration under this subparagraph. If, at the end of ninety days after the date of delivery to the original data submitter of the offer to compensate, the original data submitter and the applicant have neither agreed on the amount and terms of compensation nor on a procedure for reaching an agreement on the amount and terms of compensation, either person may initiate binding arbitration proceedings by requesting the Federal Mediation and Conciliation Service to appoint an arbitrator from the roster of arbitrators maintained by such Service. The procedure and rules of the Service shall be applicable to the selection of such arbitrator and to such arbitration proceedings, and the findings and determination of the arbitrator shall be final and conclusive, and no official or court of the United States shall have power or jurisdiction to review any such findings and determination, except for fraud, misrepresentation, or other misconduct by one of the parties to the arbitration or the arbitrator where there is a verified complaint with supporting affidavits attesting to specific instances of such fraud, misrepresentation, or other misconduct. The parties to the arbitration shall share equally in the payment of the fee and expenses of the arbitrator. If the Administrator determines that an original data submitter has failed to participate in a procedure for reaching an agreement or in an arbitration proceeding as required by this subparagraph, or failed to comply with the terms of an agreement or arbitration decision concerning compensation under this subparagraph, the original data submitter shall forfeit the right to compensation for the use of the data in support of the application. Notwithstanding any other provision of this subchapter, if the Administrator determines that an applicant has failed to participate in a procedure for reaching an agreement or in an arbitration proceeding as required by this subparagraph, or failed to comply with the terms of an agreement or arbitration decision concerning compensation under this subparagraph, the Administrator shall deny the application or cancel the registration of the pesticide in support of which the data were used without further hearing. Before the Administrator takes action under either of the preceding two sentences, the Administrator shall furnish to the affected person, by certified mail, notice of intent to take action and allow fifteen days from the date of delivery of the notice for the affected person to respond. If a registration is denied or canceled under this subparagraph, the Administrator may make such order as the Administrator deems appropriate concerning the continued sale and use of existing stocks of such pesticide. Registration action by the Administrator shall not be delayed pending the fixing of compensation;

(iii) after expiration of any period of exclusive use and any period for which compensation is required for the use of an item of data under clauses (i) and (ii), the Administrator may consider such item of data in support of an application by any other applicant without the permission of the original data submitter and without an offer having been received to compensate the original data submitter for the use of such item of data;

(E) the complete formula of the pesticide; and
(F) a request that the pesticide be classified for general use, for restricted use, or for both.

(2) Data in support of registration

(A) The Administrator shall publish guidelines specifying the kinds of information which will be required to support the registration of a pesticide and shall revise such guidelines from time to time. If thereafter he requires any additional kind of information under subparagraph (B) of this paragraph, he shall permit sufficient time for applicants to obtain such additional information. The Administrator, in establishing standards for data requirements for the registration of pesticides with respect to minor uses, shall make such standards commensurate with the anticipated extent of use, pattern of use, and the level and degree of potential exposure of man and the environment to the pesticide. *The Administrator shall not require a person to submit, in relation to a registration or reregistration of a pesticide for minor agricultural use under this Act, any field residue data from a geographic area where the pesticide will not be registered for such use.*[4] In the development of these standards, the Administrator shall consider the economic factors of potential national volume of use, extent of distribution, and the impact of the cost of meeting the requirements on the incentives for any potential registrant to undertake the development of the required data. Except as provided by FIFRA § 10 [136h] of this title, within 30 days after the Administrator registers a pesticide under this subchapter he shall make available to the public the data called for in the registration statement together with such other scientific information as he deems relevant to his decision.

(B)(i) If the Administrator determines that additional data are required to maintain in effect an existing registration of a pesticide, the Administrator shall notify all existing registrants of the pesticide to which the determination relates and provide a list of such registrants to any interested person.

(ii) Each registrant of such pesticide shall provide evidence within ninety days after receipt of notification that it is taking appropriate steps to secure the additional data that are required. Two or more registrants may agree to develop jointly, or to share in the cost of developing, such data if they agree and advise the Administrator of their intent within ninety days after notification. Any registrant who agrees to share in the cost of producing the data shall be entitled to examine and rely upon such data in support of maintenance of such registration. The Administrator shall issue a notice of intent to suspend the registration of a pesticide in accordance with the procedures prescribed by clause (iv) if a registrant fails to comply with this clause.

(iii) If, at the end of sixty days after advising the Administrator of their agreement to develop jointly, or share in the cost of developing, data, the registrants have not further agreed on the terms of the data development arrangement or on a procedure for reaching such agreement, any of such registrants may initiate binding arbitration proceedings by requesting the Federal Mediation and Conciliation Service to appoint an arbitrator from the roster of arbitrators maintained by such Service. The procedure and rules of the Service shall be applicable to the selection of such arbitrator and to such arbitration proceedings, and the findings and determination of the arbitrator shall be final and conclusive, and no official or court of the United States shall have power or jurisdiction to review any such findings and determination, except for fraud, misrepresentation, or other misconduct by one of the parties to the arbitration or the arbitrator where there is a verified complaint with supporting affidavits attesting to specific instances of such fraud, misrepresentation, or other misconduct. All

---

[4] Italicized text signifies changes to Statute or new language added by Pub. L. 101-624.

parties to the arbitration shall share equally in the payment of the fee and expenses of the arbitrator. The Administrator shall issue a notice of intent to suspend the registration of a pesticide in accordance with the procedures prescribed by clause (iv) if a registrant fails to comply with this clause.

(iv) Notwithstanding any other provision of this subchapter, if the Administrator determines that a registrant, within the time required by the Administrator, has failed to take appropriate steps to secure the data required under this subparagraph, to participate in a procedure for reaching agreement concerning a joint data development arrangement under this subparagraph or in an arbitration proceeding as required by this subparagraph, or to comply with the terms of an agreement or arbitration decision concerning a joint data development arrangement under this subparagraph, the Administrator may issue a notice of intent to suspend such registrant's registration of the pesticide for which additional data is required. The Administrator may include in the notice of intent to suspend such provisions as the Administrator deems appropriate concerning the continued sale and use of existing stocks of such pesticide. Any suspension proposed under this subparagraph shall become final and effective at the end of thirty days from receipt by the registrant of the notice of intent to suspend, unless during that time a request for hearing is made by a person adversely affected by the notice or the registrant has satisfied the Administrator that the registrant has complied fully with the requirements that served as a basis for the notice of intent to suspend. If a hearing is requested, a hearing shall be conducted under FIFRA § 6(d) [136d(d)] of this title. The only matters for resolution at that hearing shall be whether the registrant has failed to take the action that served as the basis for the notice of intent to suspend the registration of the pesticide for which additional data is required, and whether the Administrator's determination with respect to the disposition of existing stocks is consistent with this subchapter. If a hearing is held, a decision after completion of such hearing shall be final. Notwithstanding any other provision of this subchapter, a hearing shall be held and a determination made within seventy-five days after receipt of a request for such hearing. Any registration suspended under this subparagraph shall be reinstated by the Administrator if the Administrator determines that the registrant has complied fully with the requirements that served as a basis for the suspension of the registration.

(v) Any data submitted under this subparagraph shall be subject to the provisions of paragraph (1)(D) of this section. Whenever such data are submitted jointly by two or more registrants, an agent shall be agreed on at the time of the joint submission to handle any subsequent data compensation matters for the joint submitters of such data.

(C) Within nine months after September 30, 1978, the Administrator shall, by regulation, prescribe simplified procedures for the registration of pesticides, which shall include the provisions of subparagraph (D) of this paragraph.

(D) Exemption — No applicant for registration of a pesticide who proposes to purchase a registered pesticide from another producer in order to formulate such purchased pesticide into the pesticide that is the subject of the application shall be required to—

(i) submit or cite data pertaining to such purchased product;
or
(ii) offer to pay reasonable compensation otherwise required
by paragraph (1)(D) of this subsection for the use of any such data.

(3) **Time for acting with respect to application —**

(A) The Administrator shall review the data after receipt of the application and shall, as expeditiously as possible, either register the pesticide in accordance with paragraph (5), or notify the applicant of his determination that it does not comply with the provisions of the subchapter in accordance with paragraph (6).

(B)(i) The Administrator shall, as expeditiously as possible, review and act on any application received by the Administrator that —

(I) proposes the initial or amended registration of an end-use pesticide that, if registered as proposed, would be identical or substantially similar in composition and labeling to a currently-registered pesticide identified in the application, or that would differ in composition and labeling from such currently-registered pesticide only in ways that would not significantly increase the risk of unreasonable adverse effects on the environment; or

(II) proposes an amendment to the registration of a registered pesticide that does not require scientific review of data.

(ii) In expediting the review of an application for an action described in clause (i), the Administrator shall —

(I) within 45 days after receiving the application, notify the registrant whether or not the application is complete and, if the application is found to be incomplete, reject the application;

(II) within 90 days after receiving a complete application, notify the registrant if the application has been granted or denied; and

(III) if the application is denied, notify the registrant in writing of the specific reasons for the denial of the application.

(4) **Nctice of application** — The Administrator shall publish in the **Federal Register,** promptly after receipt of the statement and other data required pursuant to paragraphs (1) and (2), a notice of each application for registration on any pesticide if it contains any new active ingredient or if it would entail a changed use pattern. The notice shall provide for a period of 30 days in which any Federal agency or any other interested person may comment.

(5) **Approval of registration** — The Administrator shall register a pesticide if he determines that, when considered with any restrictions imposed under subsection (d) of this section—

(A) its composition is such as to warrant the proposed claims for it;

(B) its labeling and other material required to be submitted comply with the requirements of this subchapter;

(C) it will perform its intended function without unreasonable adverse effects on the environment; and

(D) when used in accordance with widespread and commonly recognized practice it will not generally cause unreasonable adverse effects on the environment.

The Administrator shall not make any lack of essentiality a criterion for denying registration of any pesticide. Where two pesticides meet the requirements of this paragraph, one should not be registered in preference to the other. In considering an application for the registration of a pesticide, the Administrator may waive data requirements pertaining to efficacy, in which event the Administrator may register the pesticide without determining that the pesticide's composition is such as to warrant proposed claims of efficacy. If a pesticide is found to be efficacious by any State under FIFRA § 24(c) [136v(c)] of this title, a presumption is established that the Administrator shall waive data requirements pertaining to efficacy for use of the pesticide in such State.

(6) Denial of registration — If the Administrator determines that the requirements of paragraph (5) for registration are not satisfied, he shall notify the applicant for registration of his determination and of his reasons (including the factual basis) therefor, and that, unless the applicant corrects the conditions and notifies the Administrator thereof during the 30-day period beginning with the day after the date on which the applicant receives the notice, the Administrator may refuse to register the pesticide. Whenever the Administrator refuses to register a pesticide, he shall notify the applicant of his decision and of his reasons (including the factual basis) therefor. The Administrator shall promptly publish in the **Federal Register** notice of such denial of registration and the reasons therefor. Upon such notification, the applicant for registration or other interested person with the concurrence of the applicant shall have the same remedies as provided for in FIFRA § 6 [136d] of this title.

(7) Registration under special circumstances — Notwithstanding the provisions of paragraph (5) —

(A) The Administrator may conditionally register or amend the registration of a pesticide if the Administrator determines that (i) the pesticide and proposed use are identical or substantially similar to any currently registered pesticide and use thereof, or differ only in ways that would not significantly increase the risk of unreasonable adverse effects on the environment, and (ii) approving the registration or amendment in the manner proposed by the applicant would not significantly increase the risk of any unreasonable adverse effect on the environment. An applicant seeking conditional registration or amended registration under this subparagraph shall submit such data as would be required to obtain registration of a similar pesticide under paragraph (5). If the applicant is unable to submit an item of data because it has not yet been generated, the Administrator may register or amend the registration of the pesticide under such conditions as will require the submission of such data not later than the time such data are required to be submitted with respect to similar pesticides already registered under this subchapter.

(B) The Administrator may conditionally amend the registration of a pesticide to permit additional uses of such pesticide notwithstanding that data concerning the pesticide may be insufficient to support an unconditional amendment, if the Administrator determines that (i) the applicant has submitted satisfactory data pertaining to the proposed additional use, and (ii) amending the registration in the manner proposed by the applicant would not significantly increase the risk of any unreasonable adverse effect on the environment. Notwithstanding the foregoing provisions of this subparagraph, no registration of a pesticide may be amended to permit an additional use of such pesticide if the Administrator has issued a notice stating that such pesticide, or any ingredient thereof, meets or exceeds risk criteria associated in whole or in part with human dietary exposure enumerated in regulations issued under this subchapter, and during the pendency of any risk-benefit evaluation initiated by such notice, if (I) the additional use of such pesticide involves a major food or feed crop, or (II) the additional use of such pesticide involves a minor food or feed crop and the Administrator determines, with the concurrence of the Secretary of Agriculture, there is available an effective alternative pesticide that does not meet or exceed such risk criteria. An applicant seeking amended registration under this subparagraph shall submit such data as would be required to obtain registration of a similar pesticide under paragraph (5). If the applicant is unable to submit an item of data (other than data pertaining to the proposed additional use) because it has not yet been generated, the Administrator may amend the registration under such conditions as will require the submission of such data not later than the time such data are required to be submitted with respect to similar pesticides already registered under this subchapter.

(C) The Administrator may conditionally register a pesticide containing an active ingredient not contained in any currently registered pesticide for a period reasonably sufficient for the generation and submission of required data (which are lacking because a period reasonably sufficient for generation of the data has not elapsed since the Administrator first imposed the data requirement) on the condition that by the end of such period the Administrator receives such data and the data do not meet or exceed risk criteria enumerated in regulations issued under this subchapter, and on such other conditions as the Administrator may prescribe. A conditional registration under this subparagraph shall be granted only if the Administrator determines that use of the pesticide during such period will not cause any unreasonable adverse effect on the environment, and that use of the pesticide is in the public interest.

(8) **Interim administrative review** — Notwithstanding any other provision of this subchapter, the Administrator may not initiate a public interim administrative review process to develop a risk-benefit evaluation of the ingredients of a pesticide or any of its uses prior to initiating a formal action to cancel, suspend, or deny registration of such pesticide, required under this subchapter, unless such interim administrative process is based on a validated test or other significant evidence raising prudent concerns of unreasonable adverse risk to man or to the environment. Notice of the definition of the terms "validated test" and "other significant evidence" as used herein shall be published by the Administrator in the **Federal Register**.

(d) Classification of pesticides

(1) Classification for general use, restricted use, or both

(A) As a part of the registration of a pesticide the Administrator shall classify it as being for general use or for restricted use. If the Administrator determines that some of the uses for which the pesticide is registered should be for general use and that other uses for which it is registered should be for restricted use, he shall classify it for both general use and restricted use. Pesticide uses may be classified by regulation on the initial classification, and registered pesticides may be classified prior to reregistration. If some of the uses of the pesticide are classified for general use, and other uses are classified for restricted use, the directions relating to its general uses shall be clearly separated and distinguished from those directions relating to its restricted uses. The Administrator may require that its packaging and labeling for restricted uses shall be clearly distinguishable from its packaging and labeling for general uses.

(B) If the Administrator determines that the pesticide, when applied in accordance with its directions for use, warnings and cautions and for the uses for which it is registered, or for one or more of such uses, or in accordance with a widespread and commonly recognized practice, will not generally cause unreasonable adverse effects on the environment, he will classify the pesticide, or the particular use or uses of the pesticide to which the determination applies, for general use.

(C) If the Administrator determines that the pesticide, when applied in accordance with its directions for use, warnings and cautions and for the uses for which it is registered, or for one or more of such uses, or in accordance with a widespread and commonly recognized practice, may generally cause, without additional regulatory restrictions, unreasonable adverse effects on the environment, including injury to the applicator, he shall classify the pesticide, or the particular use or uses to which the determination applies, for restricted use:

(i) If the Administrator classifies a pesticide, or one or more uses of such pesticide, for restricted use because of a determination that the acute dermal or inhalation toxicity of the pesticide presents a hazard to the applicator or other persons, the pesticide shall be applied for any use to which the restricted classification applies only by or under the direct supervision of a certified applicator.

(ii) If the Administrator classifies a pesticide, or one or more uses of such pesticide, for restricted use because of a determination that its use without additional regulatory restriction may cause unreasonable adverse effects on the environment, the pesticide shall be applied for any use to which the determination applies only by or under the direct supervision of a certified applicator, or subject to such other restrictions as the Administrator may provide by regulation. Any such regulation shall be reviewable in the appropriate court of appeals upon petition of a person adversely affected filed within 60 days of the publication of the regulation in final form.

(2) Change in classification — If the Administrator determines that a change in the classification of any use of a pesticide from general use to restricted use is necessary to prevent unreasonable adverse effects on the environment, he shall notify the registrant of such pesticide of such determination at least forty-five days before making the change and shall publish the proposed change in the **Federal Register.** The registrant, or other interested person with the concurrence of the registrant, may seek relief from such determination under FIFRA § 6(b) [136d(b)] of this title.

(3) Change in classification from restricted use to general use — The registrant of any pesticide with one or more uses classified for restricted use may petition the Administrator to change any such classification from restricted to general use. Such petition shall set out the basis for the registrant's position that restricted use classification is unnecessary because classification of the pesticide for general use would not cause unreasonable adverse effects on the environment. The Administrator, within sixty days after receiving such petition, shall notify the registrant whether the petition has been granted or denied. Any denial shall contain an explanation therefor and any such denial shall be subject to judicial review under FIFRA § 16 [136n] of this title.

(e) **Products with same formulation and claims** — Products which have the same formulation, are manufactured by the same person, the labeling of which contains the same claims, and the labels of which bear a designation identifying the product as the same pesticide may be registered as a single pesticide; and additional names and labels shall be added to the registration by supplemental statements.

(f) **Miscellaneous** —

(1) Effect of change of labeling or formulation
If the labeling or formulation for a pesticide is changed, the registration shall be amended to reflect such change if the Administrator determines that the change will not violate any provision of this subchapter.

(2) Registration not a defense
In no event shall registration of an article be construed as a defense for the commission of any offense under this subchapter. As long as no cancellation proceedings are in effect registration of a pesticide shall be prima facie evidence that the pesticide, its labeling and packaging comply with the registration provisions of the subchapter.

(3) Authority to consult other Federal agencies
In connection with consideration of any registration or application for registration under this section, the Administrator may consult with any other Federal agency.

FIFRA § 4. REREGISTRATION OF REGISTERED PESTICIDES. [7 U.S.C. § 136b.]

(a) General rule — The Administrator shall reregister, in accordance with this section, each registered pesticide containing any active ingredient contained in any pesticide first registered before November 1,

1984, except for any pesticide as to which the Administrator has determined, after November 1, 1984, and before the effective date of this section, that —

(1) there are no outstanding data requirements; and

(2) the requirements of FIFRA § 3(c)(5) [136a(c)(5)] have been satisfied.

(b) Reregistration phases
Reregistrations of pesticides under this section shall be carried out in the following phases:

(1) The first phase shall include the listing under subsection (c) of the active ingredients of the pesticides that will be reregistered.

(2) The second phase shall include the submission to the Administrator under subsection (d) of notices by registrants respecting their intention to seek reregistration, identification by registrants of missing and inadequate data for such pesticides, and commitments by registrants to replace such missing or inadequate data within the applicable time period.

(3) The third phase shall include submission to the Administrator by registrants of the information required under subsection (e).

(4) The fourth phase shall include an independent, initial review by the Administrator under subsection (f) of submissions under phases two and three, identification of outstanding data requirements, and the issuance, as necessary, of requests for additional data.

(5) The fifth phase shall include the review by the Administrator under subsection (g) of data submitted for reregistration and appropriate regulatory action by the Administrator.

(c) Phase One

(1) Priority for reregistration
For purposes of the reregistration of the pesticides described in subsection (a), the Administrator shall list the active ingredients of pesticides and shall give priority to, among others, active ingredients (other than active ingredients for which registration standards have been issued before the effective date of this section) that —

(A) are in use on or in food or feed and may result in postharvestresidues;

(B) may result in residues of potential toxicological concern in potable ground water, edible fish, or shellfish;

(C) have been determined by the Administrator before the effective date of this section to have significant outstanding data requirements; or

(D) are used on crops, including in greenhouses and nurseries, where worker exposure is most likely to occur.

(2) Reregistration lists
For purposes of reregistration under this section, the Administrator shall by order —

(A) not later than 70 days after the effective date of this section, list pesticide active ingredients for which registration standards have been issued before such effective date;

(B) not later than 4 months after such effective date, list the first 150 pesticide active ingredients, as determined under paragraph (1);

(C) not later than 7 months after such effective date, list the second 150 pesticide active ingredients, as determined under paragraph (1); and

(D) not later than 10 months after such effective date, list the remainder of the pesticide active ingredients, as determined under paragraph (1).

Each list shall be published in the **Federal Register.**

(3) Judicial review — The content of a list issued by the Administrator under paragraph (2) shall not be subject to judicial review.

(4) Notice to registrants — On the publication of a list of pesticide active ingredients under paragraph (2), the Administrator shall send by certified mail to the registrants of the pesticides containing such active ingredients a notice of the time by which the registrants are to notify the Administrator under subsection (d) whether the registrants intend to seek or not to seek reregistration of such pesticides.

(d) Phase Two

(1) In general

The registrant of a pesticide that contains an active ingredient listed under subparagraph (B), (C), or (D) of subsection (c)(2) shall submit to the Administrator, within the time period prescribed by paragraph (4), the notice described in paragraph (2) and any information, commitment, or offer described in paragraph (3).

(2) Notice of intent to seek or not to seek reregistration

(A) The registrant of a pesticide containing an active ingredient listed under subparagraph (B), (C), or (D) of subsection (c)(2) shall notify the Administrator by certified mail whether the registrant intends to seek or does not intend to seek reregistration of the pesticide.

(B) If a registrant submits a notice under subparagraph (A) of an intention not to seek reregistration of a pesticide, the Administrator shall publish a notice in the **Federal Register** stating that such a notice has been submitted.

(3) Missing or inadequate data — Each registrant of a pesticide that contains an active ingredient listed under subparagraph (B), (C), or (D) of subsection (c)(2) and for which the registrant submitted a notice under paragraph (2) of an intention to seek reregistration of such pesticide shall submit to the Administrator —

(A) in accordance with regulations issued by the Administrator under FIFRA § 3 [136a], an identification of —

(i) all data that are required by regulation to support the registration of the pesticide with respect to such active ingredient;

(ii) data that were submitted by the registrant previously in support of the registration of the pesticide that are inadequate to meet such regulations; and

(iii) data identified under clause (i) that have not been submitted to the Administrator; and

(B) either —

(i) a commitment to replace the data identified under subparagraph (A)(ii) and submit the data identified under subparagraph (A)(iii) within the applicable time period prescribed by paragraph (4)(B); or

(ii) an offer to share in the cost to be incurred by a person who has made a commitment under clause (i) to replace or submit the data and an offer to submit to arbitration as described by FIFRA § 3(c)(2)(B) [136a(c)(2)(B)] with regard to such cost sharing.

For purposes of a submission by a registrant under subparagraph (A)(ii), data are inadequate if the data are derived from a study with respect to which the registrant is unable to make the certification prescribed by subsection (e)(1)(G) that the registrant possesses or has access to the raw data used in or generated by such study. For purposes of a submission by a registrant under such subparagraph, data shall be considered to be inadequate if the data are derived from a study submitted before January 1, 1970, unless it is demonstrated to the satisfaction of the Administrator that such data should be considered to support the registration of the pesticide that is to be reregistered.

(4) Time periods

(A) A submission under paragraph (2) or (3) shall be made —

(i) in the case of a pesticide containing an active ingredient listed under subsection (c)(2)(B), not later than 3 months after the date of publication of the listing of such active ingredient;

(ii) in the case of a pesticide containing an active ingredient listed under subsection (c)(2)(C), not later than 3 months after the date of publication of the listing of such active ingredient; and

(iii) in the case of a pesticide containing an active ingredient listed under subsection (c)(2)(D), not later than 3 months after the date of publication of the listing of such active ingredient.

On application, the Administrator may extend a time period prescribed by this subparagraph if the Administrator determines that factors beyond the control of the registrant prevent the registrant from complying with such period.

(B) A registrant shall submit data in accordance with a commitment entered into under paragraph (3)(B) within a reasonable period of time, as determined by the Administrator, but not more than 48 months after the date the registrant submitted the commitment. The Administrator, on application of a registrant, may extend the period prescribed by the preceding sentence by no more than 2 years if extraordinary circumstances beyond the control of the registrant prevent the registrant from submitting data within such prescribed period.

(5) Cancellation and removal —

(A) If the registrant of a pesticide does not submit a notice under paragraph (2) or (3) within the time prescribed by paragraph (4)(A), the Administrator shall issue a notice of intent to cancel the registration of such registrant for such pesticide and shall publish the notice in the **Federal Register** and allow 60 days for the submission of comments on the notice. On expiration of such 60 days, the Administrator, by order and without a hearing, may cancel the registration or take such other action, including extension of applicable time periods, as may be necessary to enable reregistration of such pesticide by another person.

(B)(i)If —

(I) no registrant of a pesticide containing an active ingredient listed under subsection (c)(2) notifies the Administrator under paragraph (2) that the registrant intends to seek reregistration of any pesticide containing that active ingredient;

(II) no such registrant complies with paragraph (3)(A); or

(III) no such registrant makes a commitment under paragraph (3)(B) to replace or submit all data described in clauses (ii) and (iii) of paragraph (3)(A);

the Administrator shall publish in the **Federal Register** a notice of intent to remove the active ingredient from the list established under subsection (c)(2) and a notice of intent to cancel the registrations of all pesticides containing such active ingredient and shall provide 60 days for comment on such notice.

(ii) After the 60-day period has expired, the Administrator, by order, may cancel any such registration without hearing, except that the Administrator shall not cancel a registration under this subparagraph if —

(I) during the comment period a person acquires the rights of the registrant in that registration;
(II) during the comment period that person furnishes a notice of intent to reregister the pesticide in accordance with paragraph (2); and
(III) not later than 120 days after the publication of the notice under this subparagraph, that person has complied with paragraph (3) and the fee prescribed by subsection (i)(1) has been paid.

(6) Suspensions and penalties — The Administrator shall issue a notice of intent to suspend the registration of a pesticide in accordance with the procedures prescribed by FIFRA § 3(c)(2)(B)(iv) [136a(c)(2)(B)(iv)] if the Administrator determines that (A) progress is insufficient to ensure the submission of the data required for such pesticide under a commitment made under paragraph (3)(B) within the time period prescribed by paragraph (4)(B) or (B) the registrant has not submitted such data to the Administrator within such time period.

(e) **Phase Three** —

(1) Information about studies — Each registrant of a pesticide that contains an active ingredient listed under subparagraph (B), (C), or (D) of subsection (c)(2) who has submitted a notice under subsection (d)(2) of an intent to seek the reregistration of such pesticide shall submit, in accordance with the guidelines issued under paragraph (4), to the Administrator —

(A) a summary of each study concerning the active ingredient previously submitted by the registrant in support of the registration of a pesticide containing such active ingredient and considered by the registrant to be adequate to meet the requirements of FIFRA § 3 [136a] and the regulations issued under such section;
(B) a summary of each study concerning the active ingredient previously submitted by the registrant in support of the registration of a pesticide containing such active ingredient that may not comply with the requirements of FIFRA § 3 [136a] and the regulations issued under such section but which the registrant asserts should be deemed to comply with such requirements and regulations;
(C) a reformat of the data from each study summarized under subparagraph (A) or (B) by the registrant concerning chronic dosing, oncogenicity, reproductive effects, mutagenicity, neurotoxicity, teratogenicity, or residue chemistry of the active ingredient that were submitted to the Administrator before January 1, 1982;
(D) where data described in subparagraph (C) are not required for the active ingredient by regulations issued under FIFRA § 3 [136a], a reformat of acute and subchronic dosing data submitted by the registrant to the Administrator before January 1, 1982, that the registrant considers to be adequate to meet the requirements of FIFRA § 3 [136a] and the regulations issued under such section;
(E) an identification of data that are required to be submitted to the Administrator under FIFRA § 6(a)(2) [136d(a)(2)] indicating an adverse effect of the pesticide;
(F) an identification of any other information available that in the view of the registrant supports the registration;
(G) a certification that the registrant or the Administrator possesses or has access to the raw data used in or generated by the studies that the registrant summarized under subparagraph (A) or (B);
(H) either —

(i) a commitment to submit data to fill each outstanding data requirement identified by the registrant; or

(ii) an offer to share in the cost of developing such data to be incurred by a person who has made a commitment under clause (i) to submit such data, and an offer to submit to arbitration as described by FIFRA § 3(c)(2)(B) [136a(c)(2)(B)] with regard to such cost sharing; and

(I) evidence of compliance with FIFRA § 3(c)(1)(D)(ii) [136a(c)(1)(D)(ii)] and regulations issued thereunder with regard to previously submitted data as if the registrant were now seeking the original registration of the pesticide.

A registrant who submits a certification under subparagraph (G) that is false shall be considered to have violated this Act and shall be subject to the penalties prescribed by FIFRA § 14 [136l].

(2) Time periods

(A) The information required by paragraph (1) shall be submitted to the Administrator

(i) in the case of a pesticide containing an active ingredient listed under subsection (c)(2)(B), not later than 12 months after the date of publication of the listing of such active ingredient;

(ii) in the case of a pesticide containing an active ingredient listed under subsection (c)(2)(C), not later than 12 months after the date of publication of the listing of such active ingredient; and

(iii) in the case of a pesticide containing an active ingredient listed under subsection (c)(2)(D), not later than 12 months after the date of publication of the listing of such active ingredient.

(B) A registrant shall submit data in accordance with a commitment entered into under paragraph (1)(H) within a reasonable period of time, as determined by the Administrator, but not more than 48 months after the date the registrant submitted the commitment under such paragraph. The Administrator, on application of a registrant, may extend the period prescribed by the preceding sentence by no more than 2 years if extraordinary circumstances beyond the control of the registrant prevent the registrant from submitting data within such prescribed period.

(3) Cancellation

(A) If the registrant of a pesticide fails to submit the information required by paragraph (1) within the time prescribed by paragraph (2), the Administrator, by order and without hearing, shall cancel the registration of such pesticide.

(B)

(i) If the registrant of a pesticide submits the information required by paragraph (1) within the time prescribed by paragraph (2) and such information does not conform to the guidelines for submissions established by the Administrator, the Administrator shall determine whether the registrant made a good faith attempt to conform its submission to such guidelines.

(ii) If the Administrator determines that the registrant made a good faith attempt to conform its submission to such guidelines, the Administrator shall provide the registrant a reasonable period of time to make any necessary changes or corrections.

(iii)

(I) If the Administrator determines that the registrant did not make a good faith attempt to conform its submission to such guidelines, the Administrator may issue a notice of intent to cancel the registration. Such a notice shall be sent to the registrant by certified mail.

(II) The registration shall be canceled without a hearing or further notice at the end of 30 days after receipt by the registrant of the notice unless during that time a request for a hearing is made by the registrant.

(III) If a hearing is requested, a hearing shall be conducted under FIFRA § 6(d) [136d(d)], except that the only matter for resolution at the hearing shall be whether the registrant made a good faith attempt to conform its submission to such guidelines. The hearing shall be held and a determination made within 75 days after receipt of a request for hearing.

(4) Guidelines —
(A) Not later than 1 year after the effective date of this section, the Administrator, by order, shall issue guidelines to be followed by registrants in —
(i) summarizing studies;
(ii) reformatting studies;
(iii) identifying adverse information; and
(iv) identifying studies that have been submitted previously that may not meet the requirements of FIFRA § 3 [136a] or regulations issued under such section, under paragraph (1).

(B) Guidelines issued under subparagraph (A) shall not be subject to judicial review.

(5) Monitoring —The Administrator shall monitor the progress of registrants in acquiring and submitting the data required under paragraph (1).

(f) Phase Four —

(1) Independent review and identification of outstanding data requirements —
(A) The Administrator shall review the submissions of all registrants of pesticides containing a particular active ingredient under subsections (d)(3) and (e)(1) to determine if such submissions identified all the data that are missing or inadequate for such active ingredient. To assist the review of the Administrator under this subparagraph, the Administrator may require a registrant seeking reregistration to submit complete copies of studies summarized under subsection (e)(1).
(B) The Administrator shall independently identify and publish in the **Federal Register** the outstanding data requirements for each active ingredient that is listed under subparagraph (B), (C), or (D) of subsection (c)(2) and that is contained in a pesticide to be reregistered under this section. The Administrator, at the same time, shall issue a notice under FIFRA § 3(c)(2)(B) [136a(c)(2)(B)] for the submission of the additional data that are required to meet such requirements.

(2) Time periods —
(A) The Administrator shall take the action required by paragraph (1) —
(i) in the case of a pesticide containing an active ingredient listed under subsection (c)(2)(B), not later than 18 months after the date of the listing of such active ingredient;
(ii) in the case of a pesticide containing an active ingredient listed under subsection (c)(2)(C), not later than 24 months after the date of the listing of such active ingredient;
(iii) in the case of a pesticide containing an active ingredient listed under subsection (c)(2)(D), not later than 33 months after the date of the listing of such active ingredient.
(B) If the Administrator issues a notice to a registrant under paragraph (1)(B) for the submission of additional data, the registrant shall submit such data within a reasonable period of time, as determined by the Administrator, but not to exceed 48 months after the issuance of such notice. The Administrator, on application of a registrant, may extend the period prescribed by the preceding sentence by no more than 2 years if extraordinary circumstances beyond the control of the registrant prevent the registrant from submitting data within such prescribed period.

(3) **Suspensions and penalties** — The Administrator shall issue a notice of intent to suspend the registration of a pesticide in accordance with the procedures prescribed by FIFRA § 3(c)(2)(B)(iv) [136a(c)(2)(B)(iv)] if the Administrator determines that (A) tests necessary to fill an outstanding data requirement for such pesticide have not been initiated within 1 year after the issuance of a notice under paragraph (1)(B), or (B) progress is insufficient to ensure submission of the data referred to in clause (A) within the time period prescribed by paragraph (2)(B) or the required data have not been submitted to the Administrator within such time period.

(g) **Phase Five** —

(1) **Data review**
The Administrator shall conduct a thorough examination of all data submitted under this section concerning an active ingredient listed under subsection (c)(2) and of all other available data found by the Administrator to be relevant.

(2) **Reregistration and other actions** —

(A) Within 1 year after the submission of all data concerning an active ingredient of a pesticide under subsection (f), the Administrator shall determine whether pesticides containing such active ingredient are eligible for reregistration. For extraordinary circumstances, the Administrator may extend such period for not more than 1 additional year.

(B) Before reregistering a pesticide, the Administrator shall obtain any needed product-specific data regarding the pesticide by use of FIFRA § 3(c)(2)(B) [136a(c)(2)(B)] and shall review such data within 90 days after its submission. The Administrator shall require that data under this subparagraph be submitted to the Administrator not later than 8 months after a determination of eligibility under subparagraph (A) has been made for each active ingredient of the pesticide, unless the Administrator determines that a longer period is required for the generation of the data.

(C) After conducting the review required by paragraph (1) for each active ingredient of a pesticide and the review required by subparagraph (B) of this paragraph, the Administrator shall determine whether to reregister a pesticide by determining whether such pesticide meets the requirements of FIFRA § 3(c)(5) [136a(c)(5).] If the Administrator determines that a pesticide is eligible to be reregistered, the Administrator shall reregister such pesticide within 6 months after the submission of the data concerning such pesticide under subparagraph (B).

(D) If after conducting a review under paragraph (1) or subparagraph (B) of this paragraph the Administrator determines that a pesticide should not be reregistered, the Administrator shall take appropriate regulatory action.

(h) **Compensation of data submitter** — If data that are submitted by a registrant under subsection (d), (e), (f), or (g) are used to support the application of another person under FIFRA § 3 [136a], the registrant who submitted such data shall be entitled to compensation for the use of such data as prescribed by FIFRA § 3(c)(1)(D) [136a(c)(1)(D)]. In determining the amount of such compensation, the fees paid by the registrant under this section shall be taken into account.

(i) Fees —

(1)**Initial fee for food or feed use pesticide active ingredients** — The registrants of pesticides that contain an active ingredient that is listed under subparagraph (B), (C), or (D) of subsection (c)(2) and that is an active ingredient of any pesticide registered for a major food or feed use shall collectively pay a fee of $50,000 on submission of information under paragraphs (2) and (3) of subsection (d) for such ingredient.

(2) **Final fee for food or feed use pesticide active ingredients —**

(A) The registrants of pesticides that contain an active ingredient that is listed under subparagraph (B), (C), or (D) of subsection (c)(2) and that is an active ingredient of any pesticide registered for a major food or feed use shall collectively pay a fee of $100,000 —

(i) on submission of information for such ingredient under subsection (e)(1) if data are reformatted under subsection (e)(1)(C); or

(ii) on submission of information for such ingredient under subsection (e)(2)(B) if data are reformatted under subsection (e)(1)(C).

(B) The registrants of pesticides that contain an active ingredient that is listed under subsection (c)(2)(A) and that is an active ingredient of any pesticide registered for a major food or feed use shall collectively pay a fee of $150,000 at such time as the Administrator shall prescribe.

(3) **Fees for other pesticide active ingredients —**

(A) The registrants of pesticides that contain an active ingredient that is listed under subparagraph (B), (C), or (D) of subsection (c)(2) and that is not an active ingredient of any pesticide registered for a major food or feed use shall collectively pay fees in amounts determined by the Administrator. Such fees may not be less than one-half of, nor greater than, the fees required by paragraphs (1) and (2). A registrant shall pay such fees at the times corresponding to the times fees prescribed by paragraphs (1) and (2) are to be paid.

(B) The registrants of pesticides that contain an active ingredient that is listed under subsection (c)(2)(A) and that is not an active ingredient of any pesticide that is registered for a major food or feed use shall collectively pay a fee of not more than $100,000 and not less than $50,000 at such time as the Administrator shall prescribe.

(4) **Reduction or waiver of fees for minor use and other pesticides**

(A) An active ingredient that is contained only in pesticides that are registered solely for agricultural or nonagricultural minor uses, or a pesticide the value or volume of use of which is small, shall be exempt from the fees prescribed by paragraph (3).

(B) An antimicrobial active ingredient, the production level of which does not exceed 1,000,000 pounds per year, shall be exempt from the fees prescribed by paragraph (3). For purposes of this subparagraph, the term "antimicrobial active ingredient" means any active ingredient that is contained only in pesticides that are not registered for any food or feed use and that are —

(i) sanitizers intended to reduce the number of living bacteria or viable virus particles on inanimate surface or in water or air;

(ii) bacteriostats intended to inhibit the growth of bacteria in the presence of moisture;

(iii) disinfectants intended to destroy or irreversibly inactivate bacteria, fungi, or viruses on surfaces or inanimate objects;

(iv) sterilizers intended to destroy viruses and all living bacteria, fungi, and their spores on inanimate surfaces; or

(v) fungicides or fungistats.

(C)(i) Notwithstanding any other provision of this subsection, in the case of a small business registrant of a pesticide, the registrant shall pay a fee for the reregistration of each active ingredient of the pesticide that does not exceed an amount determined in accordance with this subparagraph.

(ii) If during the 3-year period prior to reregistration the average annual gross revenue of the registrant from pesticides containing such active ingredient is —

(I) less than $5,000,000, the registrant shall pay 0.5 percent of such revenue;

(II) $5,000,000 or more but less than $10,000,000, the registrant shall pay 1 percent of such revenue; or

(III) $10,000,000 or more, the registrant shall pay 1.5 percent of such revenue, but not more than $150,000.

(iii) For the purpose of this subparagraph, a small business registrant is a corporation, partnership, or unincorporated business that —

(I) has 150 or fewer employees; and

(II) during the 3-year period prior to reregistration, had an average annual gross revenue from chemicals that did not exceed $40,000,000.

(5) Maintenance fee

(A) Subject to other provisions of this paragraph, each registrant of a pesticide shall pay an annual fee by March 1 of each year of —

(i) in the case of a registrant holding not more than 50 pesticide registrations, $425 for each registration; and

(ii) in the case of a registrant holding more than 50 pesticide registrations —

(I) $425 for each registration up to 50 registrations; and

(II) $100 for each registration over 50 registrations,

except that no fee shall be charged for more than 200 registrations held by any registrant. In the case of a pesticide that is registered for a minor agricultural use, the Administrator may reduce or waive the payment of the fee imposed under this subparagraph if the Administrator determines that the fee would significantly reduce the availability of the pesticide for the use.

(B) The amount of each fee prescribed under subparagraph (A) shall be adjusted by the Administrator to a level that will result in the collection under this paragraph of, to the extent practicable, an aggregate amount of $14,000,000 each fiscal year.

(C)

(i) The maximum annual fee payable by a registrant under clause (i) of subparagraph (A) (as adjusted under subparagraph (B)) shall be $20,000.

(ii) The maximum annual fee payable by a registrant under clause (ii) of subparagraph (A) (as adjusted under subparagraph (B)) shall be $35,000.

(D) If any fee prescribed by this paragraph with respect to the registration of a pesticide is not paid by a registrant by the time prescribed, the Administrator, by order and without hearing, may cancel the registration.

(E) The authority provided under this paragraph shall terminate on September 30, 1997.

(6) Other fees — During the period beginning on the date of enactment of this section and ending on September 30, 1997, the Administrator may not levy any other fees for the registration of a pesticide under this Act except as provided in paragraphs (1) through (5).

(7) **Apportionment —**

(A) If two or more registrants are required to pay any fee prescribed by paragraph (1), (2), or (3) with respect to a particular active ingredient, the fees for such active ingredient shall be apportioned among such registrants on the basis of the market share in United States sales of the active ingredient for the 3 calendar years preceding the date of payment of such fee, except that —

(i) small business registrants that produce the active ingredient shall pay fees in accordance with paragraph (4)(C); and

(ii) registrants who have no market share but who choose to reregister a pesticide containing such active ingredient shall pay the lesser of —

(I) 15 percent of the reregistration fee; or

(II) a proportionate amount of such fee based on the lowest percentage market share held by any registrant active in the marketplace.

In no event shall registrants who have no market share but who choose to reregister a pesticide containing such active ingredient collectively pay more than 25 percent of the total active ingredient reregistration fee.

(B) The Administrator, by order, may require any registrant to submit such reports as the Administrator determines to be necessary to allow the Administrator to determine and apportion fees under this subsection or to determine the registrant's eligibility for a reduction or waiver of a fee.

(C) If any such report is not submitted by a registrant after receiving notice of such report requirement, or if any fee prescribed by this subsection (other than paragraph (5)) for an active ingredient is not paid by a registrant to the Administrator by the time prescribed under this subsection, the Administrator, by order and without hearing, may cancel each registration held by such registrant of a pesticide containing the active ingredient with respect to which the fee is imposed. The Administrator shall reapportion the fee among the remaining registrants and notify the registrants that the registrants are required to pay to the Administrator any unpaid balance of the fee within 30 days after receipt of such notice.

(j) **Exemption of certain registrants** — The requirements of subsections (d), (e), (f), and (i) (other than subjection (i)(5)) regarding data concerning an active ingredient and fees for review of such data shall not apply to any person who is the registrant of a pesticide to the extent that, under FIFRA § 3(c)(2)(D) [136a(c)(2)(D)], the person would not be required to submit or cite such data to obtain an initial registration of such pesticide.

(k) **Reregistration and expedited processing fund —**

(1) **Establishment** — There shall be established in the Treasury of the United States a reregistration and expedited processing fund.

(2) **Source and use** — All fees collected by the Administrator under subsection (i) shall be deposited into the fund and shall be available to the Administrator, without fiscal year limitation, to carry out reregistration and expedited processing of similar applications.

(3) **Expedited processing of similar applications —**

(A) The Administrator shall use each fiscal year not more than $2,000,000 of the amounts in the fund to obtain sufficient personnel and resources to assure the expedited processing and review of any application that —

(i) proposes the initial or amended registration of an end-use pesticide that, if registered as proposed, would be identical or substantially similar in composition and labeling to a currently-registered pesticide identified in the application, or that would differ in composition and labeling from any such currently-registered pesticide only in ways that would not significantly increase the risk of unreasonable adverse effects on the environment; or

(ii) proposes an amendment to the registration of a registered pesticide that does not require scientific review of data.

(B) Any amounts made available under subparagraph (A) shall be used to obtain sufficient personnel and resources to carry out the activities described in such subparagraph that are in addition to the personnel and resources available to carry out such activities on the date of enactment of this section.

(4) Unused funds — Money in the fund not currently needed to carry out this section shall be —

(A) maintained on hand or on deposit;

(B) invested in obligations of the United States or guaranteed thereby;

(C) invested in obligations, participations, or other instruments that are lawful investments for fiduciary, trust, or public funds.

(5) Accounting — The Administrator shall —

(A) provide an annual accounting of the fees collected and disbursed from the fund; and

(B) take all steps necessary to ensure that expenditures from such funds are used only to carry out this section.

(l) Judicial review — Any failure of the Administrator to take any action required by this section shall be subject to judicial review under the procedures prescribed by FIFRA § 16(b) [136n(b)].

FIFRA § 5. EXPERIMENTAL USE PERMITS. [7 U.S.C. § 136c.]

(a) Issuance — Any person may apply to the Administrator for an experimental use permit for a pesticide. The Administrator shall review the application. After completion of the review, but not later than one hundred and twenty days after receipt of the application and all required supporting data, the Administrator shall either issue the permit or notify the applicant of the Administrator's determination not to issue the permit and the reasons therefor. The applicant may correct the application or request a waiver of the conditions for such permit within thirty days of receipt by the applicant of such notification. The Administrator may issue an experimental use permit only if the Administrator determines that the applicant needs such permit in order to accumulate information necessary to register a pesticide under FIFRA § 3 [136a] of this title. An application for an experimental use permit may be filed at any time.

(b) Temporary tolerance level — If the Administrator determines that the use of a pesticide may reasonably be expected to result in any residue on or in food or feed, he may establish a temporary tolerance level for the residue of the pesticide before issuing the experimental use permit.

(c) Use under permit — Use of a pesticide under an experimental use permit shall be under the supervision of the Administrator, and shall be subject to such terms and conditions and be for such period of time as the Administrator may prescribe in the permit.

(d) Studies — When any experimental use permit is issued for a pesticide containing any chemical or combination of chemicals which has not been included in any previously registered pesticide, the Administrator may specify that studies be conducted to detect whether the use of the pesticide under the permit may cause unreasonable adverse effects on the environment. All results of such studies shall be reported to the Administrator before such pesticide may be registered under FIFRA § 3 [136a] of this title.

(e) Revocation — The Administrator may revoke any experimental use permit, at any time, if he finds that its terms or conditions are being violated, or that its terms and conditions are inadequate to avoid unreasonable adverse effects on the environment.

(f) State issuance of permits — Notwithstanding the foregoing provisions of this section, the Administrator shall, under such terms and conditions as he may by regulations prescribe, authorize any State to issue an experimental use permit for a pesticide. All provisions of FIFRA § 11 [136i] of this title relating to State plans shall apply with equal force to a State plan for the issuance of experimental use permits under this section.

(g) Exemption for agricultural research agencies — Notwithstanding the foregoing provisions of this section, the Administrator may issue an experimental use permit for a pesticide to any public or private agricultural research agency or educational institution which applies for such permit. Each permit shall not exceed more than a one-year period or such other specific time as the Administrator may prescribe. Such permit shall be issued under such terms and conditions restricting the use of the pesticide as the Administrator may require. Such pesticide may be used only by such research agency or educational institution for purposes of experimentation.

FIFRA § 6. ADMINISTRATIVE REVIEW; SUSPENSION. [7 U.S.C. § 136d.]

(a) Cancellation after five years —

(1) Procedure — The Administrator shall cancel the registration of any pesticide at the end of the five-year period which begins on the date of its registration (or at the end of any five-year period thereafter) unless the registrant, or other interested person with the concurrence of the registrant, before the end of such period, requests in accordance with regulations prescribed by the Administrator that the registration be continued in effect. The Administrator may permit the continued sale and use of existing stocks of a pesticide whose registration is canceled under this subsection or subsection (b) of this section to such extent, under such conditions, and for such uses as he may specify if he determines that such sale or use is not inconsistent with the purposes of this subchapter and will not have unreasonable adverse effects on the environment. The Administrator shall publish in the **Federal Register**, at least 30 days prior to the expiration of such five-year period, notice that the registration will be canceled if the registrant or other interested person with the concurrence of the registrant does not request that the registration be continued in effect.

(2) Information — If at any time after the registration of a pesticide the registrant has additional factual information regarding unreasonable adverse effects on the environment of the pesticide, he shall submit, such information to the Administrator.

(b) Cancellation and change in classification — If it appears to the Administrator that a pesticide or its labeling or other material required to be submitted does not comply with the provisions of this subchapter or, when used in accordance with widespread and commonly recognized practice, generally causes unreasonable adverse effects on the environment, the Administrator may issue a notice of his intent either —

(1) to cancel its registration or to change its classification together with the reasons (including the factual basis) for his action, or

(2) to hold a hearing to determine whether or not its registration should be canceled or its classification changed.

Such notice shall be sent to the registrant and made public. In determining whether to issue any such notice, the Administrator shall include among those factors to be taken into account the impact of the action proposed in such notice on production and prices of agricultural commodities, retail food prices, and otherwise on the agricultural economy. At least 60 days prior to sending such notice to the registrant or making public such notice, whichever occurs first, the Administrator shall provide the Secretary of Agriculture with a copy of such notice and an analysis of such impact on the agricultural economy. If the Secretary comments in writing to the Administrator regarding the notice and analysis within 30 days after receiving them, the Administrator shall publish in the **Federal Register** (with the notice) the comments of the Secretary and the response of the Administrator with regard to the Secretary's comments. If the Secretary does not comment in writing to the Administrator regarding the notice and analysis within 30 days after receiving them, the Administrator may notify the registrant and make public the notice at any time after such 30-day period notwithstanding the foregoing 60-day time requirement. The time requirements imposed by the preceding 3 sentences may be waived or modified to the extent agreed upon by the Administrator and the Secretary. Notwithstanding any other provisions of this subsection and FIFRA § 25(d) [136w(d)] of this title, in the event that the Administrator determines that suspension of a pesticide registration is necessary to prevent an imminent hazard to human health, then upon such a finding the Administrator may waive the requirement of notice to and consultation with the Secretary of Agriculture pursuant to this subsection and of submission to the Scientific Advisory Panel pursuant to FIFRA § 25(d) [136w(d)] of this title and proceed in accordance with subsection (c) of this section. The proposed action shall become final and effective at the end of 30 days from receipt by the registrant, or publication, of a notice issued under paragraph (1), whichever occurs later, unless within that time either (i) the registrant makes the necessary corrections, if possible, or (ii) a request for a hearing is made by a person adversely affected by the notice. In the event a hearing is held pursuant to such a request or to the Administrator's determination under paragraph (2), a decision pertaining to registration or classification issued after completion of such hearing shall be final. In taking any final action under this subsection, the Administrator shall consider restricting a pesticide's use or uses as an alternative to cancellation and shall fully explain the reasons for these restrictions, and shall include among those factors to be taken into account the impact of such final action on production and prices of agricultural commodities, retail food prices, and otherwise on the agricultural economy, and he shall publish in the **Federal Register** an analysis of such impact.

(c) Suspension —

(1) Order — If the Administrator determines that action is necessary to prevent an imminent hazard during the time required for cancellation or change in classification proceedings, he may, by order, suspend the registration of the pesticide immediately. No order of suspension may be issued unless the Administrator has issued or at the same time issues notice of his intention to cancel the registration or change the classification of the pesticide. Except as provided in paragraph (3), the Administrator shall notify the registrant prior to issuing any suspension order. Such notice shall include findings pertaining to the question of "imminent hazard". The registrant shall then have an opportunity, in accordance with the provisions of paragraph (2), for an expedited hearing before the Administrator on the question of whether an imminent hazard exists.

(2) Expedite hearing — If no request for a hearing is submitted to the Administrator within five days of the registrant's receipt of the notification provided for by paragraph (1), the suspension order may be issued and shall take effect and shall not be reviewable by a court. If a hearing is requested, it shall commence within five days of the receipt of the request for such hearing unless the registrant and the Administrator agree that it shall commence at a later time. The hearing shall be held in accordance with the provisions of subchapter II of chapter 5 of title 5, except that the presiding officer need not be a certified administrative law judge. The presiding officer shall have ten days from the conclusion of the presentation of evidence to submit recommended findings and conclusions to the Administrator, who shall then have seven days to render a final order on the issue of suspension.

(3) Emergency order — Whenever the Administrator determines that an emergency exists that does not permit him to hold a hearing before suspending, he may issue a suspension order in advance of notification to the registrant. In that case, paragraph (2) shall apply except that (A) the order of suspension shall be in effect pending the expeditious completion of the remedies provided by that paragraph and the issuance of a final order on suspension, and (B) no party other than the registrant and the Administrator shall participate except that any person adversely affected may file briefs within the time allotted by the Agency's rules. Any person so filing briefs shall be considered a party to such proceeding for the purposes of FIFRA § 16(b) [136n(b)] of this title.

(4) Judicial review — A final order on the question of suspension following a hearing shall be reviewable in accordance with FIFRA § 16 [136n] of this title, notwithstanding the fact that any related cancellation proceedings have not been completed. Any order of suspension entered prior to a hearing before the Administrator shall be subject to immediate review in an action by the registrant or other interested person with the concurrence of the registrant in an appropriate district court, solely to determine whether the order of suspension was arbitrary, capricious or an abuse of discretion, or whether the order was issued in accordance with the procedures established by law. The effect of any order of the court will be only to stay the effectiveness of the suspension order, pending the Administrator's final decision with respect to cancellation or change in classification. This action may be maintained simultaneously with any administrative review proceedings under this section. The commencement of proceedings under this paragraph shall not operate as a stay of order, unless ordered by the court.

(d) Public hearings and scientific review — In the event a hearing is requested pursuant to subsection (b) of this section or determined upon by the Administrator pursuant to subsection (b) of this section, such hearing shall be held after due notice for the purpose of receiving evidence relevant and material to the issues raised by the objections filed by the applicant or other interested parties, or to the issues stated by the Administrator, if the hearing is called by the Administrator rather than by the filing of objections. Upon a showing of relevance and reasonable scope of evidence sought by any party to a public hearing, the Hearing Examiner shall issue a subpoena to compel testimony or production of documents from any person. The Hearing Examiner shall be guided by the principles of the Federal Rules of Civil Procedure in making any order for the protection of the witness or the content of documents produced and shall order the payment of reasonable fees and expenses as a condition to requiring testimony of the witness. On contest, the subpoena may be enforced by an appropriate United States district court in accordance with the principles stated herein. Upon the request of any party to a public hearing and when in the Hearing Examiner's judgment it is necessary or desirable, the Hearing Examiner shall at any time before the hearing record is closed refer to a Committee of the National Academy of Sciences the relevant questions of scientific fact involved in the public hearing. No member of any committee of the National Academy of Sciences established to carry out the functions of this section shall have a financial or other conflict of interest with respect to any matter considered by such committee. The Committee of the National Academy of Sciences shall report in writing to the Hearing

Examiner within 60 days after such referral on these questions of scientific fact. The report shall be made public and shall be considered as part of the hearing record. The Administrator shall enter into appropriate arrangements with the National Academy of Sciences to assure an objective and competent scientific review of the questions presented to Committees of the Academy and to provide such other scientific advisory services as may be required by the Administrator for carrying out the purposes of this subchapter. As soon as practicable after completion of the hearing (including the report of the Academy) but not later than 90 days thereafter, the Administrator shall evaluate the data and reports before him and issue an order either revoking his notice of intention issued pursuant to this section, or shall issue an order either canceling the registration, changing the classification, denying the registration, or requiring modification of the labeling or packaging of the article. Such order shall be based only on substantial evidence of record of such hearing and shall set forth detailed findings of fact upon which the order is based.

(e) Conditional registration —

(1) The Administrator shall issue a notice of intent to cancel a registration issued under FIFRA § 3(c)(7) [136a(c)(7)] of this title if (A) the Administrator, at any time during the period provided for satisfaction of any condition imposed, determines that the registrant has failed to initiate and pursue appropriate action toward fulfilling any condition imposed, or (B) at the end of the period provided for satisfaction of any condition imposed, that condition has not been met. The Administrator may permit the continued sale and use of existing stocks of a pesticide whose conditional registration has been canceled under this subsection to such extent, under such conditions, and for such uses as the Administrator may specify if the Administrator determines that such sale or use is not inconsistent with the purposes of this subchapter and will not have unreasonable adverse effects on the environment.

(2) A cancellation proposed under this subsection shall become final and effective at the end of thirty days from receipt by the registrant of the notice of intent to cancel unless during that time a request for hearing is made by a person adversely affected by the notice. If a hearing is requested, a hearing shall be conducted under subsection (d) of this section. The only matters for resolution at that hearing shall be whether the registrant has initiated and pursued appropriate action to comply with the condition or conditions within the time provided or whether the condition or conditions have been satisfied within the time provided, and whether the Administrator's determination with respect to the disposition of existing stocks is consistent with this subchapter. A decision after completion of such hearing shall be final. Notwithstanding any other provision of this section, a hearing shall be held and a determination made within seventy-five days after receipt of a request for such hearing.

(f) General provisions —

(1) Voluntary cancellation —

(A) A registrant may, at any time, request that a pesticide registration of the registrant be canceled or amended to terminate one or more pesticide uses.

(B) Before acting on a request under subparagraph (A), the Administrator shall publish in the **Federal Register** a notice of the receipt of the request and provide for a 30-day period in which the public may comment.

(C) In the case of a pesticide that is registered for a minor agricultural use, if the Administrator determines that the cancellation or termination of uses would adversely affect the availability of the pesticide for use, the Administrator —

(i) shall publish in the **Federal Register** a notice of the receipt of the request and make reasonable efforts to inform persons who so use the pesticide of the request; and

(ii) may not approve or reject the request until the termination of the 90-day period beginning on the date of publication of the notice in the **Federal Register**, except that the Administrator may waive the 90-day period upon the request of the registrant or if the Administrator determines that the continued use of the pesticide would pose an unreasonable adverse effect on the environment.

(D) Subject to paragraph (3)(B), after complying with this paragraph, the Administrator may approve or deny the request"; and

(2) Publication of notice — A notice of denial of registration, intent to cancel, suspension, or intent to suspend issued under this Act or a notice issued under subsection (c)(4) or (d)(5)(A) of FIFRA § 4 [136b] shall be published in the **Federal Register** and shall be sent by certified mail, return receipt requested, to the registrant's or applicant's address of record on file with the Administrator. If the mailed notice is returned to the Administrator as undeliverable at that address, if delivery is refused, or if the Administrator otherwise is unable to accomplish delivery of the notice to the registrant or applicant after making reasonable efforts to do so, the notice shall be deemed to have been received by the registrant or applicant on the date the notice was published in the **Federal Register**.

(3) Transfer of Registration of Pesticides Registered for Minor Agricultural Uses In the case of a pesticide that is registered for a minor agricultural use:

(A) During the 90-day period referred to in paragraph (1)(C)(ii), the registrant of the pesticide may notify the Administrator of an agreement between the registrant and a person or persons (including persons who so use the pesticide) to transfer the registration of the pesticide, in lieu of canceling or amending the registration to terminate the use.

(B) An application for transfer of registration, in conformance with any regulations the Administrator may adopt with respect to the transfer of the pesticide registrations, must be submitted to the Administrator within 30 days of the date of notification provided pursuant to subparagraph (A). If such an application is submitted, the Administrator shall approve the transfer and shall not approve the request for voluntary cancellation or amendment to terminate use unless the Administrator determines that the continued use of the pesticide would cause an unreasonable adverse effect on the environment.

(C) If the Administrator approves the transfer and the registrant transfers the registration of the pesticide, the Administrator shall not cancel or amend the registration to delete the use or rescind the transfer of the registration, during the 180-day period beginning on the date of the approval of the transfer unless the Administrator determines that the continued use of the pesticide would cause an unreasonable adverse effect on the environment.

(D) The new registrant of the pesticide shall assume the outstanding data and other requirements for the pesticide that are pending at the time of the transfer".

(g) Notice for stored pesticides with canceled or suspended registrations —

(1) In general — Any producer or exporter of pesticides, registrant of a pesticide, applicant for registration of a pesticide, applicant for or holder of an experimental use permit, commercial applicator, or any person who distributes or sells any pesticide, who possesses any pesticide which has had its registration canceled or suspended under this section shall notify the Administrator and appropriate State and local officials of —

(A) such possession,

(B) the quantity of such pesticide such person possesses, and

(C) the place at which such pesticide is stored.

(2) **Copies** — The Administrator shall transmit a copy of each notice submitted under this subsection to the regional office of the Environmental Protection Agency which has jurisdiction over the place of pesticide storage identified in the notice.

(h) **Judicial review** — Final orders of the Administrator under this section shall be subject to judicial review pursuant to FIFRA § 16 [136n] of this title.

FIFRA § 7. REGISTRATION OF ESTABLISHMENTS. [7 U.S.C. § 136e.]

(a) **Requirement** — No person shall produce any pesticide subject to this subchapter or active ingredient used in producing a pesticide subject to this subchapter in any State unless the establishment in which it is produced is registered with the Administrator. The application for registration of any establishment shall include the name and address of the establishment and of the producer who operates such establishment.

(b) **Registration** — Whenever the Administrator receives an application under subsection (a) of this section, he shall register the establishment and assign it an establishment number.

(c) **Information required** —

(1) Any producer operating an establishment registered under this section shall inform the Administrator within 30 days after it is registered of the types and amounts of pesticides and, if applicable, active ingredients used in producing pesticides —
(A) which he is currently producing;
(B) which he has produced during the past year; and
(C) which he has sold or distributed during the past year.
The information required by this paragraph shall be kept current and submitted to the Administrator annually as required under such regulations as the Administrator may prescribe.

(2) Any such producer shall, upon the request of the Administrator for the purpose of issuing a stop sale order pursuant to FIFRA § 13 [136k] of this title, inform him of the name and address of any recipient of any pesticide produced in any registered establishment which he operates.

(d) **Confidential records and information** — Any information submitted to the Administrator pursuant to subsection (c) of this section other than the names of the pesticides or active ingredients used in producing pesticides produced, sold, or distributed at an establishment shall be considered confidential and shall be subject to the provisions of FIFRA § 10 [136h] of this title.

FIFRA § 8. RECORDS. [7 U.S.C. § 136f.]

(a) **Requirements** — The Administrator may prescribe regulations requiring producers, registrants, and applicants for registration to maintain such records with respect to their operations and the pesticides and devices produced as he determines are necessary for the effective enforcement of this subchapter and to make the records available for inspection and copying in the same manner as provided in subsection (b). No records required under this subsection shall extend to financial data, sales data other than shipment data, pricing data, personnel data, and research data (other than data relating to registered pesticides or to a pesticide for which an application for registration has been filed).

(b) Inspection — For the purposes of enforcing the provisions of this subchapter, any producer, distributor, carrier, dealer, or any other person who sells or offers for sale, delivers or offers for delivery any pesticide or device subject to this subchapter, shall, upon request of any officer or employee of the Environmental Protection Agency or of any State or political subdivision, duly designated by the Administrator, furnish or permit such person at all reasonable times to have access to, and to copy: (1) all records showing the delivery, movement, or holding of such pesticide or device, including the quantity, the date of shipment and receipt, and the name of the consignor and consignee; or (2) in the event of the inability of any person to produce records containing such information, all other records and information relating to such delivery, movement, or holding of the pesticide or device. Any inspection with respect to any records and information referred to in this subsection shall not extend to financial data, sales data other than shipment data, pricing data, personnel data; and research data (other than data relating to registered pesticides or to a pesticide for which an application for registration has been filed). Before undertaking an inspection under this subsection, the officer or employee must present to the owner, operator, or agent in charge of the establishment or other place where pesticides or devices are held for distribution or sale, appropriate credentials and a written statement as to the reason for the inspection, including a statement as to whether a violation of the law is suspected. If no violation is suspected, an alternate and sufficient reason shall be given in writing. Each such inspection shall be commenced and completed with reasonable promptness.

FIFRA § 9. INSPECTION OF ESTABLISHMENTS, ETC. [7 U.S.C. § 136g.]

(a) In general —

(l) For purposes of enforcing the provisions of this subchapter, officers or employees of the Environmental Protection Agency or of any State duly designated by the Administrator are authorized to enter at reasonable times, (A) any establishment or other place where pesticides or devices are held for distribution or sale for the purpose of inspecting and obtaining samples of any pesticides or devices, packaged, labeled, and released for shipment, and samples of any containers or labeling for such pesticides or devices, (B) any place where there is being held any pesticide the registration of which has been suspended or canceled for the purpose of determining compliance with FIFRA § 19 [136q].

(2) Before undertaking such inspection, the officers or employees must present to the owner, operator, or agent in charge of the establishment or other place where pesticides or devices are held for distribution or sale, appropriate credentials and a written statement as to the reason for the inspection, including a statement as to whether a violation of the law is suspected. If no violation is suspected, an alternate and sufficient reason shall be given in writing. Each such inspection shall be commenced and completed with reasonable promptness. If the officer or employee obtains any samples, prior to leaving the premises, he shall give to the owner, operator, or agent in charge a receipt describing the samples obtained and, if requested, a portion of each such sample equal in volume or weight to the portion retained. If an analysis is made of such samples, a copy of the results of such analysis shall be furnished promptly to the owner, operator, or agent in charge.

(b) Warrants — For purposes of enforcing the provisions of this subchapter and upon a showing to an officer or court of competent jurisdiction that there is reason to believe that the provisions of this subchapter have been violated, officers or employees duly designated by the Administrator are empowered to obtain and to execute warrants authorizing —

(1) entry, inspection, and copying of records for purposes of this section or FIFRA § 8 [136f];

(2) inspection and reproduction of all records showing the quantity, date of shipment, and the name of consignor and consignee of any pesticide or device found in the establishment which is adulterated, misbranded, not registered (in the case of a pesticide) or otherwise in violation of this subchapter and in the event of the inability of any person to produce records containing such information, all other records and information relating to such delivery, movement, or holding of the pesticide or device; and

(3) the seizure of any pesticide or device which is in violation of this subchapter.

(c) **Enforcement —**

(1) **Certification of facts to Attorney General —** The examination of pesticides or devices shall be made in the Environmental Protection Agency or elsewhere as the Administrator may designate for the purpose of determining from such examinations whether they comply with the requirements of this subchapter. If it shall appear from any such examination that they fail to comply with the requirements of this subchapter, the Administrator shall cause notice to be given to the person against whom criminal or civil proceedings are contemplated. Any person so notified shall be given an opportunity to present his views, either orally or in writing, with regard to such contemplated proceedings, and if in the opinion of the Administrator it appears that the provisions of this subchapter have been violated by such person, then the Administrator shall certify the facts to the Attorney General, with a copy of the results of the analysis or the examination of such pesticide for the institution of a criminal proceeding pursuant to FIFRA § 14(b) [136l(b)] of this title or a civil proceeding under FIFRA § 14(a) [136l(a)] of this title, when the Administrator determines that such action will be sufficient to effectuate the purposes of this subchapter.

(2) **Notice not required —** The notice of contemplated proceedings and opportunity to present views set forth in this subsection are not prerequisites to the institution of any proceeding by the Attorney General.

(3) **Warning notices —** Nothing in this subchapter shall be construed as requiring the Administrator to institute proceedings for prosecution of minor violations of this subchapter whenever he believes that the public interest will be adequately served by a suitable written notice of warning.

FIFRA § 10. PROTECTION OF TRADE SECRETS AND OTHER INFORMATION. [7 U.S.C. § 136h.]

(a) **In general —** In submitting data required by this subchapter, the applicant may

(1) clearly mark any portions thereof which in his opinion are trade secrets or commercial or financial information and

(2) submit such market material separately from other material required to be submitted under this subchapter.

(b) **Disclosure —** Notwithstanding any other provision of this subchapter and subject to the limitations in subsections (d) and (e) of this section, the Administrator shall not make public information which in his judgment contains or relates to trade secrets or commercial or financial information obtained from a person and privileged or confidential, except that, when necessary to carry out the provisions of this subchapter, information relating to formulas of products acquired by authorization of this subchapter may be revealed to any Federal agency consulted and may be revealed at a public hearing or in findings of fact issued by the Administrator.

(c) **Disputes** — If the Administrator proposes to release for inspection information which the applicant or registrant believes to be protected from disclosure under subsection (b) of this section, he shall notify the applicant or registrant, in writing, by certified mail. The Administrator shall not thereafter make available for inspection such data until thirty days after receipt of the notice by the applicant or registrant. During this period, the applicant or registrant may institute an action in an appropriate district court for a declaratory judgment as to whether such information is subject to protection under subsection (b) of this section.

(d) **Limitations** —

(1) All information concerning the objectives, methodology, results, or significance of any test or experiment performed on or with a registered or previously registered pesticide or its separate ingredients, impurities, or degradation products, and any information concerning the effects of such pesticide on any organism or the behavior of such pesticide in the environment, including, but not limited to, data on safety to fish and wildlife, humans and other mammals, plants, animals, and soil, and studies on persistence, translocation and fate in the environment, and metabolism, shall be available for disclosure to the public. The use of such data for any registration purpose shall be governed by FIFRA § 3 [136a] of this title. This paragraph does not authorize the disclosure of any information that—

(A) discloses manufacturing or quality control processes,

(B) discloses the details of any methods for testing, detecting, or measuring the quantity of any deliberately added inert ingredient of a pesticide, or

(C) discloses the identity or percentage quantity of any deliberately added inert ingredient of a pesticide, unless the Administrator has first determined that disclosure is necessary to protect against an unreasonable risk of injury to health or the environment.

(2) Information concerning production, distribution, sale, or inventories of a pesticide that is otherwise entitled to confidential treatment under subsection (b) of this section may be publicly disclosed in connection with a public proceeding to determine whether a pesticide, or any ingredient of a pesticide, causes unreasonable adverse effects on health or the environment, if the Administrator determines that such disclosure is necessary in the public interest.

(3) If the Administrator proposes to disclose information described in clause (A), (B), or (C) of paragraph (1) or in paragraph (2) of this subsection, the Administrator shall notify by certified mail the submitter of such information of the intent to release such information. The Administrator may not release such information, without the submitter's consent, until thirty days after the submitter has been furnished such notice. Where the Administrator finds that disclosure of information described in clause (A), (B), or (C) of paragraph (1) of this subsection is necessary to avoid or lessen an imminent and substantial risk of injury to the public health, the Administrator may set such shorter period of notice (but not less than ten days) and such method of notice as the Administrator finds appropriate. During such period the data submitter may institute an action in an appropriate district court to enjoin or limit the proposed disclosure. The court may enjoin disclosure, or limit the disclosure or the parties to whom disclosure shall be made, to the extent that —

(A) in the case of information described in clause (A), (B), or (C) of paragraph (1) of this subsection, the proposed disclosure is not required to protect against an unreasonable risk of injury to health or the environment; or

(B) in the case of information described in paragraph (2) of this subsection, the public interest in availability of the information in the public proceeding does not outweigh the interests in preserving the confidentiality of the information.

(e) **Disclosure to Contractors** — Information otherwise protected from disclosure to the public under subsection (b) of this section may be disclosed to contractors with the United States and employees of such contractors if, in the opinion of the Administrator, such disclosure is necessary for the satisfactory performance by the contractor of a contract with the United States for the performance of work in connection with this subchapter and under such conditions as the Administrator may specify. The Administrator shall require as a condition to the disclosure of information under this subsection that the person receiving it take such security precautions respecting the information as the Administrator shall by regulation prescribe.

(f) **Penalty for disclosure by Federal employees** —

(1) Any officer or employee of the United States or former officer or employee of the United States who, by virtue of such employment or official position, has obtained possession of, or has access to, material the disclosure of which is prohibited by subsection (b) of this section, and who, knowing that disclosure of such material is prohibited by such subsection, willfully discloses the material in any manner to any person not entitled to receive it, shall be fined not more than $10,000 or imprisoned for not more than one year, or both. Section 1905 of title 18 shall not apply with respect to the publishing, divulging, disclosure, or making known of, or making available, information reported or otherwise obtained under this subchapter. Nothing in this subchapter shall preempt any civil remedy under State or Federal law or wrongful disclosure of trade secrets.

(2) For the purposes of this section, any contractor with the United States who is furnished information as authorized by subsection (e) of this section, or any employee of any such contractor, shall be considered to be an employee of the United States.

(g) **Disclosure to foreign and multinational pesticide producers** —

(1) The Administrator shall not knowingly disclose information submitted by an applicant or registrant under this subchapter to any employee or agent of any business or other entity engaged in the production, sale, or distribution of pesticides in countries other than the United States or in addition to the United States or to any other person who intends to deliver such data to such foreign or multinational business or entity unless the applicant or registrant has consented to such disclosure. The Administrator shall require an affirmation from any person who intends to inspect data that such person does not seek access to the data for purposes of delivering it or offering it for sale to any such business or entity or its agents or employees and will not purposefully deliver or negligently cause the data to be delivered to such business or entity or its agents or employees. Notwithstanding any other provision of this subsection, the Administrator may disclose information to any person in connection with a public proceeding under law or regulation, subject to restrictions on the availability of information contained elsewhere in this subchapter, which information is relevant to a determination by the Administrator with respect to whether a pesticide, or any ingredient of a pesticide, causes unreasonable adverse effects on health or the environment.

(2) The Administrator shall maintain records of the names of persons to whom data are disclosed under this subsection and the persons or organizations they represent and shall inform the applicant or registrant of the names and affiliations of such persons.

(3) Section 1001 of title 18 shall apply to any affirmation made under paragraph (1) of this subsection.

FIFRA § 11. RESTRICTED USE PESTICIDES; APPLICATORS. [7 U.S.C. § 136i.]

(a) Certification procedure

(1) Federal certification — In any State for which a State plan for applicator certification has not been approved by the Administrator, the Administrator, in consultation with the Governor of such State, shall conduct a program for the certification of applicators of pesticides. Such program shall conform to the requirements imposed upon the States under the provisions of subsection (a)(2) of this section and shall not require private applicators to take any examination to establish competency in the use of pesticides. Prior to the implementation of the program, the Administrator shall publish in the **Federal Register** for review and comment a summary of the Federal plan for applicator certification and shall make generally available within the State copies of the plan. The Administrator shall hold public hearings at one or more locations within the State if so requested by the Governor of such State during the thirty days following publication of the **Federal Register** notice inviting comment on the Federal plan. The hearings shall be held within thirty days following receipt of the request from the Governor. In any State in which the Administrator conducts a certification program, the Administrator may require any person engaging in the commercial application, sale, offering for sale, holding for sale, or distribution of any pesticide one or more uses of which have been classified for restricted use to maintain such records and submit such reports concerning the commercial application, sale, or distribution of such pesticide as the Administrator may by regulation prescribe. Subject to paragraph (2), the Administrator shall prescribe standards for the certification of applicators of pesticides. Such standards shall provide that to be certified, an individual must be determined to be competent with respect to the use and handling of the pesticides, or to the use and handling of the pesticide or class of pesticides covered by such individual's certification. The certification standard for a private applicator shall, under a State plan submitted for approval, be deemed fulfilled by his completing a certification form. The Administrator shall further assure that such form contains adequate information and affirmations to carry out the intent of this subchapter, and may include in the form an affirmation that the private applicator has completed a training program approved by the Administrator so long as the program does not require the private applicator to take, pursuant to a requirement prescribed by the Administrator, any examination to establish competency in the use of the pesticide. The Administrator may require any pesticide dealer participating in a certification program to be licensed under a State licensing program approved by him.

(2) State certification — If any State, at any time, desires to certify applicators of pesticides, the Governor of such State shall submit a State plan for such purpose. The Administrator shall approve the plan submitted by any State, or any modification thereof, if such plan in his judgment—

(A) designates a State agency as the agency responsible for administering the plan throughout the State;

(B) contains satisfactory assurances that such agency has or will have the legal authority and qualified personnel necessary to carry out the plan;

(C) gives satisfactory assurances that the State will devote adequate funds to the administration of the plan;

(D) provides that the State agency will make such reports to the Administrator in such form and containing such information as the Administrator may from time to time require; and

(E) contains satisfactory assurances that State standards for the certification of applicators of pesticides conform with those standards prescribed by the Administrator under paragraph (1).

Any State certification program under this section shall be maintained in accordance with the State plan approved under this section.

(b) **State plans** — If the Administrator rejects a plan submitted under this paragraph, he shall afford the State submitting the plan due notice and opportunity for Shearing[5] before so doing. If the Administrator approves a plan submitted under this paragraph, then such State shall certify applicators of pesticides with respect to such State. Whenever the Administrator determines that a State is not administering the certification program in accordance with the plan approved under this section, he shall so notify the State and provide for a hearing at the request of the State, and, if appropriate corrective action is not taken within a reasonable time, not to exceed ninety days, the Administrator shall withdraw approval of such plan.

(c) **Instruction in integrated pest management techniques** — Standards prescribed by the Administrator for the certification of applicators of pesticides under subsection (a) of this section, and State plans submitted to the Administrator under subsections (a) and (b) of this section, shall include provisions for making instructional materials concerning integrated pest management techniques available to individuals at their request in accordance with the provisions of FIFRA § 23(c) [136u(c)] of this title, but such plans may not require that any individual receive instruction concerning such techniques or to be shown to be competent with respect to the use of such techniques. The Administrator and States implementing such plans shall provide that all interested individuals are notified on the availability of such instructional materials.

(d) **In general** — No regulations prescribed by the Administrator for carrying out the provisions of this subchapter shall require any private applicator to maintain any records or file any reports or other documents.

(e) **Separate standards** — When establishing or approving standards for licensing or certification, the Administrator shall establish separate standards for commercial and private applicators.

FIFRA § 12. UNLAWFUL ACTS. [7 U.S.C. § 136j.]

(a) **In general** —

(1) Except as provided by subsection (b) of this section, it shall be unlawful for any person in any State to distribute or sell to any person—

(A) any pesticide that is not registered under FIFRA § 3 [136a] or whose registration has been canceled or suspended, except to the extent that distribution or sale otherwise has been authorized by the Administrator under this Act;

(B) any registered pesticide if any claims made for it as a part of its distribution or sale substantially differ from any claims made for it as a part of the statement required in connection with its registration under FIFRA § 3 [136a] of this title;

(C) any registered pesticide the composition of which differs at the time of its distribution or sale from its composition as described in the statement required in connection with its registration under FIFRA § 3 [136a] of this title;

(D) any pesticide which has not been colored or discolored pursuant to the provisions of FIFRA § 25(c)(5) [136w(c)(5)] of this title;

(E) any pesticide which is adulterated or misbranded; or

(F) any device which is misbranded.

---

5 So in original. Probably should be "hearing."

(2) It shall be unlawful for any person--

(A) to detach, alter, deface, or destroy, in whole or in part, any labeling required under this subchapter;

(B) to refuse to —

(i) prepare, maintain, or submit any records required by or under FIFRA §§ 5, 7, 8, 11, or 19 [136c, 136e, 136f, 136i, or 136q];

(ii) submit any reports required by or under FIFRA §§ 5, 6, 7, 8, 11, or 19 [136c, 136d, 136e, 136f, 136i, or 136q]; or

(iii) allow any entry, inspection, copying of records, or sampling authorized by this Act;

(C) to give a guaranty or undertaking provided for in subsection (b) of this section which is false in any particular, except that a person who receives and relies upon a guaranty authorized under subsection (b) of this section may give a guaranty to the same effect, which guaranty shall contain, in addition to his own name and address, the name and address of the person residing in the United States from whom he received the guaranty or undertaking;

(D) to use for his own advantage or to reveal, other than to the Administrator, or officials or employees of the Environmental Protection Agency or other Federal executive agencies, or to the courts, or to physicians, pharmacists, and other qualified persons, needing such information for the performance of their duties, in accordance with such directions as the Administrator may prescribe, any information acquired by authority of this subchapter which is confidential under this subchapter;

(E) who is a registrant, wholesaler, dealer, retailer, or other distributor to advertise a product registered under this subchapter for restricted use without giving the classification of the product assigned to it under FIFRA § 3 [136a] of this title;

(F) to distribute or sell, or to make available for use, or to use, any registered pesticide classified for restricted use for some or all purposes other than in accordance with FIFRA § 3(d) [136a(d)] of this title and any regulations thereunder. It shall not be unlawful to sell, under regulations issued by the Administrator, a restricted use pesticide to a person who is not a certified applicator for application by a certified applicator;

(G) to use any registered pesticide in a manner inconsistent with its labeling;

(H) to use any pesticide which is under an experimental use permit contrary to the provisions of such permit;

(I) to violate any order issued under FIFRA § 13 [136k] of this title;

(J) to violate any suspension order issued under FIFRA §§ 3(c)(2)(B), 4, or 6 [136a(c)(2)(B), 136b, or 136d];

(K) to violate any cancellation order issued under this Act or to fail to submit a notice in accordance with FIFRA § 6(g) [136d(g)];

(L) who is a producer to violate any of the provisions of FIFRA § 7 [136e] of this title;

(M) to knowingly falsify all or part of any application for registration, application for experimental use permit, any information submitted to the Administrator pursuant to FIFRA § 7 [136e] of this title, any records required to be maintained pursuant to this Act[6] of this title, any report filed under this subchapter, or any information marked as confidential and submitted to the Administrator under any provision of this subchapter;

(N) who is a registrant, wholesaler, dealer, retailer, or other distributor to fail to file reports required by this subchapter;

---

[6] So in the original.

(O) to add any substance to, or take any substance from, any pesticide in a manner that may defeat the purpose of this subchapter; or

(P) to use any pesticide in tests on human beings unless such human beings

(i) are fully informed of the nature and purposes of the test and of any physical and mental health consequences which are reasonably foreseeable therefrom, and (ii) freely volunteer to participate in the test.

(Q) to falsify all or part of any information relating to the testing of any pesticide (or any ingredient, metabolite, or degradation product thereof), including the nature of any protocol, procedure, substance, organism, or equipment used, observation made, or conclusion or opinion formed, submitted to the Administrator, or that the person knows will be furnished to the Administrator or will become a part of any records required to be maintained by this Act;

(R) to submit to the Administrator data known to be false in support of a registration; or

(S) to violate any regulation issued under FIFRA §§ 3(a) or 19 [136a(a) or 136q].

(b) Exemptions — The penalties provided for a violation of paragraph (1) of subsection (a) of this section shall not apply to —

(1) any person who establishes a guaranty signed by, and containing the name and address of, the registrant or person residing in the United States from whom he purchased or received in good faith the pesticide in the same unbroken package, to the effect that the pesticide was lawfully registered at the time of sale and delivery to him, and that it complies with the other requirements of this subchapter, and in such case the guarantor shall be subject to to the penalties which would otherwise attach to the person holding the guaranty under the provisions of this subchapter;

(2) any carrier while lawfully shipping, transporting, or delivering for shipment any pesticide or device, if such carrier upon request of any officer or employee duly designated by the Administrator shall permit such officer or employee to copy all of its records concerning such pesticide or device;

(3) any public official while engaged in the performance of his official duties;

(4) any person using or possessing any pesticide as provided by an experimental use permit in effect with respect to such pesticide and such use or possession; or

(5) any person who ships a substance or mixture of substances being put through tests in which the purpose is only to determine its value for pesticide purposes or to determine its toxicity or other properties and from which the user does not expect to receive any benefit in pest control from its use.

FIFRA § 13. STOP SALE, USE, REMOVAL, AND SEIZURE. [7 U.S.C. § 136k.]

(a) Stop sale, etc., orders — Whenever any pesticide or device is found by the Administrator in any State and there is reason to believe on the basis of inspection or tests that such pesticide or device is in violation of any of the provisions of this subchapter, or that such pesticide or device has been or is intended to be distributed or sold in violation of any such provisions, or when the registration of the pesticide has been canceled by a final order or has been suspended, the Administrator may issue a written or printed "stop sale, use, or removal" order to any person who owns, controls, or has custody of such pesticide or device, and after receipt of such order no person shall sell, use, or remove the pesticide or device described in the order except in accordance with the provisions of the order.

(b) **Seizure** — Any pesticide or device that is being transported or, having been transported, remains unsold or in original unbroken packages, or that is sold or offered for sale in any State, or that is imported from a foreign country, shall be liable to be proceeded against in any district court in the district where it is found and seized for confiscation by a process in rem for condemnation if—

(1) in the case of a pesticide—
(A) it is adulterated or misbranded;
(B) it is not registered pursuant to the provisions of FIFRA § 3 [136a] of this title;
(C) its labeling fails to bear the information required by this subchapter;
(D) it is not colored or discolored and such coloring or discoloring is required under this subchapter; or
(E) any of the claims made for it or any of the directions for its use differ in substance from the representations made in connection with its registration;

(2) in the case of a device, it is misbranded; or

(3) in the case of a pesticide or device, when used in accordance with the requirements imposed under this subchapter and as directed by the labeling, it nevertheless causes unreasonable adverse effects on the environment.

In the case of a plant regulator, defoliant, or desiccant, used in accordance with the label claims and recommendations, physical or physiological effects on plants or parts thereof shall not be deemed to be injury, when such effects are the purpose for which the plant regulator, defoliant, or desiccant was applied.

(c) **Disposition after condemnation** — If the pesticide or device is condemned it shall, after entry of the decree, be disposed of by destruction or sale as the court may direct and the proceeds, if sold, less the court costs, shall be paid into the Treasury of the United States, but the pesticide or device shall not be sold contrary to the provisions of this subchapter or the laws of the jurisdiction in which it is sold. On payment of the costs of the condemnation proceedings and the execution and delivery of a good and sufficient bond conditioned that the pesticide or device shall not be sold or otherwise disposed of contrary to the provisions of the subchapter or the laws of any jurisdiction in which sold, the court may direct that such pesticide or device be delivered to the owner thereof. The proceedings of such condemnation cases shall conform, as near as may be to the proceedings in admiralty, except that either party may demand trial by jury of any issue of fact joined in any case, and all such proceedings shall be at the suit of and in the name of the United States.

(d) **Court costs, etc.** — When a decree of condemnation is entered against the pesticide or device, court costs and fees, storage, and other proper expenses shall be awarded against the person, if any, intervening as claimant of the pesticide or device.

FIFRA § 14. PENALTIES. [7 U.S.C. § 136l.]

(a) **Civil Penalties** —

(1) **In general** — Any registrant, commercial applicator, wholesaler, dealer, retailer, or other distributor who violates any provision of this subchapter may be assessed a civil penalty by the Administrator of not more than $5,000 for each offense.

(2) **Private applicator** — Any private applicator or other person not included in paragraph (1) who violates any provision of this subchapter subsequent to receiving a written warning from the Administrator or following a citation for a prior violation, may be assessed a civil penalty by the Administrator of not more than $1,000 for each offense: **Provided**, That any applicator not included under paragraph (1) of this subsection who holds or applies registered pesticides, or use dilutions of registered pesticides, only to provide a service of controlling pests without delivering any unapplied pesticide to any person so served, and who violates any provision of this subchapter may be assessed a civil penalty by the Administrator of not more than $500 for the first offense nor more than $1,000 for each subsequent offense.

(3) **Hearing** — No civil penalty shall be assessed unless the person charged shall have been given notice and opportunity for a hearing on such charge in the county, parish, or incorporated city of the residence of the person charged.

(4) **Determination of penalty** — In determining the amount of the penalty, the Administrator shall consider the appropriateness of such penalty to the size of the business of the person charged, the effect on the person's ability to continue in business, and the gravity of the violation. Whenever the Administrator finds that the violation occurred despite the exercise of due care or did not cause significant harm to health or the environment, the Administrator may issue a warning in lieu of assessing a penalty.

(5) **References to Attorney General** — In case of inability to collect such civil penalty or failure of any person to pay all, or such portion of such civil penalty as the Administrator may determine, the Administrator shall refer the matter to the Attorney General, who shall recover such amount by action in the appropriate United States district court.

(b) **Criminal penalties** —

(1) **In general** —

(A) Any registrant, applicant for a registration, or producer who knowingly violates any provision of this Act shall be fined not more than $50,000 or imprisoned for not more than 1 year, or both.

(B) Any commercial applicator of a restricted use pesticide, or any other person not described in subparagraph (A) who distributes or sells pesticides or devices, who knowingly violates any provision of this Act shall be fined not more than $25,000 or imprisoned for not more than 1 year, or both.

(2) **Private applicator** — Any private applicator or other person not included in paragraph (1) who knowingly violates any provision of this subchapter shall be guilty of a misdemeanor and shall on conviction be fined not more than $1,000, or imprisoned for not more than 30 days, or both.

(3) **Disclosure of information** — Any person, who, with intent to defraud, uses or reveals information relative to formulas of products acquired under the authority of FIFRA § 3 [136a] of this title, shall be fined not more than $10,000, or imprisoned for not more than three years, or both.

(4) **Acts of officers, agents, etc.** — When construing and enforcing the provisions of this subchapter, the act, omission, or failure of any officer, agent, or other person acting for or employed by any person shall in every case be also deemed to be the act, omission, or failure of such person as well as that of the person employed.

FIFRA § 15.[7] INDEMNITIES. [7 U.S.C. § 136m.]

(a) General indemnification

(1) In general — Except as otherwise provided in this section, if —
(A) the Administrator notifies a registrant under FIFRA § 6(c)(1) [136d(c)(1)] that the Administrator intends to suspend a registration or that an emergency order of suspension of a registration under FIFRA § 6(c)(3) [136d(c)(3)] has been issued;
(B) the registration in question is suspended under FIFRA § 6(c) [136d(c)], and thereafter is canceled under FIFRA § 6(b), (d), or (f), [136d(b), 136d(d), or 136d(f)]; and
(C) any person who owned any quantity of the pesticide immediately before the notice to the registrant under subparagraph (A) suffered losses by reason of suspension or cancellation of the registration; the Administrator shall make an indemnity payment to the person.

(2) Exception — Paragraph (1) shall not apply if the Administrator finds that the person —
(A) had knowledge of facts that, in themselves, would have shown that the pesticide did not meet the requirements of FIFRA § 3(c)(5) [136a(c)(5)] for registration; and
(B) continued thereafter to produce the pesticide without giving timely notice of such facts to the Administrator.

(3) Report — If the Administrator takes an action under paragraph (1) that requires the payment of indemnification, the Administrator shall report to the Committee on Agriculture of the House of Representatives, the Committee on Agriculture, Nutrition, and Forestry of the Senate, and the Committees on Appropriations of the House of Representatives and the Senate on —
(A) the action taken that requires the payment of indemnification;
(B) the reasons for taking the action;
(C) the estimated cost of the payment; and
(D) a request for the appropriation of funds for the payment.

(4) Appropriation — The Administrator may not make a payment of indemnification under paragraph (1) unless a specific line item appropriation of funds has been made in advance for the payment.

---

7 The following language is contained in the statute at large; however, it does not amend the FIFRA statute:

(b) Interim Payments. --

(1) SOURCE. — Any obligation of the Administrator to pay an indemnity arising under FIFRA § 15 [136m], as it existed prior to the effective date of the amendment made by this section, shall be made from the appropriation provided under § 1304 of title 31, United States Code.

(2) ADMINISTRATIVE SETTLEMENT. — An administrative settlement of a claim for such indemnity may be made in accordance with the third paragraph of § 2414 of title 28, United States Code, and shall be regarded as if it were made under that section for purposes of § 1304 of title 31, United States Code.

(b) Indemnification of end users, dealers, and distributors

(1) End users — If —

(A) the Administrator notifies a registrant under FIFRA § 6(c)(1) [136d(c)(1)] that the Administrator intends to suspend a registration or that an emergency order of suspension of a registration under FIFRA § 6(c)(3) [136d(c)(3)] has been issued;

(B) the registration in question is suspended under FIFRA § 6(c) [136d(c)], and thereafter is canceled under FIFRA § 6(b), (d), or (f) [136d(b), 136d(d), or 136d(f)]; and

(C) any person who, immediately before the notice to the registrant under subparagraph (A), owned any quantity of the pesticide for purposes of applying or using the pesticide as an end user, rather than for purposes of distributing or selling it or further processing it for distribution or sale, suffered a loss by reason of the suspension or cancellation of the pesticide; the person shall be entitled to an indemnity payment under this subsection for such quantity of the pesticide.

(2) Dealers and distributors —

(A) Any registrant, wholesaler, dealer, or other distributor (hereinafter in this paragraph referred to as a "seller") of a registered pesticide who distributes or sells the pesticide directly to any person not described as an end user in paragraph (1)(C) shall, with respect to any quantity of the pesticide that such person cannot use or resell as a result of the suspension or cancellation of the pesticide, reimburse such person for the cost of first acquiring the pesticide from the seller (other than the cost of transportation, if any), unless the seller provided to the person at the time of distribution or sale a notice, in writing, that the pesticide is not subject to reimbursement by the seller.

(B) If —

(i) the Administrator notifies a registrant under FIFRA§ 6(c)(1) [136d(c)(1)] that the Administrator intends to suspend a registration or that an emergency order of suspension of a registration under FIFRA § 6(c)(3) [136d(c)(3)] has been issued;

(ii) the registration in question is suspended under FIFRA§ 6(c) [136d(c)], and thereafter is canceled under FIFRA § 6(b), (d), or (f) [136d(b), 136d(d) or 136d(f)];

(iii) any person who, immediately before the notice to the registrant under clause (i) —

(I) has not been notified in writing by the seller, as provided under subparagraph (A), that any quantity of the pesticide owned by such person is not subject to reimbursement by the seller in the event of suspension or cancellation of the pesticide; and

(II) owned any quantity of the pesticide for purposes of —

(aa) distributing or selling it; or

(bb) further processing it for distribution or sale directly to an end user; suffered a loss by reason of the suspension or cancellation of the pesticide; and

(iv) the Administrator determines on the basis of a claim of loss submitted to the Administrator by the person, that the seller —

(I) did not provide the notice specified in subparagraph

(A) to such person; and

(II) is and will continue to be unable to provide reimbursement to such person, as provided under subparagraph (A), for the loss referred to in clause (iii), as a result of the insolvency or bankruptcy of the seller and the seller's resulting inability to provide such reimbursement; the person shall be entitled to an indemnity payment under this subsection for such quantity of the pesticide.

(C) If an indemnity payment is made by the United States under this paragraph, the United States shall be subrogated to any right that would otherwise be held under this paragraph by a seller who is unable to make a reimbursement in accordance with this paragraph with regard to reimbursements that otherwise would have been made by the seller.

(3) **Source** — Any payment required to be made under paragraph (1) or (2) shall be made from the appropriation provided under § 1304 of title 31, United States Code.

(4) **Administrative settlement** — An administrative settlement of a claim for such indemnity may be made in accordance with the third paragraph of § 2414 of title 28, United States Code, and shall be regarded as if it were made under that section for purposes of § 1304 of title 31, United States Code.

(c) **Amount of payment** —

(1) **In general** — The amount of an indemnity payment under subsection (a) or (b) to any person shall be determined on the basis of the cost of the pesticide owned by the person (other than the cost of transportation, if any) immediately before the issuance of the notice to the registrant referred to in subsection (a)(1)(A), (b)(1)(A), or (b)(2)(B)(i), except that in no event shall an indemnity payment to any person exceed the fair market value of the pesticide owned by the person immediately before the issuance of the notice.

(2) **Special rule** — Notwithstanding any other provision of this Act, the Administrator may provide a reasonable time for use or other disposal of the pesticide. In determining the quantity of any pesticide for which indemnity shall be paid under this section, proper adjustment shall be made for any pesticide used or otherwise disposed of by the owner.

FIFRA § 16. ADMINISTRATIVE PROCEDURE; JUDICIAL REVIEW. [7 U.S.C. § 136n.]

(a) **District court review** — Except as otherwise provided in this Act, the refusal of the Administrator to cancel or suspend a registration or to change a classification not following a hearing and other final actions of the Administrator not committed to the discretion of the Administrator by law are judicially reviewable in the district courts of the United States.

(b) **Review by court of appeals** — In the case of actual controversy as to the validity of any order issued by the Administrator following a public hearing, any person who will be adversely affected by such order and who had been a party to the proceedings may obtain judicial review by filing in the United States court of appeals for the circuit wherein such person resides or has a place of business, within 60 days after the entry of such order, a petition praying that the order be set aside in whole or in part. A copy of the petition shall be forthwith transmitted by the clerk of the court to the Administrator or any officer designated by him for that purpose, and thereupon the Administrator shall file in the court the record of the proceedings on which he based his order, as provided in § 2112 of title 28. Upon the filing of such petition the court shall have exclusive jurisdiction to affirm or set aside the order complained of in whole or in part. The court shall consider all evidence of record. The order of the Administrator shall be sustained if it is supported by substantial evidence when considered on the record as a whole. The judgment of the court affirming or setting aside, in whole or in part, any order under this section shall be final, subject to review by the Supreme Court of the United States upon certiorari or certification as provided in § 1254 of title 28. The commencement of proceedings under this section shall not, unless specifically ordered by the court to the contrary, operate as a stay of an order.

(c) Jurisdiction of district courts — The district courts of the United States are vested with jurisdiction specifically to enforce, and to prevent and restrain violations of, this subchapter.

(d) Notice of judgments — The Administrator shall, by publication in such manner as he may prescribe, give notice of all judgments entered in actions instituted under the authority of this subchapter.

FIFRA § 17. IMPORTS AND EXPORTS. [7 U.S.C. § 136o.]

(a) Pesticides and devices intended for export — Notwithstanding any other provision of this subchapter, no pesticide or device or active ingredient used in producing a pesticide intended solely for export to any foreign country shall be deemed in violation of this subchapter—

(1) when prepared or packed according to the specifications or directions of the foreign purchaser, except that producers of such pesticides and devices and active ingredients used in producing pesticides shall be subject to FIFRA §§ 2(p), 2(q)(1)(A), (C), (D), (E), (G), and (H), 2(q)(2)(A), (B), (C)(i) and (iii), and (D), 7, and 8 [136(p), 136(q)(1)(A), (C), (D), (E), (G), and (H), 136(q)(2)(A), (B), (C)(i) and (iii), and (D), 136e, and 136f] of this title; and

(2) in the case of any pesticide other than a pesticide registered under FIFRA § 3 [136a] or sold under FIFRA § 6(a)(1) [136d(a)(1)] of this title, if, prior to export, the foreign purchaser has signed a statement acknowledging that the purchaser understands that such pesticide is not registered for use in the United States and cannot be sold in the United States under this subchapter. A copy of that statement shall be transmitted to an appropriate official of the government of the importing country.

(b) Cancellation notices furnished to foreign governments — Whenever a registration, or a cancellation or suspension of the registration of a pesticide becomes effective, or ceases to be effective, the Administrator shall transmit through the State Department notification thereof to the governments of other countries and to appropriate international agencies. Such notification shall, upon request, include all information related to the cancellation or suspension of the registration of the pesticide and information concerning other pesticides that are registered under FIFRA § 3 [136a] of this title and that could be used in lieu of such pesticide.

(c) Importation of pesticides and devices — The Secretary of the Treasury shall notify the Administrator of the arrival of pesticides and devices and shall deliver to the Administrator, upon his request, samples of pesticides or devices which are being imported into the United States, giving notice to the owner or consignee, who may appear before the Administrator and have the right to introduce testimony. If it appears from the examination of a sample that it is adulterated, or misbranded or otherwise violated the provisions set forth in this subchapter, or is otherwise injurious to health or the environment, the pesticide or device may be refused admission, and the Secretary of the Treasury shall refuse delivery to the consignee and shall cause the destruction of any pesticide or device refused delivery which shall not be exported by the consignee within 90 days from the date of notice of such refusal under such regulations as the Secretary of the Treasury may prescribe. The Secretary of the Treasury may deliver to the consignee such pesticide or device pending examination and decision in the matter on execution of bond for the amount of the full invoice value of such pesticide or device, together with the duty thereon, and on refusal to return such pesticide or device for any cause to the custody of the Secretary of the Treasury, when demanded, for the purpose of excluding them from the country, or for any other purpose, said consignee shall forfeit the full amount of said bond. All changes for storage, cartage, and labor on pesticides or devices which are refused admission or delivery shall be paid by the owner or consignee, and in default of such payment shall constitute a lien against any future importation made by such owner or consignee.

(d) Cooperation in international efforts — The Administrator shall, in cooperation with the Department of State and any other appropriate Federal agency, participate and cooperate in any international efforts to develop improved pesticide research and regulations.

(e) Regulations — The Secretary of the Treasury, in consultation with the Administrator, shall prescribe regulations for the enforcement of subsection (c) of this section.

FIFRA § 18. EXEMPTION OF FEDERAL AND STATE AGENCIES. [7 U.S.C. § 136p.]

The Administrator may, at his discretion, exempt any Federal or State agency from any provision of this subchapter if he determines that emergency conditions exist which require such exemption. The Administrator, in determining whether or not such emergency conditions exist, shall consult with the Secretary of Agriculture and the Governor of any State concerned if they request such determination.

FIFRA § 19. STORAGE, DISPOSAL, TRANSPORTATION, AND RECALL. [7 U.S.C. § 136q.]

(a) Storage, disposal, and transportation —

(1) Data requirements and registration of pesticides — The Administrator may require under FIFRA §§ 3 or 6 [136a or 136d] that —

(A) the registrant or applicant for registration of a pesticide submit or cite data or information regarding methods for the safe storage and disposal of excess quantities of the pesticide to support the registration or continued registration of a pesticide;

(B) the labeling of a pesticide contain requirements and procedures for the transportation, storage, and disposal of the pesticide, any container of the pesticide, any rinsate containing the pesticide, or any other material used to contain or collect excess or spilled quantities of the pesticide; and

(C) the registrant of a pesticide provide evidence of sufficient financial and other resources to carry out a recall plan under subsection (b), and provide for the disposition of the pesticide, in the event of suspension and cancellation of the pesticide.

(2) Pesticides — The Administrator may by regulation, or as part of an order issued under FIFRA § 6 [136d] or an amendment to such an order —

(A) issue requirements and procedures to be followed by any person who stores or transports a pesticide the registration of which has been suspended or canceled;

(B) issue requirements and procedures to be followed by any person who disposes of stocks of a pesticide the registration of which has been suspended; and

(C) issue requirements and procedures for the disposal of any pesticide the registration of which has been canceled.

(3) Containers, rinsates, and other materials — The Administrator may by regulation, or as part of an order issued under FIFRA § 6 [136d] or an amendment to such an order —

(A) issue requirements and procedures to be followed by any person who stores or transports any container of a pesticide the registration of which has been suspended or canceled, any rinsate

containing the pesticide, or any other material used to contain or collect excess or spilled quantities of the pesticide;
(B) issue requirements and procedures to be followed by any person who disposes of stocks of any container of a pesticide the registration of which has been suspended, any rinsate containing the pesticide, or any other material used to contain or collect excess or spilled quantities of the pesticide; and
(C) issue requirements and procedures for the disposal of any container of a pesticide the registration of which has been canceled, any rinsate containing the pesticide, or any other material used to contain or collect excess or spilled quantities of the pesticide.

(b) Recalls —

(1) In general — If the registration of a pesticide has been suspended and canceled under FIFRA § 6 [136d] and if the Administrator finds that recall of the pesticide is necessary to protect health or the environment, the Administrator shall order a recall of the pesticide in accordance with this subsection.

(2) Voluntary recall — If, after determining under paragraph (1) that a recall is necessary, the Administrator finds that voluntary recall by the registrant and others in the chain of distribution may be as safe and effective as a mandatory recall, the Administrator shall request the registrant of the pesticide to submit, within 60 days of the request, a plan for the voluntary recall of the pesticide. If such a plan is requested and submitted, the Administrator shall approve the plan and order the registrant to conduct the recall in accordance with the plan unless the Administrator determines, after an informal hearing, that the plan is inadequate to protect health or the environment.

(3) Mandatory recall — If, after determining under paragraph (1) that a recall is necessary, the Administrator does not request the submission of a plan under paragraph (2) or finds such a plan to be inadequate, the Administrator shall issue a regulation that prescribes a plan for the recall of the pesticide. A regulation issued under this paragraph may apply to any person who is or was a registrant, distributor, or seller of the pesticide, or any successor in interest to such a person.

(4) Recall procedure — A regulation issued under this subsection may require any person that is subject to the regulation to —
(A) arrange to make available one or more storage facilities to receive and store the pesticide to which the recall program applies, and inform the Administrator of the location of each such facility;
(B) accept and store at such a facility those existing stocks of such pesticide that are tendered by any other person who obtained the pesticide directly or indirectly from the person that is subject to such regulation;
(C) on the request of a person making such a tender, provide for proper transportation of the pesticide to a storage facility; and
(D) take such reasonable steps as the regulation may prescribe to inform persons who may be holders of the pesticide of the terms of the recall regulation and how those persons may tender the pesticide and arrange for transportation of the pesticide to a storage facility.

(5) Contents of recall plan — A recall plan established under this subsection shall include —
(A) the level in the distribution chain to which the recall is to extend, and a schedule for recall; and
(B) the means to be used to verify the effectiveness of the recall.

(6) Requirements or procedures — No requirement or procedure imposed in accordance with paragraph (2) of subsection (a) may require the recall of existing stocks of the pesticide except as provided by this subsection.

(c) Storage costs

(1) Submission of plan — A registrant who wishes to become eligible for reimbursement of storage costs incurred as a result of a recall prescribed under subsection (b) for a pesticide whose registration has been suspended and canceled shall, as soon as practicable after the suspension of the registration of the pesticide, submit to the Administrator a plan for the storage and disposal of the pesticide that meets criteria established by the Administrator by regulation.

(2) Reimbursement — Within a reasonable period of time after such storage costs are incurred and paid by the registrant, the Administrator shall reimburse the registrant, on request, for —

(A) none of the costs incurred by the registrant before the date of submission of the plan referred to in paragraph (1) to the Administrator;

(B) 100 percent of the costs incurred by the registrant after the date of submission of the plan to the Administrator or the date of cancellation of the registration of the pesticide, whichever is later, but before the approval of the plan by the Administrator;

(C) 50 percent of the costs incurred by the registrant during the 1-year period beginning on the date of the approval of the plan by the Administrator or the date of cancellation of the registration of the pesticide, whichever is later;

(D) none of the costs incurred by the registrant during the 3-year period beginning on the 366th day following approval of the plan by the Administrator or the date of cancellation of the registration of the pesticide; whichever is later; and

(E) 25 percent of the costs incurred by the registrant during the period beginning on the first day of the 5th year following the date of the approval of the plan by the Administrator or the date of cancellation of the registration of the pesticide, whichever is later, and ending on the date that a disposal permit for the pesticide is issued by a State or an alternative plan for disposal of the pesticide in accordance with applicable law has been developed.

(d) Administration of storage, disposal, transportation, and recall programs

(1) Voluntary agreements — Nothing in this section shall be construed as preventing or making unlawful any agreement between a seller and a buyer of any pesticide or other substance regarding the ultimate allocation of the costs of storage, transportation, or disposal of a pesticide.

(2) Rule and regulation review — FIFRA § 25(a)(4) [136w(a)(4)] shall not apply to any regulation issued under subsection (a)(2) or (b).

(3) Limitations — No registrant shall be responsible under this section for a pesticide the registration of which is held by another person. No distributor or seller shall be responsible under this section for a pesticide that the distributor or seller did not hold or sell.

(4) Seizure and penalties — If the Administrator finds that a person who is subject to a regulation or order under subsection (a)(2) or (b) has failed substantially to comply with that regulation or order, the Administrator may take action under FIFRA §§ 13 or 14 [136k or 136l] or obtain injunctive relief under FIFRA § 16(c) [136n(c)] against such person or any successor in interest of any such person.

(e) Container design —

(1) Procedures —

(A) Not later than 3 years after the effective date of this subsection, the Administrator shall, in consultation with the heads of other interested Federal agencies, promulgate regulations for the design of pesticide containers that will promote the safe storage and disposal of pesticides.

(B) The regulations shall ensure, to the fullest extent practicable, that the containers —

(i) accommodate procedures used for the removal of pesticides from the containers and the rinsing of the containers;

(ii) facilitate the safe use of the containers, including elimination of splash and leakage of pesticides from the containers;

(iii) facilitate the safe disposal of the containers; and

(iv) facilitate the safe refill and reuse of the containers.

(2) Compliance — The Administrator shall require compliance with the regulations referred to in paragraph (1) not later than 5 years after the effective date of this subsection.

(f) Pesticide residue removal —

(1) Procedures —

(A) Not later than 3 years after the effective date of this subsection, the Administrator shall, in consultation with the heads of other interested Federal agencies, promulgate regulations prescribing procedures and standards for the removal of pesticides from containers prior to disposal.

(B) The regulations may —

(i) specify, for each major type of pesticide container, procedures and standards providing for, at a minimum, triple rinsing or the equivalent degree of pesticide removal;

(ii) specify procedures that can be implemented promptly and easily in various circumstances and conditions;

(iii) provide for reuse, whenever practicable, or disposal of rinse water and residue; and

(iv) be coordinated with requirements for the rinsing of containers imposed under the Solid Waste Disposal Act (42 U.S.C. 6901 et seq.).

(C) The Administrator may, at the discretion of the Administrator, exempt products intended solely for household use from the requirements of this subsection.

(2) Compliance — Effective beginning 5 years after the effective date of this subsection, a State may not exercise primary enforcement responsibility under FIFRA § 26 [136w-1], or certify an applicator under FIFRA § 11 [136i], unless the Administrator determines that the State is carrying out an adequate program to ensure compliance with this subsection.

(3) Solid Waste Disposal Act — Nothing in this subsection shall affect the authorities or requirements concerning pesticide containers under the Solid Waste Disposal Act (42 U.S.C. 6901).

(g) Pesticide container study —

(1) Study —

(A) The Administrator shall conduct a study of options to encourage or require —

(i) the return, refill, and reuse of pesticide containers;

(ii) the development and use of pesticide formulations that facilitate the removal of pesticide residues from containers; and

(iii) the use of bulk storage facilities to reduce the number of pesticide containers requiring disposal.

(B) In conducting the study, the Administrator shall —

(i) consult with the heads of other interested Federal agencies, State agencies, industry groups, and environmental organizations; and

(ii) assess the feasibility, costs, and environmental benefits of encouraging or requiring various measures or actions.

(2) **Report** — Not later than 2 years after the effective date of this subsection, the Administrator shall submit to Congress a report describing the results of the study required under paragraph (1).

(h) **Relationship to Solid Waste Disposal Act** — Nothing in this section shall diminish the authorities or requirements of the Solid Waste Disposal Act (42 U.S.C. 6901 et seq.).

FIFRA § 20. RESEARCH AND MONITORING. [7 U.S.C. § 136r.]

(a) **Research** — The Administrator shall undertake research including research by grant or contract with other Federal agencies, universities, or others as may be necessary to carry out the purposes of this subchapter, and he shall conduct research into integrated pest management in coordination with the Secretary of Agriculture. The Administrator shall also take care to insure that such research does not duplicate research being undertaken by any other Federal agency.

(b) **National monitoring plan** — The Administrator shall formulate and periodically revise, in cooperation with other Federal, State, or local agencies, a national plan for monitoring pesticides.

(c) **Monitoring** — The Administrator shall undertake such monitoring activities, including, but not limited to monitoring in air, soil, water, man, plants, and animals, as may be necessary for the implementation of this subchapter and of the national pesticide monitoring plan. The Administrator shall establish procedures for the monitoring of man and animals and their environment for incidential[8] pesticide exposure, including, but not limited to, the quantification of incidental human and environmental pesticide pollution and the secular trends thereof, and identification of the sources of contamination and their relationship to human and environmental effects. Such activities shall be carried out in cooperation with other Federal, State, and local agencies.

FIFRA § 21. SOLICITATION OF COMMENTS; NOTICE OF PUBLIC HEARINGS. [7 U.S.C. § 136s.]

(a) **Secretary of Agriculture** — The Administrator, before publishing regulations under this subchapter, shall solicit the views of the Secretary of Agriculture in accordance with the procedure described in FIFRA § 25(a) [136w(a)] of this title.

---

[8] So in original. Probably should read "incidental".

(b) Views — In addition to any other authority relating to public hearings and solicitation of views, in connection with the suspension or cancellation of a pesticide registration or any other actions authorized under this subchapter, the Administrator may, at his discretion, solicit the views of all interested persons, either orally or in writing, and seek such advice from scientists, farmers, farm organizations, and other qualified persons as he deems proper.

(c) Notice — In connection with all public hearings under this subchapter the Administrator shall publish timely notice of such hearings in the Federal Register.

FIFRA § 22. DELEGATION AND COOPERATION. [7 U.S.C. § 136t.]

(a) Delegation — All authority vested in the Administrator by virtue of the provisions of this subchapter may with like force and effect be executed by such employees of the Environmental Protection Agency as the Administrator may designate for the purpose.

(b) Cooperation — The Administrator shall cooperate with Department of Agriculture, any other Federal agency, and any appropriate agency of any State or any political subdivision thereof, in carrying out the provisions of this subchapter, and in securing uniformity of regulations.

FIFRA § 23. STATE COOPERATION, AID, AND TRAINING. [7 U.S.C. § 136u.]

(a) Cooperative agreements — The Administrator may enter into cooperative agreements with States and Indian tribes—

(1) to delegate to any State or Indian tribe the authority to cooperate in the enforcement of this subchapter through the use of its personnel or facilities, to train personnel of the State or Indian tribe to cooperate in the enforcement of this subchapter, and to assist States and Indian tribes in implementing cooperative enforcement programs through grants-in-aid; and

(2) to assist States in developing and administering State programs, and Indian tribes that enter into cooperative agreements, to train and certify applicators consistent with the standards the Administrator prescribes.

Effective with the fiscal year beginning October 1, 1978, there are authorized to be appropriated annually such funds as may be necessary for the Administrator to provide through cooperative agreements an amount equal to 50 percent of the anticipated cost to each State or Indian tribe, as agreed to under such cooperative agreements, of conducting training and certification programs during such fiscal year. If funds sufficient to pay 50 percent of the costs for any year are not appropriated, the share of each State and Indian tribe shall be reduced in a like proportion in allocating available funds.

(b) Contracts for training — In addition, the Administrator may enter into contracts with Federal, State, or Indian tribal agencies for the purpose of encouraging the training of certified applicators.

(c) Information and education — The Administrator shall, in cooperation with the Secretary of Agriculture, use the services of the cooperative State extension services to inform and educate pesticide users about accepted uses and other regulations made under this subchapter.

FIFRA § 24. AUTHORITY OF STATES. [7 U.S.C. § 136v.]

(a) In general — A State may regulate the sale or use of any federally registered pesticide or device in the State, but only if and to the extent the regulation does not permit any sale or use prohibited by this subchapter.

(b) Uniformity — Such State shall not impose or continue in effect any requirements for labeling or packaging in addition to or different from those required under this subchapter.

(c) Additional uses —

(1) A State may provide registration for additional uses of federally registered pesticides formulated for distribution and use within that State to meet special local needs in accord with the purposes of this subchapter and if registration for such use has not previously been denied, disapproved, or canceled by the Administrator. Such registration shall be deemed registration under FIFRA § 3 [136a] of this title for all purposes of this subchapter, but shall authorize distribution and use only within such State.
(2) A registration issued by a State under this subsection shall not be effective for more than ninety days if disapproved by the Administrator within that period. Prior to disapproval, the Administrator shall, except as provided in paragraph (3) of this subsection, advise the State of the Administrator's intention to disapprove and the reasons therefor, and provide the State time to respond. The Administrator shall not prohibit or disapprove a registration issued by a State under this subsection (A) on the basis of lack of essentiality of a pesticide or (B) except as provided in paragraph (3) of this subsection, if its composition and use patterns are similar to those of a federally registered pesticide.
(3) In no instance may a State issue a registration for a food or feed use unless there exists a tolerance or exemption under the Federal Food, Drug, and Cosmetic Act [21 U.S.C. 301 et seq.] that permits the residues of the pesticides on the food or feed. If the Administrator determines that a registration issued by a State is inconsistent with the Federal Food, Drug, and Cosmetic Act, or the use of, a pesticide under a registration issued by a State constitutes an imminent hazard, the Administrator may immediately disapprove the registration.
(4) If the Administrator finds, in accordance with standards set forth in regulations issued under FIFRA § 25 [136w] of this title, that a State is not capable of exercising adequate controls to assure that State registration under this section will be in accord with the purposes of this subchapter or has failed to exercise adequate controls, the Administrator may suspend the authority of the State to register pesticides until such time as the Administrator is satisfied that the State can and will exercise adequate controls. Prior to any such suspension, the Administrator shall advise the State of the Administrator's intention to suspend and the reasons therefor and provide the State time to respond.

FIFRA § 25. AUTHORITY OF ADMINISTRATOR. [7 U.S.C. § 136w.]

(a) In general —

(1) **Regulations** — The Administrator is authorized, in accordance with the procedure described in paragraph (2), to prescribe regulations to carry out the provisions of this subchapter. Such regulations shall take into account the difference in concept and usage between various classes of pesticides and differences in environmental risk and the appropriate data for evaluating such risk between agricultural and nonagricultural pesticides.

(2) **Procedure —**

(A) Proposed regulations — At least 60 days prior to signing any proposed regulation for publication in the **Federal Register**, the Administrator shall provide the Secretary of Agriculture with a copy of such regulation. If the Secretary comments in writing to the Administrator regarding any such regulation within 30 days after receiving it, the Administrator shall publish in the **Federal Register** (with the proposed regulation) the comments of the Secretary and the response of the Administrator with regard to the Secretary's comments. If the Secretary does not comment in writing to the Administrator regarding the regulation within 30 days after receiving it, the Administrator may sign such regulation for publication in the **Federal Register** any time after such 30-day period notwithstanding the foregoing 60-day time requirement.

(B) Final regulations — At least 30 days prior to signing any regulation in final form for publication in the **Federal Register**, the Administrator shall provide the Secretary of Agriculture with a copy of such regulation. If the Secretary comments in writing to the Administrator regarding any such final regulation within 15 days after receiving it, the Administrator shall publish in the **Federal Register** (with the final regulation) the comments of the Secretary, if requested by the Secretary, and the response of the Administrator concerning the Secretary's comments. If the Secretary does not comment in writing to the Administrator regarding the regulation within 15 days after receiving it, the Administrator may sign such regulation for publication in the **Federal Register** at any time after such 15-day period notwithstanding the foregoing 30-day time requirement. In taking any final action under this subsection, the Administrator shall include among those factors to be taken into account the effect of the regulation on production and prices of agricultural commodities, retail food prices, and otherwise on the agricultural economy, and the Administrator shall publish in the **Federal Register** an analysis of such effect.

(C) Time requirements — The time requirements imposed by subparagraphs (A) and (B) may be waived or modified to the extent agreed upon by the Administrator and the Secretary.

(D) Publication in the Federal Register — The Administrator shall, simultaneously with any notification to the Secretary of Agriculture under this paragraph prior to the issuance of any proposed or final regulation, publish such notification in the **Federal Register**.

(3) **Congressional committees —** At such time as the Administrator is required under paragraph (2) of this subsection to provide the Secretary of Agriculture with a copy of proposed regulations and a copy of the final form of regulations, he shall also furnish a copy of such regulations to the Committee on Agriculture of the House of Representatives and the Committee on Agriculture, Nutrition, and Forestry of the Senate.

(4) **Rule and regulation review —**

(A) Congressional review of regulations — Simultaneously with the promulgation of any rule or regulation under this Act, the Administrator shall transmit a copy thereof to the Secretary of the Senate and the Clerk of the House of Representatives. The rule or regulation shall not become effective until the passage of 60 calendar days after the rule or regulation is so transmitted.

(B) Effective date — If at the end of 60 calendar days of continuous session of Congress after the date of promulgation of a rule or regulation, no committee of either House of Congress has reported or been discharged from further consideration of a concurrent resolution disapproving the rule or regulation, and neither House has adopted such a resolution, the rule or regulation may go into effect immediately. If, within such 60 calendar days, such a committee has reported or been discharged from further consideration of such a resolution, or either House has adopted such a resolution, the rule or regulation may go into effect not sooner than 90 calendar days of continuous session of Congress after its promulgation unless disapproved as provided in subparagraph (A).

(C) Computation of review period — For the purposes of subparagraphs (A) and (B) of this paragraph —

(i) continuity of session is broken only by an adjournment of Congress sine die; and

(ii) the days on which either House is not in session because of an adjournment of more than 3 days to a day certain are excluded in the computation of 60 and 90 calendar days of continuous session of Congress.

(D) Effect of Congressional inaction — Congressional inaction on or rejection of a resolution ofdisapproval shall not be deemed an expression of approval of such a rule.

(E) Judicial review —

(i) Any interested party, including any person who participated in the rulemaking involved, may institute such actions in the appropriate district court of the United States, including actions for declaratory judgment, as may be appropriate to construe the constitutionality of any provision of this paragraph. The district court immediately shall certify all questions of the constitutionality of this paragraph to the United states court of appeals for the circuit involved, which shall hear the matter sitting en banc.

(ii) Notwithstanding any other provision of law, any decision on a matter certified under clause (i) of this subparagraph shall be reviewable by appeal directly to the Supreme Court of the United States. Such appeal shall be brought not later than 20 days after the decision of the court of appeals.

(iii) It shall be the duty of the court of appeals and of the Supreme Court of the United States to advance on the docket and to expedite to the greatest possible extent the disposition of any matter certified under clause (i) of this subparagraph.

(b) Exemption of pesticides — The Administrator may exempt from the requirements of this subchapter by regulation any pesticide which he determines either (1) to be adequately regulated by another Federal agency, or (2) to be of a character which is unnecessary to be subject to this subchapter in order to carry out the purposes of this subchapter.

(c) Other authority — The Administrator, after notice and opportunity for hearing, is authorized —

(1) to declare a pest any form of plant or animal life (other than man and other than bacteria, virus, and other micro-organisms on or in living man or other living animals) which is injurious to health or the environment;

(2) to determine any pesticide which contains any substance or substances in quantities highly toxic to man;

(3) to establish standards (which shall be consistent with those established under the authority of the Poison Prevention Packaging Act (Public Law 91-601) [15 U.S.C. 1471 et seq.] with respect to the package, container, or wrapping in which a pesticide or device is enclosed for use or consumption, in order to protect children and adults from serious injury or illness resulting from accidental ingestion or contact with pesticides or devices regulated by this subchapter as well as to accomplish the other purposes of this subchapter;

(4) to specify those classes of devices which shall be subject to any provision of FIFRA § 2(q)(1) [136(q)(1)] or FIFRA § 7 [136e] of this title upon his determination that application of such provision is necessary to effectuate the purposes of this subchapter;

(5) to prescribe regulations requiring any pesticide to be colored or discolored if he determines that such requirement is feasible and is necessary for the protection of health and the environment; and

(6) to determine and establish suitable names to be used in the ingredient statement.

(d) **Scientific advisory panel** — The Administrator shall submit to an advisory panel for comment as to the impact on health and the environment of the action proposed in notices of intent issued under FIFRA § 6(b) [136d(b)] of this title and of the proposed and final form of regulations issued under subsection (a) of this section within the same time periods as provided for the comments of the Secretary of Agriculture under such FIFRA § 6(b) [136d(b)] and subsection (a) of this section. The time requirements for notices of intent and proposed and final forms of regulation may not be modified or waived unless in addition to meeting the requirements of FIFRA § 6(b) [136d(b)] of this title or subsection (a) of this section, as applicable, the advisory panel has failed to comment on the proposed action within the prescribed time period or has agreed to the modification or waiver. The Administrator shall also solicit from the advisory panel comments, evaluations, and recommendations for operating guidelines to improve the effectiveness and quality of scientific analyses made by personnel of the Environmental Protection Agency that lead to decisions by the Administrator in carrying out the provisions of this subchapter. The comments, evaluations, and recommendations of the advisory panel submitted under this subsection and the response of the Administrator shall be published in the **Federal Register** in the same manner as provided for publication of the comments of the Secretary of Agriculture under such sections. The chairman of the advisory panel, after consultation with the Administrator, may create temporary subpanels on specific projects to assist the full advisory panel in expediting and preparing its evaluations, comments, and recommendations. The subpanels may be composed of scientists other than members of the advisory panel, as deemed necessary for the purpose of evaluating scientific studies relied upon by the Administrator with respect to proposed action. Such additional scientists shall be selected by the advisory panel. The panel referred to in this subsection shall consist of 7 members appointed by the Administrator from a list of 12 nominees, 6 nominated by the National Institutes of Health and 6 by the National Science Foundation, utilizing a system of staggered terms of appointment. Members of the panel shall be selected on the basis of their professional qualifications to assess the effects of the impact of pesticides on health and the environment. To the extent feasible to insure multidisciplinary representation, the panel membership shall include representation from the disciplines of toxicology, pathology, environmental biology, and related sciences. If a vacancy occurs on the panel due to expiration of a term, resignation, or any other reason, each replacement shall be selected by the Administrator from a group of 4 nominees, 2 submitted by each of the nominating entities named in this subsection. The Administrator may extend the term of a panel member until the new member is appointed to fill the vacancy. If a vacancy occurs due to resignation, or reason other than expiration of a term, the Administrator shall appoint a member to serve during the unexpired term utilizing the nomination process set forth in this subsection. Should the list of nominees provided under this subsection be unsatisfactory, the Administrator may request an additional set of nominees from the nominating entities. The Administrator may require such information from the nominees to the advisory panel as he deems necessary, and he shall publish in the **Federal Register** the name, address, and professional affiliations of each nominee. Each member of the panel shall receive per diem compensation at a rate not in excess of that fixed for GS-18 of the General Schedule as may be determined by the Administrator, except that any such member who holds another office or position under the Federal Government the compensation for which exceeds such rate may elect to receive compensation at the rate provided for such other office or position in lieu of the compensation provided by this subsection. In order to assure the objectivity of the advisory panel, the Administrator shall promulgate regulations regarding conflicts of interest with respect to the members of the panel. The advisory panel established under this section shall be permanent. In performing the functions assigned by this subchapter, the panel shall consult and coordinate its activities with the Science Advisory Board established under the Environmental Research,

Development, and Demonstration Authorization Act of 1978 [42 U.S.C. 4365]. Whenever the Administrator exercises authority under FIFRA § 6(c) [136d(c)] of this title to immediately suspend the registration of any pesticide to prevent an imminent hazard, the Administrator shall promptly submit to the advisory panel for comment, as to the impact on health and the environment, the action taken to suspend the registration of such pesticide.

(e) **Peer review** — The Administrator shall, by written procedures, provide for peer review with respect to the design, protocols, and conduct of major scientific studies conducted under this subchapter by the Environmental Protection Agency or by any other Federal agency, any State or political subdivision thereof, or any institution or individual under grant, contract, or cooperative agreement from or with the Environmental Protection Agency. In such procedures, the Administrator shall also provide for peer review, using the advisory panel established under subsection (d) of this section or appropriate experts appointed by the Administrator from a current list of nominees maintained by such panel, with respect to the results of any such scientific studies relied upon by the Administrator with respect to actions the Administrator may take relating to the change in classification, suspension, or cancellation of a pesticide. Whenever the Administrator determines that circumstances do not permit the peer review of the results of any such scientific study prior to the Administrator's exercising authority under FIFRA § 6(c) [136d(c)] of this title to immediately suspend the registration of any pesticide to prevent an imminent hazard, the Administrator shall promptly thereafter provide for the conduct of peer review as provided in this sentence. The evaluations and relevant documentation constituting the peer review that relate to the proposed scientific studies and the results of the completed scientific studies shall be included in the submission for comment forwarded by the Administrator to the advisory panel as provided in subsection (d) of this section. As used in this subsection, the term "peer review" shall mean an independent evaluation by scientific experts, either within or outside the Environmental Protection Agency, in the appropriate disciplines.

FIFRA § 26. STATE PRIMARY ENFORCEMENT RESPONSIBILITY. [7 U.S.C. § 136w-1.]

(a) **In general** — For the purposes of this subchapter, a State shall have primary enforcement responsibility for pesticide use violations during any period for which the Administrator determines that such State —

(1) has adopted adequate pesticide use laws and regulations, except that the Administrator may not require a State to have pesticide use laws that are more stringent than this subchapter;

(2) has adopted and is implementing adequate procedures for the enforcement of such State laws and regulations; and

(3) will keep such records and make such reports showing compliance with paragraphs (1) and (2) of this subsection as the Administrator may require by regulation.

(b) **Special rules** — Notwithstanding the provisions of subsection (a) of this section, any State that enters into a cooperative agreement with the Administrator under FIFRA § 23 [136u] of this title for the enforcement of pesticide use restrictions shall have the primary enforcement responsibility for pesticide use violations. Any State that has a plan approved by the Administrator in accordance with the requirements of FIFRA § 11 [136i] of this title that the Administrator determines meets the criteria set out in subsection (a) of this section shall have the primary enforcement responsibility for pesticide use violations. The Administrator shall make such determinations with respect to State plans under FIFRA § 11 [136i] of this title in effect on September 30, 1978, not later than six months after that date.

(c) **Administrator** — The Administrator shall have primary enforcement responsibility for those States that do not have primary enforcement responsibility under this subchapter. Notwithstanding the provisions of FIFRA § 2(e)(1) [136(e)(1)] of this title, during any period when the Administrator has such enforcement responsibility, FIFRA § 8(b) [136f(b)] of this title shall apply to the books and records of commercial applicators and to any applicator who holds or applies pesticides, or use dilutions of pesticides, only to provide a service of controlling pests without delivering any unapplied pesticide to any person so served, and FIFRA § 9(a) [136g(a)] of this title shall apply to the establishment or other place where pesticides or devices are held for application by such persons with respect to pesticides or devices held for such application.

FIFRA § 27.FAILURE BY THE STATE TO ASSURE ENFORCEMENT OF STATE PESTICIDE USE REGULATIONS. [7 U.S.C. § 136w-2.]

(a) **Referral** — Upon receipt of any complaint or other information alleging or indicating a significant violation of the pesticide use provisions of this subchapter, the Administrator shall refer the matter to the appropriate State officials for their investigation of the matter consistent with the requirements of this subchapter. If, within thirty days, the State has not commenced appropriate enforcement action, the Administrator may act upon the complaint or information to the extent authorized under this subchapter.

(b) **Notice** — Whenever the Administrator determines that a State having primary enforcement responsibility for pesticide use violations is not carrying out (or cannot carry out due to the lack of adequate legal authority) such responsibility, the Administrator shall notify the State. Such notice shall specify those aspects of the administration of the State program that are determined to be inadequate. The State shall have ninety days after receipt of the notice to correct any deficiencies. If after that time the Administrator determines that the State program remains inadequate, the Administrator may rescind, in whole or in part, the State's primary enforcement responsibility for pesticide use violations.

(c) **Construction** — Neither FIFRA § 26 [136w-1] of this title nor this section shall limit the authority of the Administrator to enforce this subchapter, where the Administrator determines that emergency conditions exist that require immediate action on the part of the Administrator and the State authority is unwilling or unable adequately to respond to the emergency.

FIFRA § 28. IDENTIFICATION OF PESTS; COOPERATION WITH DEPARTMENT OF AGRICULTURE'S PROGRAM. [7 U.S.C. § 136w-3.]

(a) **In General** — The Administrator, in coordination with the Secretary of Agriculture, shall identify those pests that must be brought under control. The Administrator shall also coordinate and cooperate with the Secretary of Agriculture's research and implementation programs to develop and improve the safe use and effectiveness of chemical, biological, and alternative methods to combat and control pests that reduce the quality and economical production and distribution of agricultural products to domestic and foreign consumers.

(b) **Pest Control Availability** —

(1) In General. — The Administrator, in cooperation with the Secretary of Agriculture, shall identify —

(A) available methods of pest control by crop or animal;

(B) minor pest control problems, both in minor crops and minor or localized problems in major crops; and
(C) factors limiting the availability of specific pest control methods, such as resistance to control methods and regulatory actions limiting the availability of control methods.

(2) Report. — The Secretary of Agriculture shall, not later than 180 days after the date of enactment of this subsection and annually thereafter, prepare a report and send the report to the Administrator. The report shall —
(A) contain the information described in paragraph (1) and the information required by section 1651 of the Food, Agriculture, Conservation, and Trade Act of 1990;
(B) identify the crucial pest control needs where a shortage of control methods is indicated by the information described in paragraph (1); and
(C) describe in detail research and extension efforts designed to address the needs identified in subparagraph (B).

(c) **Integrated Pest Management** — The Administrator, in cooperation with the Secretary of Agriculture, shall develop approaches to the control of pests based on integrated pest management that respond to the needs of producers, with a special emphasis on minor pests.

FIFRA § 29. ANNUAL REPORT. [7 U.S.C. § 136w-4.]

The Administrator shall submit an annual report to Congress before February 16 of each year and the first report shall be due February 15, 1979. The report shall include the total number of applications for conditional registration under FIFRA §§ 3(c)(7)(B) and 3(c)(7)(C) [136a(c)(7)(B) and 136a(c)(7)(C)] of this title that were filed during the immediately preceding fiscal year, and, with respect to those applications approved, the Administrator shall report the Administrator's findings in each case, the conditions imposed and any modification of such conditions in each case, and the quantities produced of such pesticides.

FIFRA § 30. SEVERABILITY. [7 U.S.C. § 136x.]

If any provision of this subchapter or the application thereof to any person or circumstance is held invalid, the invalidity shall not affect other provisions or applications of this subchapter which can be given effect without regard to the invalid provision or application, and to this end the provisions of this subchapter are severable.

FIFRA § 31. AUTHORIZATION FOR APPROPRIATIONS. [7 U.S.C. § 136y.]

There is authorized to be appropriated to carry out this Act (other than FIFRA § 23(a) [136u(a)]) —

(1) $83,000,000 for fiscal year 1989, of which not more than $13,735,500 shall be available for research under this Act;

(2) $95,000,000 for fiscal year 1990, of which not more than $14,343,600 shall be available for research under this Act; and

(3) $95,000,000 for fiscal year 1991, of which not more than $14,978,200 shall be available for research under this Act.

EFFECTIVE DATES OF 1988 AMENDMENTS TO THE FEDERAL INSECTICIDE, FUNGICIDE, AND RODENTICIDE ACT

Except as otherwise provided, the amendments to this Act shall take effect on the expiration of 60 days after the date of enactment.

FIFRA § 15 [136m] shall take effect on the expiration of 180 days after the date of enactment.

FIFRA § 31 [136y] shall take effect on October 1, 1988.

ACRONYMS

AAA	American Arbitration Association
ADI	Acceptable Daily Intake
AFL-CIO	American Federation of Labor and Congress of Industrial Organizations
ALJ	Administrative Law Judge
AOAC	Association of Official Analytical Chemists
APA	Administrative Procedure Act
APHIS	Animal and Plant Health Inspection Service
ARC	Anticipated Residue Concentration
BAT	Best Available Technology Economically Available
BDPA	Birth Defect Prevention Act
BEAD	Biological and Economic Analysis Division
BSCC	Biotechnology Science Coordinating Committee
CAA	Clean Air Act
CAIR	Comprehensive Assessment Information Rule
CAMA	Calcium Acid Methanearsonate
CDFA	California Department of Food and Agriculture
CERCLA	Comprehensive Environmental Response, Compensation, and Liability Act
CEQ	Council on Environmental Quality
CFC	Chlorofluoroalkanes
CJO	Chief Judicial Officer
CMA	Chemical Manufacturers Association
CPSC	Consumer Product Safety Commission
CR	Conditionally Required
CRP	Child-Resistant Packaging
CSMA	Chemical Specialties Manufacturers Association
CWA	Clean Water Act
DBCP	Dibromochloropropane
DDD	Dicholorodiphenyl dichloroethane
DDT	Dicholorodiphenyl trichloroethane
DDVP	Dichlorvos
DESI	Drug Efficacy Study Implementation
DMSA	Disodium Methanearsonate
DOT	Department of Transportation
DRES	Dietary Risk and Exposure System
DRG	Data Reporting Guidelines
EBC	Environmental Biosafety Committee
EDB	Ethylene Oxide
EDF	Environmental Defense Fund
EEB	Ecological Effects Branch

EEC	Estimated Environmental Concentration
EFED	Environmental Fate and Effects Division
EIS	Environmental Impact Statement
EP	End-Use Product
EPA	Environmental Protection Agency
EPCRA	Emergency Planning and Community Right-to-Know Act
EPIA	Egg Products Inspection Act
EPN	Ethyl p-nitrophenyl thiobenzene phosphate
ESA	Endangered Species Act
ERP	Enforcement Response Policy
EtO	Ethylene Oxide
EUP	Experimental Use Permit
FDA	Food and Drug Administration
FFDCA	Federal Food, Drug, and Cosmetic Act
FEPCA	Federal Environmental Pesticide Control Act of 1972
FHSA	Federal Hazardous Substances Act
FIFRA	Federal Insecticide, Fungicide and Rodenticide Act
FMCS	Federal Mediation and Conciliation Service
FMIA	Federal Meat Inspection Act
FOD	Field Operations Division
FSIS	Food Safety and Inspection Service
GAO	General Accounting Office
GEMP	Genetically Engineered Microbial Pesticide
GLP	Good Laboratory Practices
HCS	Hazard Communication Standard
HED	Health Effects Division
HMTA	Hazardous Materials Transportation Act
IARC	International Agency for Research on Cancer
ICC	Ingredient Communications Council
ITC	Interagency Testing Committee
LEPC	Local Emergency Planning Committee
LMS	Linearized Multistage Model
MCL	Maximum Contaminant Level
MCLG	Maximum Contaminant Level Goals
MOA	Memorandum of Agreement
MOS	Margin of Safety
MOU	Memorandum of Understanding
MP	Manufacturing-Use Product
MPI	Maximum Permissible Intake
MSDS	Material Safety Data Safety
MSMA	Monosodium Methanearsonate

MTD	Maximum Theoretical Dose
NADA	New Animal Drug Approval
NAS	National Academy of Sciences
NDA	New Drug Application
NEPA	National Environmental Policy Act of 1969
NIH	National Institutes of Health
NOEL	No Observable Effect Level
NOIS	Notice of Intent to Suspend
NTP	National Toxicology Program
OBPA	Oxybisphenoxarsine
OCM	Office of Compliance Monitoring
OE	Office of Enforcement
OECM	Office of Enforcement and Compliance Monitoring
OES	Office of Endangered Species
OGC	Office of General Counsel
OMPA	Octamethylpyrophosphoramide
OPP	Office of Pesticide Programs
OPTS	Office of Pesticides and Toxic Programs
ORM	Other Regulated Materials
OSHA	Occupational Safety and Health Administration
OSH ACT	Occupational Safety and Health Act of 1970
OSTP	Office of Science and Technology Policy
OTS	Office of Toxic Substances
PAIR	Preliminary Assessment Information Rule
PCMB	Pentachloronitrobenzene
PD4	Position Document 4
PIC	Prior Informed Consent
PM	Product Manager
PMIA	Pesticide Monitoring Improvements Act
PMN	Premanufacture Notification
PMSD	Program Management and Support Division
PPIA	Poultry Products Inspection Act
PPM	Parts Per Million
PPPA	Poison Prevention Packaging Act
PR	Pesticide Registration
RAC	Raw Agricultural Commodity
RD	Registration Division
RfD	Reference Dose
R&D	Research and Development
RCRA	Resource Conversation and Recovery Act of 1976
RPAR	Rebuttable Presumption Against Registration

SACB	Science Analysis and Coordination Branch
SAP	Scientific Advisory Panel
SARA	Superfund Amendments and Reauthorization Act of 1986
SDWA	Safe Drinking Water Act
SEP	Standard Evaluation Procedures
SERC	State Emergency Response Commission
SLN	Special Local Need
SOM	Sensitivity of Method
SOP	Standard Operating Procedures
SPRD	Special Review Division
SRRD	Special Review and Reregistration Division
SSURO	Stop Sale, Use or Removal Order
SWC	Settlement with Conditions
TCP	Trichlorophenol
TMRC	Theoretical Maximum Residue Contribution
TPTH	Triphenyltin hydroxide
TSCA	Toxic Substances Control Act
USDA	United States Department of Agriculture

LIST OF REFERENCES

CASES

AFL-CIO v. Cal-OSHA Board, No. A048574
(Cal. Ct. App. 1st Dist. July 12, 1990) 310

AFL-CIO v. Deukmejian, No. 502541 (Sacramento Super. Ct.
filed May 1988) 311

AFL-CIO v. Deukmejian, 212 Cal. App. 3d 425,
260 Cal. Rptr. 479 (1989) 310

Alabama Power Co. v. Costle, 636 F.2d 323 (D.C. Cir. 1979) 154

American Cyanamid Co. v. Aceto Corp., No. 13 171 0080 85
(Am. Arb. Ass'n Feb. 7, 1989) (Juten, Foy, Mathis,
Arbs.) 160

American Wood Preservers Institute v. Ruckelshaus,
No. 842695 (W.D. La. Sept. 9, 1985) 178

Abbott Laboratories v. Agtrol Chemical Products,
No. 16 171 00536 89G (Am. Arb. Ass'n Sept. 6, 1990)
(Birch, Boyd, Charnoff, Arbs.) 161

E.I. Du Pont de Nemours & Co. v. Griffin Corp.,
No. 16 171 0080 86M (Am. Arb. Ass'n Dec. 21, 1988)
(Birch, Juten, Foy, Arbs.) 160,161

FMC Corp. v. Tricon International,
No. 16 199 0033 84G (Am. Arb. Ass'n Jan. 10, 1985)
(Fay, Krister, Morris, Arbs.) 160

American Petroleum Institute v. OSHA,
No. 80-3018 (5th Cir. Jan. 13, 1987) 395

American Textile Manufacturers Institute, Inc. v. Donovan,
452 U.S. 490 (1981) 394

Ciba-Geigy Corp. v. EPA, 801 F.2d 430 (D.C. Cir. 1986) 134,151, 183,335

Ciba-Geigy Corp. v. EPA, 874 F.2d 277 (5th Cir. 1989) 183

Community Nutrition Institute v. Young, 818 F.2d 943 (D.C. Cir. 1987) 147

COPARR, Ltd. v. City of Boulder, 735 F. Supp. 363 (D. Colo. 1989), *appeal filed*, No. 89-1341 (10th Cir. Nov. 1, 1989) 297

Cosmetic Toiletry & Fragrance Association v. Minnesota, 440 F. Supp. 1216 (D. Minn. 1977), *aff'd*, 575 F.2d 1256 (8th Cir. 1978) 298

d-Con v. Allenby, 728 F. Supp. 605 (N.D. Cal. 1989) 298,309

Defenders of Wildlife v. Administrator, 882 F.2d 1294 (8th Cir. 1989) 186

Dow Chemical Co. v. Blum, 469 F. Supp. 892 (E.D. Mich. 1979) 192,195

Dow Chemical Co. v. Ruckelhaus, 477 F.2d 1317 (8th Cir. 1973) 184

Environmental Defense Fund, Inc. v. Costle, 631 F.2d 922 (D.C. Cir. 1980) 182,184,187

Environmental Defense Fund, Inc. v. EPA, 465 F.2d 528 (D.C. Cir. 1972) 184,195

Environmental Defense Fund, Inc. v. EPA, 489 F.2d 1247 (D.C. Cir. 1973) 190,191

Environmental Defense Fund, Inc. v. EPA, 510 F.2d 1292 (D.C. Cir. 1975) 189,192,195

Environmental Defense Fund, Inc. v. EPA, 548 F.2d 998 (D.C. Cir. 1976), *cert. denied*, 431 U.S. 925 (1977) 189

FEDERAL STATUTES

FEDERAL REGULATIONS

* All references to 40 C.F.R. are to the 1990 edition.

STATE STATUTES AND REGULATIONS

MISCELLANEOUS

Restatement of Torts § 757 comment (b) 162

Ruckelshaus, *Science, Risk, and Public Policy*, 221 Science 1027 (1983) 213

S. Conf. Rep. No. 92-1540, 92nd Cong., 2d Sess., *reprinted in* 1972 U.S. Code Cong. & Admin. News 4130 296

S. Rep. No. 698, 94th Cong., 2d Sess. 5, *reprinted in* 1976 U.S. Code Cong. & Admin. News 4491 376

S. Rep. No. 838, 92d Cong., 2d Sess. 16 (1972), *reprinted in* 1972 U.S. Code Cong. & Admin. News 3993 295

S. Rep. No. 838, 92d Cong., 2d Sess. 5, *reprinted in* 1972 U.S. Code Cong. & Admin. News 3993 171

S. Rep. No. 452, 94th Cong., 1st Sess. 8 (1975) 4,207

S. Rep. No. 838 (pt. II), 92d Cong., 2d Sess. *reprinted in* 1972 U.S. Code Cong. & Admin. News 3993 165

S. Rep. No. 1188, 95th Cong., 2d Sess. 71 (1978) (Conference Report on the Federal Pesticide Act of 1978) 172,174

SRRD Reregistration Branch, Reregistration Plan for Chemicals on List A (Mar. 1990) 13

Speech by EPA Administrator William K. Reilly to the Industrial Biotechnology Association, Oct. 13, 1989, Washington, D.C. 267

State Relationship Options Paper (EPA Groundwater Task Force, 1990) (Federal/State Options Paper) 244

Wash. Post, Sept. 8, 1990, at A5 238

INDEX

A

B

C

D

E

F

G

H

I

J

K

L

M

N

O

P

Q

R

S

T

U

V

W